S0-BPI-755

GENETICS IN CLINICAL ONCOLOGY

GENETICS IN CLINICAL ONCOLOGY

R. S. K. Chaganti
James German

New York Oxford
OXFORD UNIVERSITY PRESS
1985

Oxford University Press
Oxford New York Toronto
Delhi Bombay Calcutta Madras Karachi
Kuala Lumpur Singapore Hong Kong Tokyo
Nairobi Dar es Salaam Cape Town
Melbourne Auckland

and associated companies in
Beirut Berlin Ibadan Nicosia

Published by Oxford University Press, Inc., 200 Madison Avenue,
New York, New York, 10016

Library of Congress Cataloging in Publication Data
Main entry under title:

Genetics in clinical oncology.

Includes bibliographies and index.
1. Cancer—Genetic aspects. I. Chaganti, R. S. K.
II. German, James. [DNLM: 1. Neoplasms—familial &
genetic. QZ 202 G3318]
RC268.4.G457 1985 616.99′4071 84-29452
ISBN 0-19-503609-3

Printing (last digit): 9 8 7 6 5 4 3 2 1

Printed in the United States of America

Preface

This book has been written for the clinician who takes care of cancer patients. It derives from a course given at Memorial Hospital in New York City in 1980 and again in 1982. Our reason for offering the course, and then for encouraging those who participated as faculty to convert their lectures into book chapters, is our belief that physicians' understanding of genetics as it pertains to cancer improves the quality of medicine they can provide affected persons and their families.

Genetics is central to neoplasia. The essential feature of cancer is a heritable genetic change (or changes) in a somatic cell and its progeny. This is a truism for the experimental oncologist, but for it to be of practical use, clinicians must be knowledgeable about cancer genetics. Understanding cancer genetics entails an understanding not just of the essential steps in neoplastic transformation and tumor progression but of several other rapidly expanding branches of biology that impinge directly on human medicine as well. The physician truly interested in human biology cannot fail to find intellectual stimulation and satisfaction in the rapidly unfolding story of neoplasia and, therefore, of normal biology. Moreover, today many cancer patients and their families wish to learn about the disease, its etiology, and its pathogenesis, sometimes in considerable detail. They are interested in the role of genetics (now a familiar subject to many nonscientists) in any family history of cancer and the risk it may pose to other family members. A firm understanding of cancer genetics is therefore necessary for physicians dealing with cancer, so that they may provide an understanding of the nature of the disease, and accurate genetic advice, to patients and their families.

The book is divided into two parts. Part I covers the fundamental aspects of neoplasia in a way that should help the clinician to understand the nature of the disease; heritable factors of importance in predisposing to human cancer are discussed also. Part II is devoted to certain applications of knowledge of the genetics of neoplasia in clinical medicine.

Contents

ONE. GENETICS AND CANCER 3

A. The Role of Genomic Alterations in Somatic Cells in the Etiology of Neoplasia

1. Gene-Environment Interactions in Cancer Etiology, JOHN B. LITTLE 7
2. Viral and Cellular Oncogenes in Cancer Etiology, WILLIAM S. HAYWARD 22
3. Mutational Models for Cancer Etiology, LOUISE C. STRONG 39
4. Chromosomal Mechanisms of Cancer Etiology, R. S. K. CHAGANTI AND SURESH C. JHANWAR 60

B. The Role of Heredity in Clinical Cancer

5. Heritable Conditions That Predispose to Cancer, JAMES GERMAN 80
6. Recessive Inheritance in Human Cancer, JAMES GERMAN 90
7. Dominant Inheritance in Human Cancer, R. NEIL SCHIMKE 103
8. Genetic Epidemiology of Human Cancer: Application to Familial Breast Cancer, MARY-CLAIRE KING 122
9. Familial Aggregation of Cancer, NANCY R. SCHNEIDER 133

TWO. GENETICS AND THE CLINICAL ONCOLOGIST 147

10. Genetic Indicators of Cancer Predisposition, R. S. K. CHAGANTI 149
11. Chromosome Changes in Leukemia, JOSEPH R. TESTA AND SHINICHI MISAWA 159
12. Chromosome Changes in Lymphoma and Solid Tumors, AVERY A. SANDBERG 185
13. The Significance of Identifying a Cancer-Predisposed Person: Lessons from the Chromosome-Breakage Syndromes, JAMES GERMAN 211
14. Managing Families Genetically Predisposed to Cancer: The "Cancer-Family Syndrome" as a Model, CHRISTOPHER J. WILLIAMS 222

15. Familial Cancer: Implications for Healthy Relatives,
DAVID E. ANDERSON AND WICK R. WILLIAMS 241

Contributors 257
Index 261

GENETICS IN CLINICAL ONCOLOGY

PART ONE
GENETICS AND CANCER

Cancer is a genetic disturbance which itself has a genetics. As this statement suggests, genetics pertains to the cancer problem in not one but multiple ways; the term *cancer genetics* has multiple meanings. The chapters in Part One of this book address several of these.

A. The role of genomic alterations in somatic cells in the etiology of neoplasia

Cancer is a genetic disturbance means that the genetic constitutions of cancers themselves can in most instances be shown to differ from those of their hosts, in that their karyotypes are altered from normal. Often the presence of the same marker (i.e., structurally rearranged) chromosome in each cell of a given cancer indicates that the cancer is a clone, a population of cells descended from a single cell that itself bore the particular chromosome mutation. The clone that constitutes the cancer is, by definition, genetically different from the host. The observation that the cancer clone is genetically abnormal does not in itself prove that the genomic alteration was either the initial or an essential event in transformation of the cell from normal to neoplastic. A reason for caution on this score is that one impressive feature of established cancers is an extraordinary degree of fluctuation in chromosome number and structure that often exists from cell to cell. (Genomic instability is not a feature of actively proliferating normal tissues, such as those of the developing embryo or adult hematogenous and lymphoid tissue.) Genomic fluctuation appears to be of the nature of neoplasia, at least of neoplasia that satisfies the clinical criteria of "malignant." Fluctuation in chromosome number is attributable to the malsegregation of chromatids during mitosis. The basis for the mitotic errors is unknown, as is the explanation for the appearance in occasional cells of structurally rearranged chromosomes. Regardless of whether some specific, microscopically visible chromosome change was pivotal etiologically in neoplastic transformation, the increased fluctuation in chromosome number and the frequent emergence of stable new rearrangements are believed to be important in the adaptability and evolution of cancer clones.

The step or steps by which a normal cell is transformed into a cell with the features that constitute neoplasia finally are being defined. As just implied, a mutation of some type is very probably necessary for the conversion of a normal cell to a cancerous one; possibly, multiple stepwise mutations are required. The mutations can be of different types: integration of viral DNA, base substitution, translocation with position effect, or segmental chromosome deletion arising through mechanisms such as unequal crossing-over or unequal sister-chromatid exchange. Somatic recombination can play a role, perhaps by producing homozygosity of some mutant locus. The fact that visible change in chromosomes often is seen in human cancers and, when present, exists in all the cells of the clone has long indicated that chromosome rearrangement may play some role in neoplastic transformation. The loci that, when affected by mutation, can be important in neoplastic transformation—presumably those important in the control of cellular proliferation—are being identified. For the present, they are being referred to as cellular oncogenes.

B. The role of heredity in clinical cancer

Whatever the steps in cellular transformation from normal to malignant turn out to be, the probability that they will be taken at all, and then that clinical cancer will emerge, can itself be genetically determined, or at least genetically influenced—thus the second part of the thesis: *cancer itself has a genetics.*

What permits the conclusion that transformation itself and the proliferation of a clone into a tumor can be genetically determined or influenced? First, certain strains of experimental animals readily develop cancer spontaneously, others do so following exposure to some oncogenic agent, and others develop cancer rarely if at all. Second, some human families have many more members affected with cancer, often of the same type or site, than expected on the basis of the general population incidence. Although such "cancer families" are familiar to medical genetics units and experienced oncologists, the significance of familial clustering of cancer is often unclear. Cancer is such a common occurrence in the general population that such families could represent just one end of a normal distribution. However, the increased occurrence of cancer in certain families also could have a genetic basis; in fact, the distribution of affected members in certain families quite clearly suggests dominant inheritance of the trait, a trait that can be called cancer proneness. Third, cancer emerges with increased frequency in individuals with many types of defective immunity, whether the defectiveness is determined genetically or therapeutically. This, along with the observation that cancer sometimes appears to have been restrained or in some way contained (e.g., the clinically silent cancers of the thyroid and prostate that are to be found in finely sectioned autopsy material), suggest that host factors, some of which could be genetic, determine the fate of a clone of cells that by conventional pathological criteria constitutes frank cancer. Fourth, certain human genes are known that lead, secondarily, to cancer. The transmission in a family of one of these genes results in a dominant pattern of inheritance of some recognizable clinical disorder, one feature of which is a disturbed growth pattern in some tissue; it is from this particular tissue that cancer emerges (e.g., neurofibromatosis, familial polyposis coli, medullary thyroid carcinoma). Finally, a few rare recessive genes have been recognized that in the homozygous state result not only in cancer proneness but also in increased chromosome instability ("breakage"), either spontaneously or in response to some environmental insult. With respect to cancer etiology, the genes for these rare disorders can be grouped with environmental agents that predispose to cancer because they share with such agents the ability to mutate chromosomes.

To recapitulate the thesis: The cells that collectively constitute a cancer are members of a clone of somatic cells that have undergone one or several mutations (addressed in Chapters 1 through 4). However, factors heritable from generation to generation (i.e., through the germ line) can influence the likelihood that the somatic mutations of importance for neoplastic transformation will occur or that a neoplastic cell will give rise to a clincial cancer (Chapters 5 through 9).—*J.G.*

A. THE ROLE OF GENOMIC ALTERATIONS IN SOMATIC CELLS IN THE ETIOLOGY OF NEOPLASIA

1.

Gene-environment interactions in cancer etiology

JOHN B. LITTLE

Approximately one quarter of all the people now living in the United States will eventually develop cancer, and the disease will affect two out of every three families. Although cure rates have improved over the past two decades, the overall mortality continues to rise. For this reason, major emphases in recent investigations have been causation and prevention as opposed to treatment. Such studies have attempted to identify the etiologic factors in the genesis of cancer and to remove them from the environment, as well as to develop methods of protection and prevention of the disease.

It has been said that 70 to 90% of all cancers are environmentally related. These figures have come largely from epidemiologic studies of demographic patterns of different types of cancer in different regions throughout this country and the world at large. These environmental components include diet and exposure to chemicals, radiation, and viruses. They may also include not only classic carcinogens, which may initiate the process of carcinogenesis, but tumor-promoting agents or factors. The latter area in particular has been receiving growing attention, as we realize that promoting factors are ubiquitous in our diet and environment.

In light of the observation that the overall mortality rate for cancer is rising even though the cure rate also is improving, it is of interest to examine more closely the current causes of cancer. About one third of all deaths due to cancer in Europe and North America are related to the use of cigarettes and other tobacco products; lung cancer in particular is a very lethal form of the disease. Occupational exposures to certain industrial chemicals are also related to the induction of cancer in human beings, but the number of cases is small. Another environmental agent of importance is sunlight. Sunlight is probably the most common cause of individual cancers, although these are skin cancers, which are generally highly curable. These three factors together (tobacco products, occupational exposures, and sunlight) account for perhaps 50% of human cancer in Europe and North America. The etiologic factors for the remaining cancers, including such common sites as colon, breast, and uterus, are not well understood. Some of these may or may not be environmentally related.

There is also evidence for a genetic component to cancer. Certain well-defined animal models for genetic susceptibility to cancer exist. For example, it has been well known for some years that wide variations occur in the susceptibility of strains of inbred mice to specific cancers; some strains are highly susceptible to certain specific cancers, whereas in others these cancers may not occur at all. The role of genetic predisposition in human cancer is less clear, although it has long been of interest to geneticists [8]. The strongest evidence at present comes from those few clinical syndromes in which specific cancer susceptibility is inherited in a mendelian fashion. These include some of the childhood tumors such as hereditary retinoblastoma, as well as the recessive chromosome-instability syndromes, which are mentioned elsewhere in this book. Other human cancers show some degree of familial aggregation, but the relative contributions of genetic and environmental factors in these cases are difficult to define.

To what extent is there genetic basis to environmentally induced cancer? What is the role of genetic susceptibility in the overall genesis of cancer, and to what extent may it involve an interaction between genetic and environmental factors; that is, to what extent may individuals be genetically susceptible to the induction of cancer by environmental agents? Let us take cigarette smokers as an example. Why do some very heavy cigarette smokers never develop lung cancer when other relatively light smokers do? Are there two distinct elements at play? Is there an underlying degree of genetic predisposition to lung cancer that renders some people more susceptible to its induction by cigarette smoke than others? The importance of trying to identify such gene–environment interactions is obvious. First, the identification of susceptible individuals could play a significant role in cancer control. Second, knowledge of these interactions would facilitate a better understanding of the mechanisms of carcinogenesis. It is the understanding of these mechanisms that ultimately will allow us to develop rational methods of prevention in human populations.

This chapter has two aims. The first is to review current knowledge of the mechanisms and characteristics of the process of carcinogenesis and its induction by environmental agents. The second is to indicate current human models for gene–environment interactions in the etiology of cancer. Unfortunately, these are isolated models and several of them are not fully understood. They thus serve only as examples of some of the current evidence suggesting that genetic susceptibility may play a role in at least some environmentally induced cancer.

Mechanisms of carcinogenesis: influence of noncarcinogenic secondary factors

In the introduction to this section of the book, it has been pointed out that cancer is a genetic disease that also has its own genetics. The formalization of a genetic basis for cancer has been the somatic mutation theory, which states that cancer is the result of a mutational event. By a mutational event, we mean a heritable change in the nucleotide sequence of deoxyribonucleic acid (DNA) that may be caused by any mechanism (base changes, translocations, rearrangements, deletions, etc.). A number of observations have been proposed as evidence for somatic mutation as the basis for the initial alteration in cells that leads to neoplasia. These include the following:

1. DNA is a critical target in carcinogenesis. Radiation and most chemical carcinogens produce DNA damage and induce DNA-repair processes. Such DNA damage is known to lead to mutations.
2. Most carcinogens indeed are mutagens.
3. Cancer is a heritable cellular alteration. Evidence suggests that a tumor is descended from a single cell; cancer is a growth of a single family of abnormal cells. Such evidence includes the finding of unique glucose-6-phosphate dehydrogenase variants in all of the cells from a single tumor and the fact that identical chromosomal aberrations may be found in all cells of one tumor.
4. Dominant inheritance of tumors in some families is evidence of a genetic component to the original tumors (e.g., hereditary retinoblastoma, adenomatosis of the colon and rectum).

I shall not review the findings that suggest an epigenetic basis for cancer, as recent evidence indicates that the development of cancer is a complex multistage process that cannot be explained on the basis of either a single mutational or epigenetic event. Thus, although mutations may be involved in the etiology of cancer, they are probably only one of a number of factors involved in the whole process. Evidence for the multistage nature of this process, which requires both time and cell proliferation, can be observed at two levels. The first is in the development of those cellular changes recognized in vitro by which a normal cell changes into a cancer cell. The second involves the progression of recognizable histopathological changes seen in vivo from the first early signs of hyperplasia to the development of a frankly invasive and metastatic tumor.

Perhaps the first experimental evidence for the multistage nature of carcinogenesis derived from the initiation-promotion model developed by Berenblum, Rous, and others [2, 14]. These investigators showed that the incidence of skin tumors induced in mice by exposure to low doses of a polycyclic hydrocarbon carcinogen could be markedly enhanced if followed by repeated applications of a noncarcinogenic irritant agent called croton oil. Croton oil produced no tumors by itself, but a high incidence of skin tumors resulted even when a subcarcinogenic dose of the initiating carcinogen was followed by croton oil treatments beginning either immediately afterwards or several months later. These results led to the two-stage theory of carcinogenesis. The first stage, called "initiation," involves the production of a heritable alteration in the cell, whereas the second stage, "promotion," involves the phenotypic expression of this change as a recognizable cancer. Promoting agents are those that facilitate this process of expression. Recently, in vitro models have been developed that demonstrate the same phenomena.

Let us now look more closely at some of these in vitro results. Figure 1–1 presents the general scheme used to investigate the malignant transformation of a line of mouse embryo fibroblasts in vitro [9]. By malignant transformation we mean the conversion of a cell with a nor-

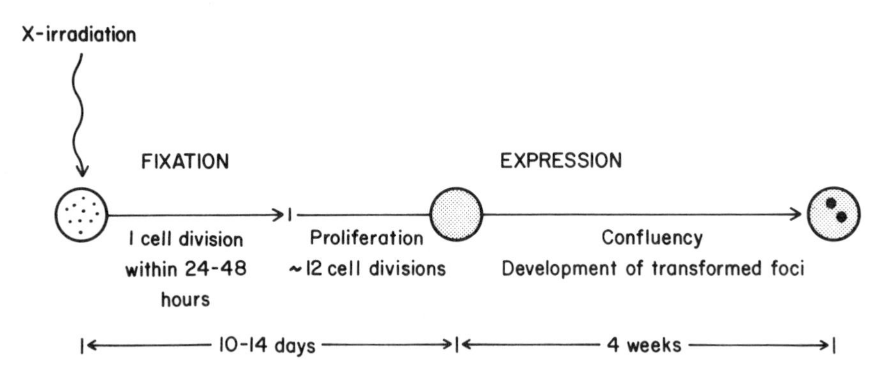

Figure 1–1 The development of malignant transformation in mouse embryo fibroblasts in vitro. Cells are treated with the carcinogen (here X-rays) at low density. One cell division must occur within 24 to 48 hours in order for the initial damage to be fixed as a heritable cellular alteration (initiation). However, additional cell proliferation and a considerable period of time (6–7 weeks) are required before this alteration is expressed in terms of morphologically transformed cells, seen on the right as densely piled up foci of cells overlying the normal monolayer. Exposure to various noncarcinogenic agents during this expression period can greatly modify—either enhance (promote) or suppress—the frequency of transformation.

mal phenotype to one with a malignant phenotype; the ultimate criterion of malignancy is the ability of the cells to form invasive tumors upon reinjection into syngeneic mice. Referring to Figure 1–1, the cells are seeded at low density in replicate petri dishes and exposed to the carcinogen. The petri dishes are then returned to the incubator and the cells allowed to proliferate until they reach confluence. At this point, when there are about 2 million cells per 100 mm petri dish, proliferation ceases as the cells have reached contact-inhibition or density-inhibition of growth. This proliferative phase requires about 10 to 14 days. The confluent monolayer of cells looks perfectly normal throughout, with no evidence of altered cell morphology. However, if these petri dishes containing the confluent monolayer of apparently normal cells are left in the incubator for 4 to 5 additional weeks with regular medium changes, discrete foci of densely piled up cells with a distinctly abnormal cytologic appearance and growth pattern develop overlying the normal monolayer. If the cells from these foci are isolated and grown up in sufficient quantities, they will form malignant tumors on subcutaneous injection into syngeneic mice.

As can be seen in Figure 1–1, the development of transformation required cell proliferation (about 12 rounds of cell division) and time (6–7 weeks). Indeed, it has been shown [16] that the yield of transformed foci induced by carcinogen exposure will be reduced markedly if the proliferative capacity of the cells prior to confluence is restricted. A number of specific phenotypic characteristics have been identified in transformed cells. These include not only changes in cytology and growth pattern but also the appearance of certain proteolytic enzymes such as plasminogen inactivator, the ability to grow in suspension (anchorage independence), and finally the ability to form tumors in animals. Careful studies of cells during the development of malignant transformation *in vitro* have shown that these markers appear in a progressive, step-wise fashion over a considerable period of time [1]. Thus, the cell progressively acquires the various characteristics of transformation; these characteristics do not appear simultaneously, as one might expect if the process resulted from a single gene mutation. Such observations on transformation in vitro suggest that the development of cancer is a multistage process even at the cellular level, which is subject to possible modification at any point.

An example of promotion at the cellular level is shown in Figure 1–2. The lower curve is the dose-response relationship for the induction of malignant transformation in mouse embryo fibroblasts by X-rays. The upper curve represents the dose-response relationship for cells incubated during the entire 6-week expression period following irradiation (Figure 1–1) with a phorbol ester tumor-promoting agent. (Phorbol esters are the active promoting agents in croton oil.) As can be seen in

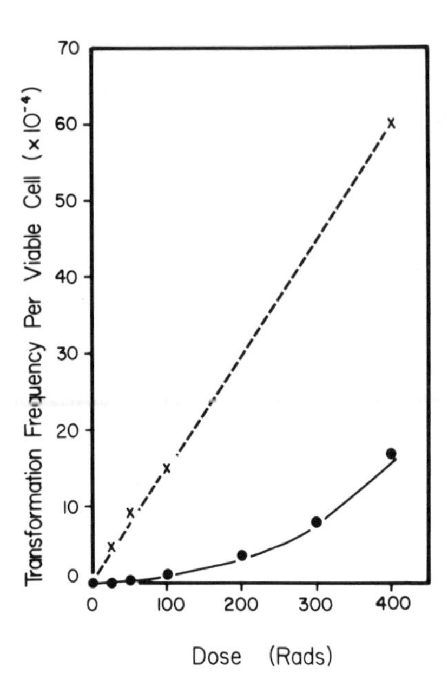

Figure 1–2 Promotion of the expression of X-ray transformation in vitro. The lower curve (●—●) represents the dose–response relationship for the induction of malignant transformation in mouse embryo fibroblasts by X-rays alone. In the upper curve (X- - - - -X), the cells were incubated for the 6-week expression period beginning 3 days after irradiation with the classic phorbol ester tumor-promoting agent 12-0-tetradecanoyl phorbol-13-acetate (TPA). TPA induced no transformation by itself but greatly enhanced transformation induced by low doses of X-rays. Reproduced from Little [10].

Figure 1–2, incubation with this noncarcinogenic agent greatly enhances the frequency of X-ray transformation in vitro, especially following low radiation doses.

It is also possible to suppress the phenotypic expression of transformation by exposure to certain agents. These include retinoic acid (vitamin A) derivatives and certain classes of inhibitors of proteolytic enzymes. The results of such an experiment are shown in Figure 1–3. Incubation with the protease inhibitor antipain for the 6-week expression period markedly suppressed the frequency of transformation induced by 600 rads of X-rays. Furthermore, antipain also suppressed

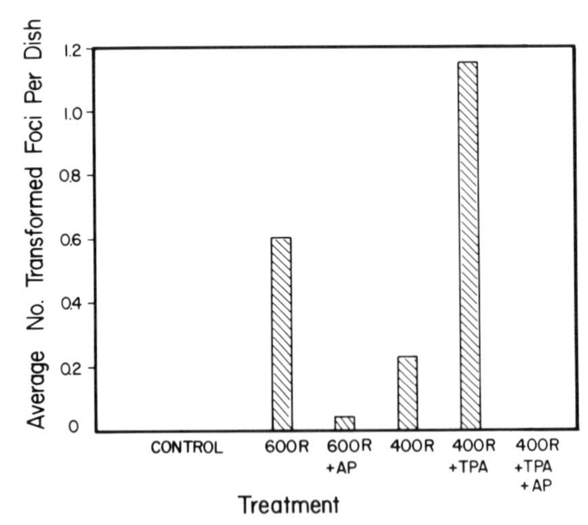

Figure 1–3 Suppression of the expression of X-ray transformation (400R or 600R) by the protease inhibitor antipain (AP). The cells were incubated with antipain during the full 6-week expression period, as in Figure 1–2. Antipain markedly suppressed both X-ray transformation (600R + AP) and the enhancement in transformation produced by the phorbol ester tumor promoter, TPA (400R + TPA, ± AP). Data are from Kennedy and Little [6].

the effects of the phorbol-ester tumor promoter. In similar experiments with soybean trypsin inhibitor, promotion was inhibited, although no effect on transformation was induced by X-rays alone [6]. This is not an unexpected finding, as this large-molecular-weight protease inhibitor probably does not enter the cell but rather acts at the level of the cell membrane, as do the phorbol-ester tumor promoters. That soybean-trypsin inhibitor is biologically effective in vitro is of interest because this inhibitor belongs to a class of naturally occurring protease inhibitors found in many foodstuffs. Thus, it has potential use in human-intervention studies.

To what extent may such noncarcinogenic modifying factors be of importance in in vivo carcinogenesis? Increasing data from animal studies indicate that the phenomenon of initiation and promotion is not restricted to the mouse skin but also occurs in other tissues. The results shown in Table 1–1 were derived from an experiment quite different from those in mouse skin, in which a noncarcinogenic secondary factor had an apparently profound effect on the ultimate tumor incidence. These experiments were initially designed to investigate the interactions between radiation and polycyclic hydrocarbon carcinogens in the genesis of lung cancer in hamsters [11]. The animals received a single intratracheal instillation of a low dose of the radioactive element polonium-210, followed 4 months later by a series of weekly instillations of

Table 1–1 Influence of low doses of benzo[*a*]pyrene or saline instillations only administered 4 months later on the induction of lung cancer in hamsters by low doses of alpha radiation.

First treatment*	Second treatment†	Experimental group‡	Number of animals with tumors / Total number of animals	Percentage
—	Saline	1	0/42	0.0
		2	0/27	
—	BP	1	0/257	0.0
		2	0/29	
^{210}Po	—	1	4/121	2.7
		2	0/29	
^{210}Po	BP	1	45/132	39
		2	16/25	
^{210}Po	Saline	1	18/82	28
		2	12/27	

*Polonium 210 (^{210}Po, 40 nCi) administered as single intratracheal instillation in 0.2 ml isotonic saline. Lifetime radiation dose is approximately 75 rads.

†Eight weekly intratracheal instillations of 0.2 ml isotonic saline beginning 15 to 18 weeks after the first treatment. In two cases, saline contained 0.3 mg of benzo[a]pyrene (BP) adsorbed onto 3.0 mg of Fe_2O_3 particles.

‡Group 1 data are from Little et al., [11]; Group 2 data are from Shami et al. [15].

a low dose of benzo[*a*]pyrene. The doses of both agents were designed to yield a very low tumor incidence by themselves. Of particular interest, however, is the comparison of the third and fifth treatment groups in Table 1–1. Both of these groups of animals received a single instillation of polonium-210. In group 5, however, this was followed 4 months later by 7 weekly instillations of 0.2 ml of isotonic saline alone. Whereas only 4 out of a total of 150 hamsters (2.7%) developed lung cancer in the radiation-only group, 30 out of 109 animals (28%) developed lung cancer in the group that also received the saline instillations. Thus, although this dose of radiation by itself had very little carcinogenic potency, when combined with so apparently innocuous a treatment as several instillations 4 months later of a very small volume of saline, a high cancer incidence resulted.

Subsequent studies [15] showed that the saline instillations induced a transient wave of cell proliferation in the mucosal cells of the lung, which may have facilitated the expression of the initial radiation-induced, heritable alteration in initiated cells as frank carcinoma. It is now widely hypothesized that cigarette smoke, because of its prolonged irritant qualities, may be acting by a similar mechanism, primarily as a promoting agent in the induction of lung cancer.

The experimental results described in this section have been presented in order to emphasize the experimental basis for the conclusion that noncarcinogenic secondary factors may greatly modify the effects of classic carcinogens such as radiation and, in some cases, may indeed be the controlling factors [10]. Genetic susceptibility might be another such modifying factor [13].

Models for genetic susceptibility to environmentally induced human cancer

One can imagine several types of models in which genetic factors could influence environmentally induced cancer. For example, certain cell populations might already be "initiated" as a heritable characteristic, and the induction of cancer in these tissues might depend only on exposure to promoting agents. This model has been proposed by Kopelovich at the Memorial Sloan-Kettering Cancer Center as the basis for the extremely high incidence of colon cancer in the adenomatosis-of-the-colon-and-rectum syndrome [7]. Increased susceptibility to the induction of cancer might also result from the hereditary lack of the production of certain protective or inhibitory substances such as the natural protease inhibitors. Finally, enhanced susceptibility to environmentally induced cancer might result from an inherited trait that leads to hypersensitivity to the induction of DNA damage or from a genetic metabolic defect in the ability of the cell to deal with or repair this damage.

A list of human disorders associated with either a high frequency of spontaneous chromosome breakage or hypersensitivity to an environmental agent is presented in Table 1–2[4]. Three of these disorders, xeroderma pigmentosum, hereditary retinoblastoma, and nevoid basal cell carcinoma syndrome, are associated with an increased susceptibility to the induction of cancer by an environmental agent [either X-rays or ultraviolet (UV) light]. Somatic cells from several of the disorders in Table 1–2 are hypersensitive to the lethal effects of specific DNA-damaging agents; though DNA-repair defects have been proposed as the mechanism for the hypersensitivity in several of these disorders, only in the case of xeroderma pigmentosum has a clear DNA-repair defect been identified. Indeed, xeroderma pigmentosum is perhaps the best human model in which a heritable genetic abnormality is associated with a marked hypersusceptibility to environmentally induced cancer.

Xeroderma pigmentosum is a rare autosomal recessive disease characterized clinically by an extreme hypersensitivity to sunlight. Patients with this disease develop multiple skin cancers in sunlight-exposed areas, often within the first few years of life. All types of skin cancers are represented, including squamous carcinomas, basal cell carcinomas, and melanomas. When somatic cells from these patients are studied in vitro, they are found to be extremely sensitive to the cytotoxic and mutagenic effects of UV light (Figure 1–4). These cells have been shown to possess a well-characterized defect in the early stages of the excision-repair process for UV-light-induced, DNA-base damage. Seven different complementation groups have been identified so far, suggesting at least seven different genes involved in this repair process. Thus, a clear genetic basis for the increased susceptibility of xeroderma pigmentosum patients to the induction of skin cancer by sunlight has been established. This involves a heritable defect in the ability to repair damage induced in the DNA of the skin cells by sunlight exposure.

There are also two autosomal dominant disorders, hereditary retinoblastoma and the nevoid basal cell carcinoma syndrome, which appear to be associated with an increased susceptibility to the induction of cancer by an environmental agent, in this case X-irradiation. Certain clinical characteristics of these disorders are discussed in detail by Dr. Strong in Chapter 3 of this volume. Patients with hereditary retinoblastoma treated by radiation therapy develop a higher-than-normal incidence of osteogenic sarcomas in the irradiated field. In the series followed by Dr. Strong, these radiation-induced tumors have developed in as many as 30 to 40% of treated patients. Somatic cells from patients with hereditary retinoblastoma appear to be slightly hypersen-

Table 1–2 Human disorders with either spontaneous chromosome breakage or susceptibility to an environmental agent.*

	Inheritance†	Cancer susceptibility	Spontaneous chromosome damage	Immune defect	Clinical hypersensitivity	In vitro hypersensitivity§
Ataxia-telangiectasia	AR	Lymphoid and others	+(clones)	+	X-rays	X-rays, bleomycin
Xeroderma pigmentosum	AR	Skin (all types)	–	–	Sunlight (skin cancer induction)	UV, certain chemicals
Franconi's anemia	AR	Leukemia	Breaks; exchanges	–	—	Cross-linking agents
Bloom's syndrome	AR	Lymphoid and others	Homologous exchanges; SCE	+	Sunlight	UV
Cockayne's syndrome	AR	—	–	–	Sunlight	UV
Progeria	AR	Sarcomas, meningiomas	–	–	—	?X-rays
Hereditary retinoblastoma	AD‡	Retinoblastoma, osteosarcoma	–	–	X-ray (sarcoma induction)	?X-rays
Nevoid basal cell carcinoma syndrome	AD	Skin (basal cell), medulloblastoma	–	–	X-ray (carcinoma induction)	?X-rays
Scleroderma	?Multifactorial	Lung	+	–	?	?Transmissible agent

*Modified from Harnden [4].

†AR = autosomal recessive; AD = autosomal dominant.

‡A small number have a constitutional deletion of chromosome 13.

§By either cell survival or chromosome damage.

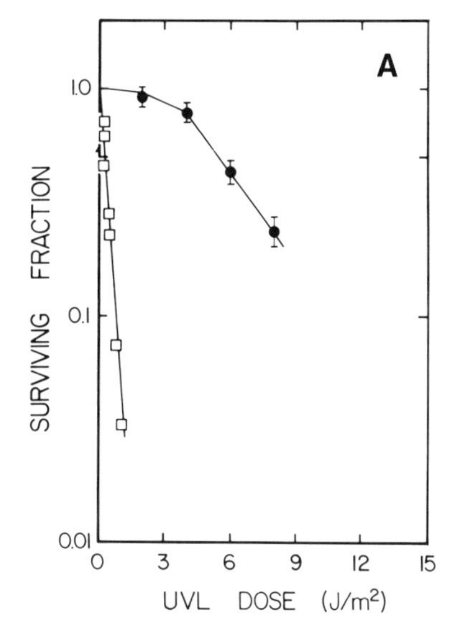

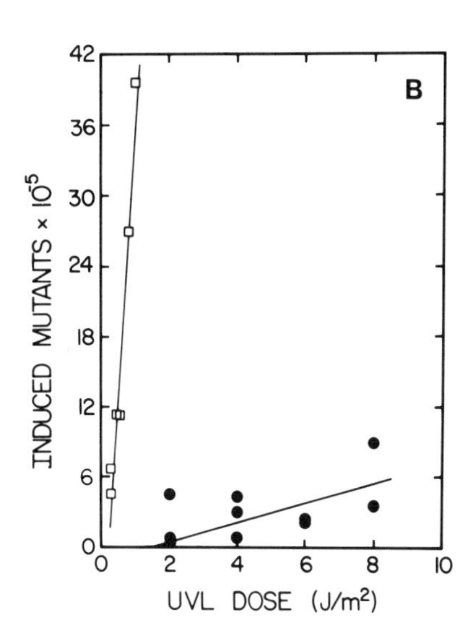

Figure 1–4 Influence of genetic factors on response to an environmental agent. The dose–response relationship for cell killing (*A*) and the induction of mutations to 6-thioguanine resistance (*B*) by UV light in normal skin fibroblasts (●—●) and skin fibroblasts from a patient with xeroderma pigmentosum (□—□). Data are from Grosovsky and Little [3].

sitive to the cytotoxic effect of X-rays [17]. Otherwise, there is as yet no evidence as to the mechanism or molecular basis for the apparent hypersusceptibility of these patients to X-ray–induced cancer. In particular, no DNA-repair defect has been identified.

Patients with the nevoid basal cell carcinoma syndrome develop multiple basal cell cancers of the skin associated with skeletal abnormalities, fibromas of the ovary, medulloblastomas, and other brain tumors. Though their skin tumors also appear more frequently in sunlight-exposed areas, this syndrome differs from xeroderma pigmentosum in one remarkable characteristic. Whereas xeroderma patients develop all types of skin tumors, nevoid basal cell carcinoma syndrome patients develop only tumors arising from the basal cells. Of particular interest has been the observation that large numbers of basal cell cancers of the skin have arisen in the irradiated field in patients with this syndrome who have been treated with radiation therapy for medulloblastoma. Again, the mechanism for this extreme hypersusceptibility is not known.

Fanconi's anemia is an autosomal recessive chromosome-breakage disorder associated with a high incidence of cancer, particularly leukemia. There is no positive evidence for an environmental component to these cancers, but recently interesting results have become available which may suggest a mechanism for this effect. Furthermore, such a mechanism could act as a model for hypersusceptibility to environmental carcinogens. Somatic cells from patients with Fanconi's anemia are particularly sensitive to the cytotoxic effects of DNA cross-linking agents such as mitomycin-C. As can be seen in Figure 1–5, however, this hypersensitivity can be significantly reduced by incubating the cells with superoxide dismutase concomitant with the mitomycin-C treatment [12]. Superoxide dismutase has no effect on the survival of normal cells treated with mitomycin-C. Other evidence has suggested that Fanconi's anemia is associated with decreased levels of intracellular superoxide dismutase. The DNA-damaging effects of mitomycin-C as well as of a number of other physical and chemical DNA-damaging agents is mediated by aqueous free radicals. The spectrum of radicals produced by these different agents may vary. The present results suggest that the superoxide anion may be important in the biological effects of DNA cross-linking agents. A genetic trait for decreased levels of specific, naturally occurring enzymes that inactivate such radicals might thus lead to an inherited hypersensitivity to these agents. The involvement of free radicals in both the initiation and promotion stages of carcinogenesis is receiving considerable current interest. Determining the role of radicals and radical inactivators in genetic susceptibility seems a fruitful line of investigation.

Two other models for gene-environment interactions are of particular interest. Kopelovich [7] has shown that somatic cells from patients with adenomatosis of the colon and rectum show a spectrum of abnor-

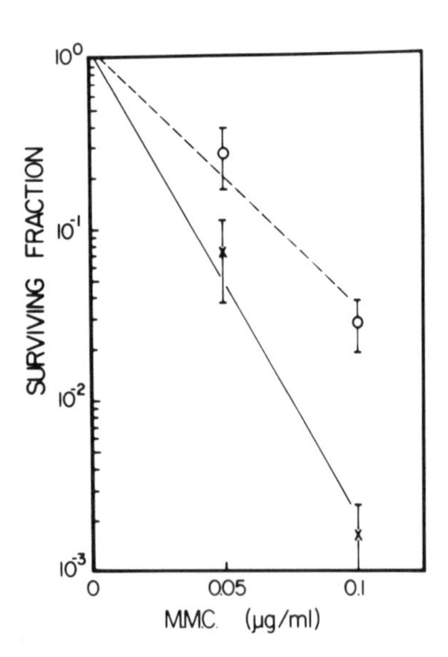

Figure 1–5 Dose–response relationship for cell killing by mitomycin-C in skin fibroblasts from a patient with Fanconi's anemia (X—X). In the upper curve (O- - - -O), the cells were incubated with superoxide dismutase simultaneously with mitomycin-C treatment. Reproduced from Nagasawa and Little [12].

malities related to growth, membrane function, and cytoskeleton that suggests they already may be initiated and require only promotional events to complete the process of carcinogenesis. Although these results require confirmation, they offer an interesting general model for genetic susceptibility to certain types of cancer and strongly suggest a role for environmental factors in the genesis of these cancers. Colon cancer in particular is one that is hypothesized to be closely related to dietary factors.

Another model involves the metabolic activation of chemical carcinogens, wherein the ability to convert chemical carcinogens to a metabolically active form may be subject to genetic variation. This idea has been given impetus by the recent elucidation of genetically determined differences among individuals in the metabolisms of various drugs. Several drugs have now been identified whose metabolic oxidation is largely determined by diallelomorphic gene loci. These polymorphisms appear to reflect genetic variations in single forms of human cytochrome P450 enzymes [5]. The cytochrome P450 system is required for the metabolism of many known carcinogens, in particular the polycyclic aromatic hydrocarbons. It exists in many forms, each differing in the pattern of metabolites produced and tissue distribution. Variability of chemical-carcinogen activation and detoxification by this system represents a possible mechanism by which the effects of carcinogens may be modulated by genetically determined host factors.

In early studies, the enzyme arylhydrocarbon hydroxylase was used as a marker enzyme. Some evidence suggested that those people who inherited the trait for synthesizing only small quantities of this enzyme might be less susceptible to the carcinogenic effect of cigarette smoke, and thus have a lower incidence of lung cancer than might be expected from their smoking habits. However, these results have not been widely confirmed. In recent studies on the polymorphic drug oxidative pathways in humans, particularly debrisoquine 4-hydroxylation, an interesting preliminary relationship to smoking and lung cancer was found [5]. In this study, light cigarette smokers who developed lung cancer were compared with very heavy smokers with no lung cancer. It was found that debrisoquine metabolism was significantly elevated in the cancer patients, even though they smoked relatively little. This metabolic pathway has been associated with the activation of certain known chemical carcinogens to their active forms. The findings are thus consistent with the hypothesis that genetic traits that control the efficiency of such metabolic pathways may significantly influence susceptibility of an individual to environmental agents. These results are at present very preliminary, but they suggest that this is a most promising area for further study of gene–environment interactions.

Conclusions

This chapter has presented several models for the interaction between genetic and environmental factors in human cancer. At present, the known interactions involve but a minute fraction of all human lung cancer. The availability of such models, however, not only yields a framework for further investigation of possible interactions but also suggests that such interactions could account for a significant fraction of the incidence of cancer in the industrialized world. Considerable effort has been expended during the past decade on identifying carcinogens in the environment. Equally important may be knowledge of the genetic factors that influence an individual's response to these agents.

Acknowledgments

The results from our laboratory described in this chapter were derived from research supported by research grants CA-11751 and CA-22704 and contract CP-33273 from the National Cancer Institute, and Center grant ES-00002 from the National Institute of Environomental Health Sciences.

Literature cited

1. Barrett JC, Ts'o POP: Evidence for the progressive nature of neoplastic transformation in vitro. *Proc. Natl. Acad. Sci. USA* 75:3761–3765, 1978.
2. Berenblum I: The cocarcinogenic action of croton resin. *Cancer Res* 1:44–48, 1941.
3. Grosovsky AG, Little JB: Mutagenesis and lethality following S phase irradiation of xeroderma pigmentosum and normal human diploid fibroblasts with ultraviolet light. *Carcinogenesis* 4:1389–1393, 1983.
4. Harnden DG: Mechanisms of Genetic Susceptibility to Cancer. *In* Carcinogenesis: Fundamental Mechanisms and Environmental Effects (Pullman B, T'so POP, Gelboin H, eds.) New York, D Reidel Publishing Co, 1980, 235–244.
5. Idle JR, Ritchie JG: Probing genetically variable carcinogen metabolism using drugs. *In* Human Carcinogenesis (Harris CC, Autrup HN, eds) New York, Academic Press, 1983, pp. 857–881.
6. Kennedy AR, Little JB: Effects of protease inhibitors on radiation transformation *in vitro. Cancer Res* 41:2103–2108, 1981.
7. Kopelovich L: Genetic forms of neoplasia in man: A model for the study of tumor promotion *in vitro. In* Carcinogenesis—A Comprehensive Survey, vol. 7, Cocarcinogenesis and Biological Effects of Tumor Promoters (Hecker E, Fusenig NE, Kunz W, Marks F, Thielmann HW, eds) New York, Raven Press, 1982, 259–271.
8. Little CC: The inheritance of a predisposition to cancer in man. *Eugen Genet Fam* 1:186–190, 1923.
9. Little JB: Radiation carcinogenesis *in vitro:* Implications for mechanisms. *In* Origins of Human Cancer (Hiatt HH, Watson JD, Winston JA, eds) *Cold Spring Harbor Conf Cell Prolif* IV: 923–939, 19⁻7.

10. Little JB: Influence of noncarcinogenic secondary factors on radiation carcinogenesis. *Radiat Res* 87:240–250, 1981.
11. Little JB, McGandy RB, Kennedy AR: Interactions between polonium-210 alpha radiation, benzo(*a*)pyrene and 0.9% NaCl solution instillations in the induction of experimental lung cancer. *Cancer Res* 38:1929–1935, 1978.
12. Nagasawa HN, Little JB: Suppression of cytotoxic effect of mitomycin-C by superoxide dismutase in Fanconi's anemia and dyskeratosis congenita fibroblasts. *Carcinogenesis* 4:795–798, 1983.
13. Ponder BAJ: Genetics and cancer. *Biochim Biophys Acta* 605:369–410, 1980.
14. Rous R, Kidd JG: Conditional neoplasms and subthreshold neoplastic states. A study of the tar tumors of rabbits. *J Exp Med* 73:365–390, 1941.
15. Shami SG, Thibodeau LA, Kennedy AR, Little JB: Proliferative and morphological changes in the pulmonary epithelium of the Syrian golden hamster during carcinogenesis initiated by 210Po-radiation. *Cancer Res* 42:1405–1411, 1982.
16. Terzaghi M, Little JB: X-radiation induced transformation in a C3H mouse embryo-derived cell line. *Cancer Res* 36:1367–1374, 1976.
17. Weichselbaum RR, Nove J, Little JB: X-ray sensitivity of diploid fibroblasts from patients with hereditary and sporadic retinoblastoma. *Proc Natl Acad Sci USA* 75:3962–3964, 1978.

2.

Viral and cellular oncogenes in cancer etiology

WILLIAM S. HAYWARD

The concept of the oncogene arose from studies of the acutely transforming retroviruses. In 1969 Huebner and Todaro [31] proposed that the retroviral genome contains two distinct classes of genes, "virogenes," encoding proteins required for virus replication, and "oncogenes," encoding transforming proteins. During the 1970s the oncogenes of a number of acutely transforming retroviruses were identified and characterized by a combination of genetic and biochemical techniques [4, 26, 72]. As a general rule, the oncogenic properties of each acute retrovirus could be attributed to a single gene (Figure 2–1). However, it became clear from these studies that the oncogenes of independently isolated acute viruses were often unrelated and that, in fact, a large number of different oncogenes exist. Another rather surprising finding, not anticipated by the oncogene hypothesis, was that viral oncogenes (v-*onc* genes) are cellular genes that have been acquired from the host by recombination [4, 64]. Retroviruses can thus be considered as transducing viruses, which occasionally capture host genes and transport them as part of their own genetic material to other cells. More than 20 different v-*onc* genes and their cellular counterparts (c-*onc* genes, or proto-oncogenes) have been identified to date (see Table 2–1).

The recognition that there exists a group of normal cellular genes with oncogenic potential led to the idea that these genes might be activated* by mechanisms other than transduction [5, 29, 34, 42, 49, 50]. The first evidence that a c-*onc* gene could be activated without being removed from its chromosomal location came from studies with avian leukosis viruses (ALVs), which lack oncogenes (see Figure 2–1). These studies demonstrated that integration of the viral DNA adjacent to a c-*onc* gene could cause a transcriptional activation of the gene [29, 42,

*The term *activation* is used throughout to indicate that a cellular gene has been genetically altered in such a way that its oncogenic potential is activated. No specific molecular mechanism is implied.

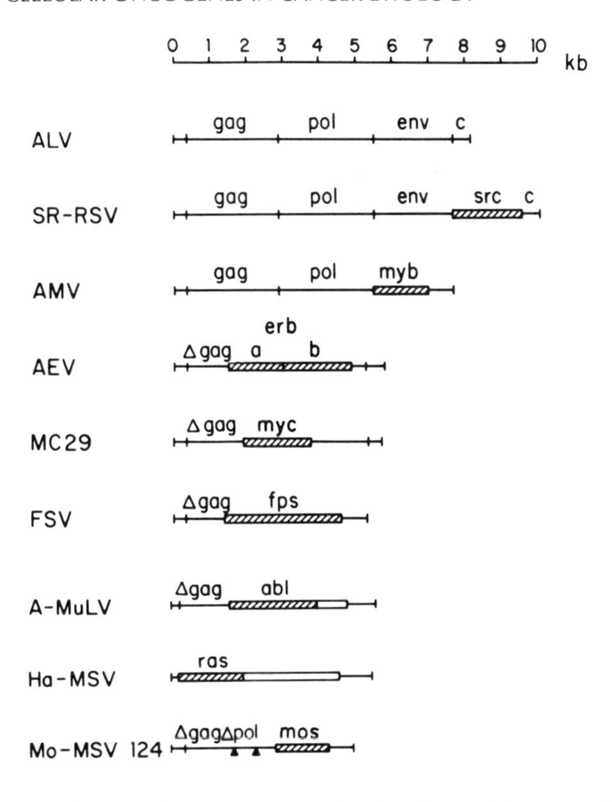

Figure 2–1 Genetic maps of some representative retroviruses. Retroviruses include avian leukosis virus (ALV), Schmidt-Ruppin strain of Rous sarcoma virus (SR-RSV), avian myeloblastosis virus (AMV), avian erythroblastosis virus (AEV), myelocytomatosis virus-29 (MC29), Fujinami sarcoma virus (FSV), Abelson murine leukemia virus (A-MuLV), Harvey murine sarcoma virus (Ha-MSV), and Moloney murine sarcoma virus (Mo-MSV). Viral genes encode the structural proteins (*gag* gene), reverse transcriptase *(pol)*, and the envelope glycoprotein *(env)*. The "c" region is apparently noncoding. Sequences that form the long terminal repeat (LTR) of proviral DNA (see Figure 2–2), are located at the extreme 5′ and 3′ ends of the RNA genomes diagrammed here. Oncogenes of the acutely transforming viruses, shown as cross-hatched boxes, are labeled according to recently adopted nomenclature [9]. These sequences are of host origin, apparently acquired by recombination between a virus such as ALV and a cellular c-*onc* gene (adapted from Reference 28).

49]. Neoplasia apparently results from the abnormal, or inappropriate, expression of the c-*onc* gene—a consequence of placing the gene under the control of viral regulatory sequences, which do not respond to the cellular signals that normally control the gene (Figure 2–2). With molecular probes corresponding to known v-*onc* genes it was possible to identify the c-*onc* gene involved in ALV-induced B-cell lymphomas as c-*myc*, the cellular counterpart of the v-*myc* oncogene of MC29 virus [29].

From the observation that the c-*myc* gene could be activated by pro-

Table 2–1 Oncogenes.*

Oncogene	Virus prototype	Species of origin (v-onc)	Neoplasms	
			v-onc	c-onc
abl	Abelson murine leukemia virus	Mouse	B-cell lymphoma	Chronic myelogenous leukemia (human)
B*lym*	—	—	—	ALV-induced B-cell lymphoma (chicken); Burkitt lymphoma (human)
erb^{A}	Avian erythroblastosis virus	Chicken	?	—
erb^{B}	Avian erythroblastosis virus	Chicken	Erythroleukemia	ALV-induced erythroleukemia (chicken)
fes/fps†	Snyder-Theilen feline sarcoma virus/ Fujinami sarcoma virus	Cat/chicken	Fibrosarcoma	—
fgr	Gardner-Rasheed feline sarcoma virus	Cat	Fibrosarcoma	—
fms	McDonough feline sarcoma virus	Cat	Fibrosarcoma	—
fos	FBJ osteosarcoma virus	Mouse	Osteosarcoma	—
kit	Hardy-Zuckerman-4 feline sarcoma virus	Cat	Fibrosarcoma	—
mos	Moloney murine leukemia virus	Mouse	Fibrosarcoma	A-particle-induced plasmacytoma (mouse)
myb	Avian myeloblastosis virus	Chicken	Myeloblastic leukemia	—

myc	Myelocytomatosis virus-29	Chicken	Myeloid leukemia; carcinoma; B-cell lymphoma	ALV-induced B-cell lymphoma (chicken); MuLV-induced B- and T-cell leukemia (mouse, rat); FeLV-induced T-cell leukemia (cat); promyelocytic leukemia (mouse); Burkitt lymphoma (human); small-cell carcinoma of the lung (human)
myc^{N}	—	—	—	Neuroblastoma (human)
raf/mil†	Murine sarcoma virus-3611/ Mill-Hill-2 virus	Mouse/chicken	Fibrosarcoma/?	—
ras^{H}	Harvey murine sarcoma virus	Rat	Fibrosarcoma	Bladder carcinoma (human)
ras^{K}	Kirsten murine sarcoma virus	Rat	Fibrosarcoma	Colon carcinoma (human); lung carcinoma (human)
ras^{N}	—	—	—	Neuroblastoma (human); T-cell leukemia (human)
rel	Reticuloendotheliosis virus-T	Turkey	Lymphatic leukemia	—
ros	UR2	Chicken	Fibrosarcoma	—
sis	Simian sarcoma virus	Woolly monkey	Fibrosarcoma	—
ski	SK770	Chicken	?	—
src	Rous sarcoma virus	Chicken	Fibrosarcoma	—
yes	Y73	Chicken	Fibrosarcoma	—

*For primary references see text and references 9 and 72.

†In two cases (*fes/fps; raf/mil*), the same v-*onc* gene was identified independently in acutely transforming retroviruses isolated from different hosts. Although subsequent analyses demonstrated the identity of the gene pairs, the original name for each has been retained.

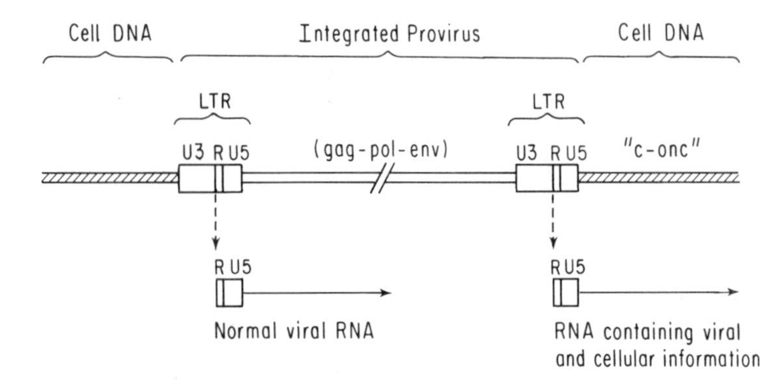

Figure 2–2 Insertional activation of a host c-*onc* gene. The integrated retroviral DNA (provirus) is shown schematically with LTRs enlarged. Synthesis of normal viral RNA initiates within the left LTR. Initiation from the right LTR, however, can cause elevated transcription of adjacent cellular sequences. When, as a rare event, the provirus integrates adjacent to a potentially oncogenic cellular gene (c-*onc*), the resulting transcriptional activation of this gene leads to neoplastic transformation [29, 42, 49]. In some cases, viral control sequences can cause "enhancement" of c-*onc* gene expression, without providing the transcriptional initiation site [50] (from Reference 29).

viral insertion, together with cytogenetic studies demonstrating specific chromosomal rearrangements associated with certain types of neoplasms [7], came the suggestion that translocations or other chromosomal rearrangements might lead to neoplastic transformation by joining the coding sequences of a c-*onc* gene to the regulatory sequences of a second gene that is transcriptionally active in the target cell [29, 34]. Support for this idea comes from experiments in a number of laboratories, which demonstrated that the c-*myc* gene in both mouse plasmacytomas [58, 68] and human Burkitt lymphomas [14, 68] is joined to one of the immunoglobulin (Ig) loci. The precise mechanism by which c-*myc* is activated is unknown, but it seems likely that the normal control of c-*myc* expression is somehow altered as a consequence of the translocation to the Ig locus (see later section).

Recent studies have demonstrated that c-*onc* genes can be activated by other types of genetic changes, including gene amplification and point mutations. The c-*myc* gene has been found to be amplified 20 to 30-fold in lines derived from a variety of neoplasms, including a promyelocytic leukemia [10, 15], colon carcinoma [2], and small-cell carcinoma of the lung [38]. A corresponding increase in c-*myc* expression was found to accompany the gene amplification.

In each of the cases described here the activated c-*onc* gene was identified by molecular hybridization, using radiolabeled probes corresponding to known c-*onc* or v-*onc* genes. Detection is generally based on identification of a c-*onc* gene that has sustained a substantial rearrangement at the DNA level or that demonstrates a significant change in expression [at the mRNA level] compared with an appropriate normal cell control. Although the preparation of c-*onc* probes was at one time quite time-consuming, probes needed for such a screening have now been molecularly cloned in many laboratories and are generally available to all investigators.

Another experimental approach used to identify activated oncogenes is the transfection assay. DNA isolated from tumor cells, or cell lines, is applied to tissue culture cells (usually NIH/3T3 cells in earlier studies), which can take up and express a small fraction of the added DNA. The presence of an activated oncogene is recognized by the appearance of focal areas of morphologically transformed cells, which are not observed when DNA from normal cells is used. This approach has the distinct advantage of not being dependent on the availability of a molecular probe for the specific oncogene involved and thus possibly revealing previously unrecognized oncogenes. Furthermore, because it is a functional assay, it can be used to detect c-*onc* genes that have sustained subtle genetic changes (e.g., point mutations) that cannot be detected by the molecular assays described. A disadvantage of this approach is that many known oncogenes will not score on the transfection assays available at the present time. Refinements of this technique, however, should widen its applicability to a broader range of tumors and oncogenes (see later section).

The transfection assay was employed successfully to identify activated oncogenes in breast, colon, and lung carcinoma cell lines [11, 71]. The oncogenes involved were subsequently shown to be cellular homologs of the known viral oncogenes v-ras^{H} and v-ras^{K} [17, 48]. Interestingly, activation of members of the c-*ras* gene family appears to occur primarily by point mutations within the coding regions of the genes, thus causing a change in the gene product (protein), rather than in the regulation of the gene [55, 66, 67].

In this chapter, I shall not attempt to describe all that is known about oncogenes, or all of the systems that have been studied. Instead, I shall present examples that illustrate the varied mechanisms by which the oncogenic potential of a c-*onc* gene can be activated and point out some of the general features that appear to be common to both viral and nonviral neoplasia.

Mechanisms of c-*onc* activation

Transduction (the acutely transforming retroviruses)

The acutely transforming retroviruses were derived by recombination between a nonacute virus such as ALV (see Figure 2–1) and a host c-

onc gene. In most cases, the oncogene sequences have been inserted into the viral genome at the expense of viral genes required for replication. Thus, with the exception of certain strains of Rous sarcoma virus (RSV), the acutely transforming retroviruses are defective (Figure 2–1), requiring a helper virus for replication. These viruses are probably the most potent carcinogenic agents known. They are probably not a major cause of cancer in nature, however, presumably because of their defectiveness, which limits their ability to spread rapidly from one individual to another. (The nonacutely transforming retroviruses, to be discussed in the next section, play a more important role in naturally occurring cancers.)

Two types of genetic change generally accompany the conversion of a c-*onc* gene to a v-*onc* gene. First, the c-*onc* gene is removed from its normal control sequences and is placed under the control of viral regulatory elements, located in the long terminal repeat (LTR) of the viral genome. The gene thus no longer responds to the normal cellular signals that turn the gene on and off at appropriate times for the normal, carefully regulated growth of the cell. Second, the coding sequences of the c-*onc* gene often are altered. In many cases, the resulting changes in the gene product contribute to the oncogenic properties of the v-*onc* gene [4, 30].

The first acutely transforming retrovirus identified, and the one most extensively studied to date, is RSV, a virus that was isolated from a chicken sarcoma in 1911 [56]. Since that time, nearly a hundred different acutely transforming retroviruses have been isolated from hosts as diverse as chickens, mice, and monkeys, and more than 20 different v-*onc* genes have been identified. (Examples are illustrated in Figure 2–1 and listed in Table 2–1). The oncogene of RSV, v-*src*, encodes a protein of approximately 60,000 daltons, which possesses a protein-kinase activity specific for tyrosine residues [23, 37, 53]. Several other oncogenes, including *abl*, *ros*, and *yes*, encode proteins with similar protein-kinase activities. These proteins are associated primarily with the cell membrane [72]. Other oncogene proteins (e.g., the *myc* and *myb* products) are located in the nucleus and do not appear to be protein kinases [1, 19]. Thus, the products of the different v-*onc* genes may function in quite different ways to induce neoplastic disease. Interestingly, the different acutely transforming retroviruses generally are associated with rather specific pathogenic properties (see Table 2–1). RSV, for example, induces fibrosarcomas almost exclusively, whereas MC29 virus induces primarily myeloid leukemias and carcinomas [25, 26]. In most cases it is not known whether the pathology associated with the virus can be attributed solely to the oncogene involved or whether other properties of the virus play a contributory role.

Certain DNA tumor viruses can be rather loosely classified as "transducing" viruses. Transforming proteins with properties analogous to those of certain retroviral oncogene proteins have been identified for SV40 and polyoma virus (the T antigens) and for adenoviruses (e.g., the E1A protein) [69]. The viral genes encoding these proteins, however, are probably not of recent host origin, as they do not demonstrate the almost complete homology with c-*onc* genes that is found in the acute retroviruses. The Epstein-Barr virus (EBV) has been implicated in human Burkitt lymphoma [22]; however, the role played by this virus is unknown.

Insertional activation (the nonacutely transforming retroviruses)

Nonacute retroviruses, such as ALV, lack oncogenes but nevertheless induce neoplastic disease after long latent periods (usually 4–12 months) [30, 52]. Experiments with ALV-infected birds demonstrated that B-cell lymphomas induced by these viruses contain viral sequences integrated adjacent to a single cellular gene—c-*myc* [29, 30]. As a consequence of this integration, the c-*myc* gene is displaced from its normal control sequences and placed under the control of the viral regulatory sequences (see Figure 2–2). Thus, c-*myc* is expressed constitutively, and at high levels, in the B-cell lymphoma. Integration of retroviral DNA is a normal part of the viral life cycle and appears to occur at more or less random sites in the host chromosome [28, 72]. Thus, integration adjacent to c-*myc* is presumably an extremely rare event (perhaps 10^{-7}). This explains in part the long latency associated with ALV-induced neoplasia. Other factors that might contribute to the long latency are (1) the time required for a single cell and its progeny to pass through enough generations to produce a tumor mass (the tumors are clonal, apparently arising from a single, initially infected cell) and (2) the possible requirement for subsequent activation of a second gene in order to induce a fully malignant state (see later section).

Insertional activation is not restricted to ALV or to the c-*myc* gene. Recent studies have demonstrated activation of c-*myc* by other retroviruses such as reticuloendotheliosis virus [46], murine leukemia virus [13, 63], and feline leukemia virus [44]. Likewise, other c-*onc* genes have been implicated in such neoplasms as ALV-induced erythroleukemias (c-*erb*^{B}) [24], MMTV (murine mammary tumor virus)-induced mammary carcinomas (*int*-1 and *int*-2) [47, 51], ring-necked pheasant virus-induced fibrosarcomas (as yet unnamed) [61], and an A-particle-induced murine plasmacytoma (c-*mos*) [6].

Transformation of tissue culture cells has been demonstrated following transfection with certain c-*onc* genes (e.g., c-*mos* [5]) that have been linked to viral LTR sequences. The artificially constructed DNA molecules used in these transfections are similar to those that occur naturally in tumors induced by nonacutely transforming retroviruses.

Translocations

It has been known for many years that nonrandom chromosomal translocations are associated with certain types of neoplasms [7]. Notable among these are the 9;22 translocation associated with chronic myelogenous leukemia (CML) and the 8;14 (or 2;8 or 8;22) translocations associated with Burkitt lymphoma [7, 57]. Recent studies have demonstrated that the human c-*myc* gene maps to a site on chromosome 8 (q24) that corresponds to the breakpoint in Burkitt lymphoma [43, 68]. The Ig lambda, heavy, and kappa genes, respectively, map at the breakpoints on the other chromosomes involved (2, 14, and 22) [8].

Direct evidence that the c-*myc* gene is closely linked to an Ig locus in Burkitt lymphoma cells derives from Southern [62] restriction analyses, molecular cloning, and nucleotide sequence determinations of the rearranged genes [14, 68]. Two examples of rearranged c-*myc* genes molecularly cloned from Burkitt lymphoma lines are illustrated in Figure 2–3. As can be seen from just these two examples, the breakpoints in both the c-*myc* gene and the Ig locus are quite variable in different tumors, making it difficult to propose a single mechanism for c-*myc* activation. In a minority of cases, the translocation is such that c-*myc* is joined to the recently identified enhancer element of the Ig heavy chain locus (e.g., the Manca cell line, Figure 2–3) [27, 73]. This is perhaps the easiest case to explain, because the Ig enhancer would be expected to induce elevated and constitutive expression of the translocated c-*myc* gene. In most cases, however, the enhancer is not linked to c-*myc* (e.g., the AW-Ramos cell line shown in Figure 2–3) [3, 14, 27, 40, 54, 68, 73]. Models that have been invoked to explain specific cases include removal of c-*myc* control sequences upstream (5′) of the gene [73], removal of the first exon of c-*myc* (which is noncoding, and therefore might play some regulatory role) [40, 58, 59], and mutation of either regulatory [3, 73] or coding [54] sequences. The latter model is consistent with the known genetic instability of the Ig locus in lymphoid cells—a property that is normally involved in the generation of antibody diversity. It seems clear, however, that mutations that result in an altered gene product are not an essential feature of c-*myc* activation. In at least two cases, the nucleotide sequence of the coding

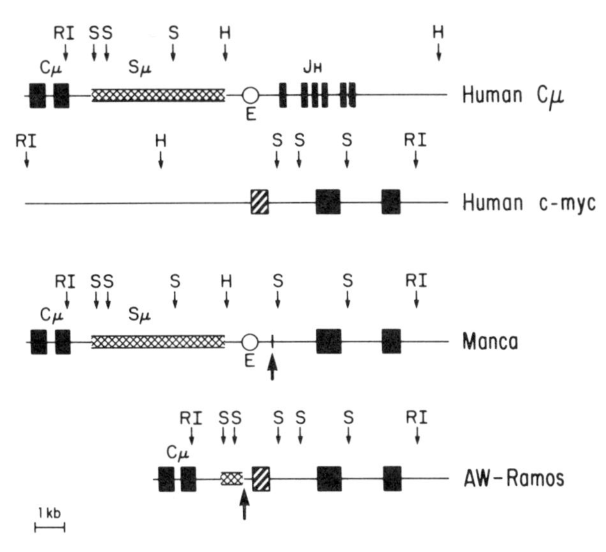

Figure 2–3 Maps of part of the human Ig heavy chain locus on chromosome 14 (Cμ), the normal human c-*myc* gene on chromosome 8, and the translocated c-*myc* genes of the Manca and AW-Ramos B-cell lymphoma lines. The heavy chain constant domain (Cμ), the switch region (Sμ), and the six functional human J segments (J_H) of the Ig locus are indicated. E denotes the position of a recently identified transcriptional enhancer element [27]. The first (noncoding) exon of the human c-*myc* gene is represented by the hatched box; the black boxes indicate coding exons. The large arrows indicate chromosomal junction points in Manca and AW-Ramos. Restriction enzyme cleavage sites are EcoRI (RI), HindIII (H), and SacI (S). Transcription of the Ig locus is from right to left; transcription of c-*myc* is from left to right (from Reference 73).

region of the translocated c-*myc* gene has been shown to be identical to that of the normal *c-myc* gene [3, 73].

Two c-*onc* genes, c-*abl* and c-*sis*, map to chromosomes involved in the specific translocation (9;22) associated with CML [32]. The map position of c-*abl* [32] corresponds to the breakpoint on chromosome 9. This gene has been shown by molecular analyses to be translocated to chromosome 22 in CML cells, and is located close to the breakpoint [16]. The c-*sis* gene, on the other hand, maps to a point distal to the breakpoint on chromosome 22 [32], and thus probably plays no direct role in the induction of CML.

Gene amplification

In several cases, increased copies of oncogenes have been observed in tumor cell lines and in tumors. In cases in which mRNA levels have been analyzed, a corresponding increase in gene expression was found.

Examples include a promyelocytic leukemia cell line, HL60 (c-*myc* is amplified) [10, 14], a colon carcinoma cell line (c-*myc*) [2], small-cell carcinomas of the lung (c-*myc*) [38], and a neuroblastoma cell line (myc^{N}) [35]. Presumably, in these cases the important factor is the increased level of expression resulting from the increased copy number of genes, but the possibility that other genetic alterations have occurred in one or more of the amplified gene copies has not been ruled out.

Point mutations

Oncogenes in lung, bladder, and colon carcinomas were identified by transfection of NIH/3T3 cells [12, 71]. Subsequently it was shown that these were related to the v-ras^{H} and v-ras^{K} genes [17, 48], first identified as the oncogenes of Harvey and Kirsten murine sarcoma viruses. A third member of the *ras* gene family, ras^{N}, was subsequently identified in neuroblastomas [60].

Activation of these genes appears to result from alterations in the oncogene product rather than from changes in regulation. Single point mutations at one of two specific domains within the coding region of the *ras* gene have been found in all cases that have been analyzed in detail [55, 66, 67]. No substantial changes in the level of expression of the *ras* genes have been observed in these tumor cell lines. Thus, the profound changes in the cell attributable to the activated *ras* genes appear to result from single amino acid changes in the *ras* proteins.

Deletions

In certain tumors, neoplasia correlates with losses of genetic material rather than from insertions, rearrangements, or point mutations [8]. Deletions have been well documented for at least two diseases (Wilms' tumor and retinoblastoma) by both cytogenetic and molecular biological techniques. These characteristic deletions can be viewed in several ways. First, it could be envisioned that the coding sequences of a c-*onc* gene might become linked to the control sequences of a second gene that is actively expressed by deletion of the DNA sequences separating the two genes. This type of mutation would be dominant, as are the mutations described in the previous sections, and would result in abnormal expression of the c-*onc* gene. However, recessive mutations might also be expected. For example, deletion of a regulatory gene, encoding a trans-acting regulatory protein that normally represses the expression of the normal c-*onc* gene, would also result in abnormal expression of the c-*onc* gene, but this would be a recessive trait.

Multistage development of neoplasia

Pathologists have long recognized that tumor development progresses through several stages, from relatively benign to highly malignant growth [45]. Several lines of evidence suggest that these stages reflect the participation of more than one oncogene in the induction of neoplasia. Cooper and Neiman [12] have reported the presence of a second oncogene (in addition to c-*myc*) that is activated in ALV-induced B-cell lymphomas. These authors suggest that c-*myc* activation by proviral insertion, as described earlier, is an early step in tumor progression and that activation of the second gene, which they call B*lym*, might be a late event resulting from a second mutational event. Several other cases are known in which more than one activated oncogene has been implicated. For example, both c-*myc* and c-*ras* have been implicated in the promyelocytic leukemia cell line HL60 [10, 15, 41], and c-*myc*, c-*ras*, and B*lym* have all been implicated in Burkitt lymphomas (although it is not clear in the latter case that all three genes would be activated in the same tumor) [14, 18, 41, 43, 68]. Consistent with this, Land et al. [36] have recently provided evidence that two different oncogenes are required to induce a fully malignant state in primary rat embryo cells.

It seems likely that multiple mutations in a single c-*onc* gene or in different c-*onc* genes, or both, will prove to be a requirement for neoplasia in many cases. An early event, for example, activation of c-*myc*, might confer a slight selective advantage on a single cell. Clonal expansion would eventually result in a large population of cells carrying the activated c-*myc* gene, thus providing a larger "target" population. A second mutational event, affecting another c-*onc* gene, might occur in a single cell within this population. Further clonal selection would then generate a subpopulation of cells that carry mutations at both loci. In this way a series of very rare events becomes statistically possible because at each step the affected cell has acquired a selective advantage that results in a clonal expansion and increased target size.

Functions of the oncogene products

The oncogene proteins appear to fall into several different classes, based on their intracellular locations and the presence or absence of tyrosine-specific protein-kinase activity [4, 23, 72]. Membrane proteins include *src* and *ras*. Nuclear proteins include the *myc* gene product, which has also been shown to bind to DNA [1, 19]. Recent reports have described an interaction of the *src* and *ros* proteins with phospholipid pathways [39, 65]. Whether or not these biochemical functions are relevant to the oncogenic properties of these proteins, however, is not known.

Because of the phenotypes associated with neoplastic disease, it has generally been assumed that the normal *c-onc* genes play some role in cell growth control or differentiation, or both [4, 28, 29]. Recent studies have lent credence to this idea, at least for two oncogenes (*sis* and *erb*^{B}) whose proteins have been shown to be closely related to the platelet-derived growth factor (PDGF) and the epidermal growth factor (EGF) receptor, respectively [20, 21, 70]. It is believed that the v-*sis* and v-*erb*^{B} genes were, in fact, derived from the cellular genes encoding PDGF and the EGF receptor. Also of interest is a recent study demonstrating that the c-*myc* gene in normal cells can be induced by addition of PDGF [33].

Conclusions

Rapid progress has been made during the past several years in identifying cellular genes that play a causative role in the induction of neoplastic disease. Much is known about the various types of mutational events that can activate the oncogenic potential of a c-*onc* gene. Although a wide variety of mutagens (e.g., nonacutely transforming retroviruses, chemical mutagens, X-irradiation) are carcinogenic, and although the mutational events range from gross rearrangements to single base changes, all of the known lesions that lead to cancer can be placed into one of two categories: (1) mutations that alter the regulation of expression of a c-*onc* gene and (2) mutations in coding regions that result in an altered c-*onc* gene product.

Relatively little is known about the biochemical functions of the oncogene proteins, but recent studies have revealed important relationships between oncogenes and host genes known to be involved in the control of cell growth. It can be expected that further progress will be made in this area in the near future.

Acknowledgments

The author would like to thank Lauren O'Connor and Ginger Black for help in preparation of this chapter. Studies conducted in the author's laboratory were supported by NIH grant CA 34502 and the Flora E. Griffin Fund.

Literature cited

1. Abrams HD, Rohrschneider LR, Eisenman RN: Nuclear location of the putative transforming protein of avian myelocytomatosis virus. *Cell* 29:427–439, 1982.
2. Alitalo K, Schwab M, Lin CC, Varmus HE, Bishop JM: Homogeneously staining chromosomal regions contain amplified copies of an abundantly expressed cellular oncogene (c-*myc*) in malignant neuroendocrine cells from a human colon carcinoma. *Proc Natl Acad Sci USA* 80:1707–1711, 1983.
3. Battey J, Moulding C, Taub R, Murphy W, Steward T, Potter H, Lenoir G, Leder P: The human c-*myc* oncogene: Structural consequences of translocation into the IgH locus in Burkitt lymphoma. *Cell* 34:779–789, 1983.
4. Bishop JM: Cellular oncogenes and retroviruses. *Annu Rev Biochem* 52:301–354, 1983.
5. Blair DG, Oskarsson M, Wood TG, McClements WL, Fischinger PJ, Vande Woude GF: Activation of the transforming potential of a normal cell sequence: A molecular model for oncogenesis. *Science* 212:941–943, 1981.
6. Canaani E, Drazen O, Klar A, Rechair G, Ram D, Cohen JB, Givol D: Activation of the c-*mos* oncogene in a mouse plasmacytoma by insertion of an endogenous intracisternal A-particle genome. *Proc Natl Acad Sci USA* 80:7118–7122, 1983.
7. Chaganti RSK: The significance of chromosome change to neoplastic development. *In* Chromosome Mutation and Neoplasia (German JL, ed) New York, Alan R Liss, 1983, 359–396.
8. Chaganti RSK, Jhanwar SC: Chromosomal mechanisms of cancer etiology. Chapter 4 in this volume.
9. Coffin JM, Varmus HE, Bishop JM, Essex M, Hardy WD, Martin GS, Rosenberg NE, Scolnick EM, Weinberg RA, Vogt PK: A proposal for naming host cell-derived inserts in retrovirus genomes. *J Virol* 49:953–957, 1981.
10. Collins S, Groudine M: Amplification of endogenous *myc*-related DNA sequences in a human myeloid leukemia cell line. *Nature* 298:679–681, 1982.
11. Cooper GM: Cellular transforming genes. *Science* 217:801–806, 1982.
12. Cooper, GM, Neiman PE: Two distinct candidate transforming genes of lymphoid leukosis virus induced neoplasms. *Nature* 292:857–858, 1981.
13. Corcoran LM, Adams JM, Dunor AR, Cory S: Murine T lymphomas in which the cellular *myc* oncogene has been activated by retroviral insertion. *Cell* 37:113–122, 1984.
14. Dalla-Favera R, Bregni M, Erickson J, Patterson D, Gallo R, Croce CM: Human c-*myc* oncogene is located on the region of chromosome 8 that is translocated in Burkitt lymphoma cells. *Proc Natl Acad Sci USA* 79:7824–7827, 1982.
15. Dalla-Favera R, Wong-Staal F, Gallo RC: *Onc* gene amplification in promyelocytic leukemia cell line HL-60 and primary leukemic cells of the same patient. *Nature* 299:61–63, 1982.
16. deKlein A, Geurts van Kessel A, Grosweld G, Bartram CR, Hagemeijer A, Bootsma D, Spurr NK, Heisterkamp N, Groffen J, Stephenson JR: A cellular oncogene (c-*abl*) is translocated to the Philadelphia chromosome in chronic myelocytic leukemia. *Nature* 300:765–767, 1982.
17. Der CJ, Krontiris TG, Cooper GM: Transforming genes of human bladder and lung carcinoma cell lines are homologous to the *ras* gene of Harvey and Kirsten sarcoma viruses. *Proc Natl Acad Sci USA* 79:3637–3640, 1983.
18. Diamond A, Cooper GM, Ritz J, Lane MA: Identification and molecular cloning of the human B*lym* transforming gene activated in Burkitt's lymphomas. *Nature* 305:112–116, 1983.
19. Donner P, Greiser-Wilke I, Moelling K: Nuclear localization and DNA binding of the transforming gene product of avian myelocytomatosis virus. *Nature* 296:262–266, 1982.

20. Doolittle RF, Hunkapiller MW, Hood LE, Devare SG, Robbins KC, Aaronson SA, Antoniades HN: Simian sarcoma virus oncogene, v-*sis*, is derived from the gene (or genes) encoding a platelet-derived growth factor. *Science* 221:275–277, 1983.
21. Downward J, Yarden Y, Mayes E, Scrace G, Totty N, Stockwell P, Ullrich A, Schlessinger J, Waterfield MD: Close similarity of epidermal growth factor receptor and v-*erb*-B oncogene protein sequences. *Nature* 307:521–527, 1984.
22. Epstein MA, Achong BG, Barr YM: Virus particles in cultured lymphoblasts from Burkitt's lymphoma. *Lancet* i:702–703, 1964.
23. Erikson RL: The transforming protein of avian sarcoma viruses and its homologue in the normal cell. *Curr Top Microbiol Immunol* 91:25–40, 1981.
24. Fung YK, Lewis WG, Crittenden LB, Kung HJ: Activation of the cellular oncogene c-*erb*^{B} by LTR insertion: Molecular basis for induction of erythroblastosis by avian leukosis viruses. *Cell* 33:357–368, 1983.
25. Graf T, Beug H: Avian leukemia viruses. Interaction with their target cell *in vivo* and *in vitro*. *Biochem. Biophys Acta Rev Cancer* 516:269–299, 1978.
26. Hanafusa H: Cell transformation by RNA tumor viruses, *In* Comprehensive Virology (Fraenkel-Conrat H, Wagner RR, ed) vol 10, New York, Plenum Press, 1977, 401–481.
27. Hayday AC, Gillies SD, Saito J, Wood C, Wiman K, Hayward W, Tonegawa S: Activation of translocated human c-*myc* gene transcription in a non-Hodgkin's lymphoma by an immunoglobulin gene-associated transcriptional enhancer element. *Nature* 307:334–340, 1984.
28. Hayward WS, Neel BG: Retroviral gene expression. *Curr Top Microbiol Immunol* 91:217–276, 1981.
29. Hayward WS, Neel BG, Astrin SM: Activation of a cellular *onc* gene by promoter insertion in ALV-induced lymphoid leukosis. *Nature* 290:475–480, 1981.
30. Hayward WS, Neel BG, Astrin SM: Avian leukosis viruses: Activation of cellular "oncogenes." *In* Advances in Viral Oncology, (Klein G, ed) vol 1, New York, Raven Press, 1982.
31. Huebner RJ, Todaro GJ: Oncogenes of RNA tumor viruses as determinants of cancer. *Proc Natl Acad Sci USA* 64:1087–1094, 1969.
32. Jhanwar SC, Neel BG, Hayward WS, Chaganti RSK: Localization of the cellular oncogenes ABL, SIS, and FES on human germ line chromosomes. *Cytogenet Cell Genet* 38:73–75, 1984.
33. Kelly K, Cochran BH, Stiles CD, Leder P: Cell-specific regulation of the c-*myc* gene by lymphocyte mitogens and platelet-derived growth factor. *Cell* 35:603–610, 1983.
34. Klein G: Changes in gene dosage and gene expression: A common denominator in the tumorigenic action of viral oncogenes and non-random chromosomal changes. *Nature* 294:313–318, 1981.
35. Kohl NE, Kanda N, Schreck RR, Bruns G, Latt SA, Gilbert F, Alt FW: Transposition and amplification of oncogene related sequences in human neuroblastomas. *Cell* 35:359–367, 1983.
36. Land H, Parada LF, Weinberg RA: Tumorigenic conversion of primary embryo fibroblasts requires at least two cooperating oncogenes. *Nature* 304:596–602, 1983.
37. Levinson AD, Opperman H, Levintow L, Varmus HE, Bishop JM: Evidence that the transforming gene of avian sarcoma virus encodes a protein kinase associated with a phosphoprotein. *Cell* 15:561–572, 1978.
38. Little CD, Nau MM, Carney DN, Gazdar AF, Minna JD: Amplification and expression of the c-*myc* oncogene in human lung cancer cell lines. *Nature* (London) 306:194–196, 1983.
39. Macara IG, Marinetti GV, Balduzzi PC: Transforming protein of avian sarcoma virus UR2 is associated with phosphatidylinositol kinase activity: Possible role in tumorigenesis. *Proc Natl Acad Sci USA* 81:2728–2732, 1984.

40. Marcu KB, Harris LJ, Stanton LW, Erickson J, Watt, R, Croce CM: Transcriptionally active c-*myc* oncogene is contained within NIARD, a DNA sequence associated with chromosome translocations in B-cell neoplasia. *Proc Natl Acad Sci USA* 80:519–523, 1983.
41. Murray, MJ, Cunningham JM, Parada LF, Dautry F, Lebowitz P, Weinberg RA: The HL-60 transforming sequence: A *ras* oncogene coexisting with altered *myc* genes in hematopoietic tumors. *Cell* 33:749–757, 1983.
42. Neel BG, Hayward WS, Robinson HL, Fang J, Astrin SM: Avian leukosis virus-induced tumors have common proviral integration sites and synthesize discrete new RNAs: Oncogenesis by promoter insertion. *Cell* 23:323–334, 1981.
43. Neel BG, Jhanwar SC, Chaganti RSK, Hayward WS: Two human c-*onc* genes are located on the long arm of chromosome 8. *Proc Natl Acad Sci USA* 79:7842–7846, 1982.
44. Neil JC, Hughes D, McFarlane R, Wilkie NM, Onions DE, Lees G, Jarrett O: Transduction and rearrangement of the *myc* gene by feline leukemia virus in naturally occurring T-cell leukemias. *Nature* 308:814–820, 1984.
45. Neiman PE, Jordan L, Weiss R, Payne LN: Malignant lymphoma of the bursa of Fabricius and analysis of early transformation. *Cold Spring Harbor Conf Cell Prolif* 7:519–528, 1980.
46. Noori-Daloii MR, Swift RA, Kung H-J, Crittenden LB, Witter RL: Specific integration of REV proviruses in avian bursal lymphomas. *Nature* 294:574–575, 1981.
47. Nusse R, Varmus HE: Many tumors induced by the mouse mammary tumor virus contain a provirus integrated in the same region of the host genome. *Cell* 31:99–109, 1982.
48. Parada LF, Tabin CJ, Shih C, Weinberg RA: Human EJ bladder carcinoma oncogene is homologue of Harvey sarcoma virus *ras* gene. *Nature* 297:474–478, 1982.
49. Payne GS, Courtneidge SA, Crittenden LB, Fadly AM, Bishop JM, Varmus HE: Analyses of avian leukosis virus DNA and RNA in bursal tumors: Viral gene expression is not required for maintenance of the tumor state. *Cell* 23:311–322, 1981.
50. Payne GS, Bishop JM, Varmus HE: Multiple arrangements of viral DNA and an activated host oncogene (c-*myc*) in bursal lymphomas. *Nature* 295:209–213, 1982.
51. Peters G, Brooke S, Smith R, Dickson C: Tumorigenesis by mouse mammary tumor virus: Evidence for a common region for provirus integration in mammary tumors. *Cell* 33:369–377, 1983.
52. Purchase HG, Okazaki W, Vogt PK, Hanafusa H, Burmester BR, Crittenden LB: Oncogenicity of avian leukosis viruses of different subgroups and of mutants of sarcoma viruses. *Infect Immun* 15:423–428, 1977.
53. Purchio AF, Erikson E, Brugge JS, Erikson RL: Identification of a polypeptide encoded by the avian sarcoma virus *src* gene. *Proc Natl Acad Sci USA* 75:1567–1571, 1978.
54. Rabbitts TH, Hamlyn PH, Baer R: The sequence of a translocated c-*myc* gene in Burkitt lymphoma indicates multiple amino acid differences from the normal gene product. *Nature* 306:760–765, 1984.
55. Reddy E, Reynolds R, Santos E, Barbacid M: A point mutation is responsible for the acquisition of transforming properties by the T24 human bladder carcinoma oncogene. *Nature* 300:149–152, 1982.
56. Rous P: A sarcoma of the fowl transmissible by an agent separable from the tumor cells. *J Exp Med* 13:397–411, 1911.
57. Rowley JD: Chromosome abnormalities in human leukemia. *Annu Rev Genet* 14:17–39, 1980.
58. Saito H, Hayday A, Wiman K, Hayward WS, Tonegawa S: Activation of c-*myc* gene by translocation: A model for translational control. *Proc Natl Acad Sci USA* 80:7476–7480, 1983.

59. Shen-Ong GLC, Keath EJ, Piccoli SP, Cole MD: Novel *myc* oncogene RNA from abortive immunoglobulin-gene recombination in mouse plasmacytomas. *Cell* 31:443–452, 1982.
60. Shimizu K, Goldfarb M, Suard Y, Perucho M, Li Y, Kamata T, Feramisco J, Stavnezer E, Fogh J, Wigler MH: Three human transforming genes are related to the viral *ras* oncogenes. *Proc Natl Acad Sci USA* 80:2112–2116, 1983.
61. Simon MC, Smith RE, Hayward WS: Mechanisms of oncogenesis by subgroup F avian leukosis viruses. *J Virol* 52:1–8, 1984.
62. Southern EM: Detection of specific sequences among DNA fragments separated by gel electrophoresis. *J Mol Biol* 98:503–517, 1975.
63. Steffen D: Proviruses are adjacent to c-*myc* in some murine leukemia virus-induced lymphomas. *Proc Natl Acad Sci USA* 81:2097–2101, 1984.
64. Stehelin D, Varmus HE, Bishop JM, Vogt PK: DNA related to the transforming gene(s) of avian sarcoma virus is present in normal avian DNA. *Nature* 260:170–173, 1976.
65. Sugimoto Y, Whitman M, Cantleyu LC, Erikson RL: Evidence that the Rous sarcoma virus transforming gene product phosphorylates phosphotidylinositol and diacylglycerol. *Proc Natl Acad Sci USA* 81:2117–2121, 1984.
66. Tabin CJ, Bradly SM, Bargmann CI, Weinberg RA, Papageorge AG, Scolnick EM, Dhar R, Lowy DL, Change EH: Mechanism of activation of a human oncogene. *Nature* 300:143–149, 1982.
67. Taparowsky E, Suard Y, Fasano O, Shimizu K, Goldfarb M, Wigler M: Activation of the T24 bladder transforming gene is linked to a single amino acid change. *Nature* 300:762–765, 1982.
68. Taub R, Kirsch I, Morton C, Lenoir G, Swan D, Tronick S, Aaronson S, Leder P: Translocation of the c-*myc* gene into the immunoglobulin heavy chain locus in human Burkitt lymphoma and murine plasmacytoma cells. *Proc Natl Acad Sci USA* 79:7837–7841, 1982.
69. Tooze J: DNA Tumor Viruses. New York, Cold Spring Harbor Laboratory, 1980.
70. Waterfield MD, Scrace GT, Wittle N, Stroobant P, Johnson A, Wasteson A, Westermark B, Heldin CH, Huang JS, Deuel TF: Platelet-derived growth factor is structurally related to the putative transforming protein $p28^{sis}$ of simian sarcoma virus. *Nature* 304:35–39, 1983.
71. Weinberg RA: Oncogenes of spontaneous and chemically induced tumors. *Adv Cancer Res* 36:149–163, 1982.
72. Weiss R, Teich N, Varmus H, Coffin J: RNA Tumor Viruses. New York, Cold Spring Harbor Laboratory, 1982.
73. Wiman KG, Clarkson B, Hayday AC, Saito H, Tonegawa S, Hayward WS: Activation of a translocated c-*myc* gene: Role of structural alterations in the upstream region. *Proc Natl Acad Sci USA* 81:6798–6802, 1984.

3.

Mutational models for cancer etiology

LOUISE C. STRONG

Development of models for cancer has been approached from numerous experimental as well as epidemiologic perspectives. To date, findings from different approaches have demonstrated certain principles that must be incorporated into any model for human cancer. These include the following:

1. Cancer as a multistage disease.

This conclusion is based on age-specific incidence patterns of human cancer, latent periods from exposure time to cancer development, experimental observation of initiation and promotion phases of carcinogenesis, and, more recently, in vitro demonstration of discrete cellular events involving immortalization and transformation and the timing of events in radiation-induced transformation [39, 45, 58, 62, 68].

2. Mutation as a critical event.

As described in the chapter by Little in this volume, data from experimental carcinogenesis as well as naturally occurring human and animal tumors strongly support the notion that one or more of the stages in cancer development involve mutation in the broad sense, including recombination, duplication, amplification, deletion, inversion, insertion of genetic material, point mutation, or any other genomic alteration that affects the translational or transcriptional properties of the gene. Cooperation between specific genetic changes in the transformation process have been described recently [32, 46].

Epidemiologists in particular have long derived multistage models for human cancer using statistical indicators to estimate the number of stages [53]; however, such approaches do not provide insight as to the nature of the cellular changes. Experimentalists have been instrumental in confirming the notion that mutation might be a critical step in carcinogenesis and that most carcinogens were indeed mutagens. Until very recently, however, the experimental approach also failed to define the specific genetic changes critical to human tumor development. With advances in technology at the cytogenetic, biochemical, and especially the recombinant DNA levels, we now are in a position to begin to describe in genetic terms the previously suggested "stages" in carcinogenesis and to describe some of the specific genetic "mutations" that may be critical to the development of at least some human cancers. Determination of those critical steps may be an important advance toward cancer control.

The specific model for human cancer that I have investigated involves the hypothesis that specific mutations involved in cancer development may be inherited, and hence present in all cells of an individual, or acquired as a somatic mutation in a given cell lineage. Under this model, the autosomal dominant genes that predispose to a given cancer may also be the genes that undergo somatic mutation in nonhereditary cancers of the same site. As an inherited predisposition does not lead to tumor in every cell of the target organ, some subsequent event, perhaps somatic mutation(s), must occur.

Evidence for a two-mutation model at the cellular level

Retinoblastoma: a specific "mutation" in cancer predisposition

The rationale and support for this model is best demonstrated by a childhood cancer, retinoblastoma. Retinoblastoma has been a model among human tumors for family studies because of the relatively successful treatment over many decades that has provided an opportunity to observe the offspring of affected individuals. In addition, the consistent early age at diagnosis of retinoblastoma permits family studies that are relatively free of the age constraints common to studies of other cancers, because an individual can be considered "unaffected" with confidence if he or she has not developed the tumor by the age of 5 years [15].

Reports of familial retinoblastoma date back to the 1800s and consistently reveal a correlation between the occurrence of familial retinoblastoma and of bilateral tumors [89]. Systematic follow-up of offspring of retinoblastoma patients has demonstrated that the risk of retinoblastoma in the offspring of bilateral retinoblastoma patients, whether sporadic or familial, is near 50%, consistent with an autosomal dominant gene with high penetrance. Among affected offspring, the frequency of bilateral tumors is very high, near 90%. However, among offspring of sporadic, unilateral retinoblastoma patients, the risk of retinoblastoma is small (5.5%), suggesting that only a small fraction of unilateral retinoblastoma is of the hereditary type. More recently, segregation analysis based on the affection status of the parent has confirmed that all bilateral retinoblastoma appears to be hereditary in an autosomal dominant manner, with high penetrance of nearly 90% and also nearly 90% probability of bilateral tumors in affected individuals. Unilateral retinoblastoma, however, may be more heterogeneous and include a subgroup of "hereditary" or "familial" cases with reduced gene penetrance and expressivity or a reduced fraction of bilateral tumors in affected individuals. The inconsistent transmission of retinoblastoma in families with unilateral retinoblastoma has been attributed to host resistance genes, germinal mosaicism, or a delayed mutation [6, 8, 33, 49, 85].

In addition to autosomal dominant retinoblastoma, another form of hereditary retinoblastoma is observed in 1 to 2% of patients, those who have a constitutional deletion of chromosome 13 involving the 13q14 region. Although the deletion most often appears de novo, the chromosome deletion also may be transmitted by a parental balanced chromosomal rearrangement. Parental karyotypes, therefore, are essential to provide counseling to families [83]. Occasionally, patients mosaic for a normal and a 13q deletion cell line have been observed, with apparent loss of the deletion cell line over time [54]. The frequency of bilateral tumors in chromosome 13–deletion retinoblastoma patients appears to be somewhat less than in the classic autosomal dominant form, that is, 55 to 60% compared with 80 to 90% [50]. Pedigree analysis in the absence of knowledge of the cytogenetics in familial chromosome 13–deletion retinoblastoma transmitted through a balanced chromosomal rearrangement might suggest a pattern of reduced penetrance and reduced expressivity.

The size of the deletion and the specific breakpoints on chromosome 13 vary in retinoblastoma patients; however, the 13q14.11 region is deleted consistently [87]. In addition, cells from most patients with chromosome 13–deletion retinoblastoma show one half the normal level of esterase D activity. Cells from patients with partial triplication of this region have three times the haploid level of esterase D activity, which is consistent with the mapping of esterase D activity to the 13q14 region. Hence, esterase D may serve as a marker for the region that, when deleted, predisposes to retinoblastoma.

One patient with a normal-appearing constitutional karyotype had only half the normal level of red blood cell esterase D activity [4]. As both parents had normal levels of esterase D activity, and there was no evidence for a familial "null allele," it was concluded that she had a submicroscopic 13q14 deletion involving esterase D and the presumed retinoblastoma region.

Initially, it was suggested that the mechanism by which a constitutional deletion might predispose to a specific tumor might be that of unmasking a mutant gene on the homologous chromosome by loss of the normal gene. The predisposition to tumor, however, is unlikely to be due to a mutation on the normal chromosome 13 as retinoblastoma occurs very consistently in patients with a constitutional 13q14 deletion from many different kindreds or different sibships within the same kindred. It would be most unlikely that all the "normal" chromosomes 13 from unrelated individuals would carry such a mutation [83].

The consistent development of tumor in patients with different chromosome 13 deletions and the absence of tumor in individuals who carry the deleted chromosome 13 balanced by translocation to some other chromosome indicate that it is not the repositioning of genes from chromosome 13 but the actual loss of genetic material that probably is critical to retinoblastoma predisposition. Additional support for the loss of gene function at 13q14 being the critical predisposing event comes from the study of two patients with constitutional chromosome 13;X rearrangements [18, 59]. In both cases the 13q14 region appears by cytogenetic techniques to be intact but translocated to an X chromosome. However, the X chromosome with the chromosome 13 translocation was shown to be the late-replicating X in fibroblasts and the late replication phenomenon included the 13q14 region in both patients. Although it has not been demonstrated that the tumor arose from cells with an inactivated region of chromosome 13, that seems the most likely explanation. These findings would imply that absence of normal gene function in the 13q14 region predisposes to retinoblastoma.

Only 1 to 2% of retinoblastoma patients have a detectable constitutional chromosome 13 deletion or translocation, or a constitutional submicroscopic deletion detectable by hemizygosity for esterase D. Yet, 20 to 40% of patients may have apparent autosomal dominant hereditary retinoblastoma. To determine whether the same gene might be involved in the more common form of hereditary retinoblastoma, Sparkes et al. [76] and Connolly et al. [14] identified families in which retinoblastoma had occurred over two to four generations and in which there were electrophoretic variants of esterase D. Using genetic linkage, the investigators demonstrated that the retinoblastoma phenotype segregated consistently with a single esterase D type in any given family, implying that the retinoblastoma gene was closely linked to the esterase D gene. No instances of recombination have been observed. The linkage findings suggest that there is a "mutation" or a critical region common to hereditary retinoblastoma in patients with demonstrable chromosomal deletions and normal karyotypes. In addition, in rare patients from families in which esterase D is polymorphic and the allele segregating with retinoblastoma can be identified, it may provide a clinical indicator for genetic counseling and prenatal diagnosis.

An additional "mutational" mechanism by which familial retinoblastoma may arise is described by Ledbetter et al. [47] in a report of familial retinoblastoma with an extreme degree of constitutional chromosomal instability. In that family, two female siblings with failure to thrive developed unilateral retinoblastoma within the first year of life, and one developed Wilms' tumor as well. Peripheral blood and fibroblast karyotypes revealed a basic 46,XX chromosome pattern but, consistently, an extraordinary number of chromosomal rearrangements. The spontaneous chromosomal breakage rate was 0.91 to 0.92 breaks per cell, compared with 0.0 to 0.12 breaks per cell observed in a blind study including parents and controls. Further analysis of the peripheral blood and fibroblasts from one patient revealed an additional cell line with a 13q14 rearrangement—t(9;13) (q13;q14)—which also had a high rate of spontaneous chromosomal breaks and rearrangements. Peripheral blood karyotypes from both parents were normal.

This particular chromosomal breakage syndrome is different from others previously recognized, not only in the high frequency of chromosome breaks and rearrangements per cell but also in the predisposition to early-onset cancer. The consistent development of retinoblastoma may indicate the chromosomal breakage is not random or that the specific site for retinoblastoma represents a "hot spot" that is especially likely to undergo breakage. This family might possibly have a new autosomal recessive syndrome of chromosomal instability that could further account for some of the genetic heterogeneity observed in familial, and especially unilateral, retinoblastoma.

In summary, we presume that hereditary retinoblastoma involves transmission of a gross chromosomal deletion or some other type of mutation, submicroscopic deletion, or point mutation in the 13q14 area, or transmission of such a high rate of chromosomal breakage and rearrangement that such a deletion or mutation in a somatic cell line would be almost inevitable.

Retinoblastoma: a subsequent somatic "mutation" at the homologous locus

In patients who inherit a genetic predisposition to retinoblastoma, whether of the chromosome-deletion type, autosomal-dominant type, or chromosomal-instability type, not all retinal cells develop into tumor. Although bilateral retinoblastoma is most common, some individuals develop tumor only in one eye, and rare individuals may not develop retinoblastoma at all. Some individuals develop an apparent benign variant, retinoma, that may represent a cell undergoing the critical genetic changes at a time near terminal differentiation [23]. Presumably, some additional critical step or steps must occur in the predisposed cell to lead to tumor. Knudson suggested that all retinoblastoma might be attributable at the cell level to two mutations and that the second mutation, a somatic mutation, might occur at the locus homologous to that of the germinal mutation [40]. Although this has long been an attractive theory, supportive findings have emerged only recently.

The initial approach to the study of tumor-specific genetic changes involved examination of chromosomes of tumors themselves. It was first noted that the group D chromosomes in retinoblastomas appeared unusually "fragile" [30]. Subsequent studies have revealed that chromosome 13 was often involved in breakage and rearrangment in the tumors. Nevertheless, deletion or rearrangement of the 13q14 region has been observed in fewer than half of the tumors studied and does not appear to be a consistent finding in retinoblastomas. Other nonrandom chromosomal abnormalities observed in retinoblastomas include the presence of an i(6p), trisomy lq, and double minutes, cytogenetic findings commonly found in other malignant tumors [3, 24, 55].

Godbout et al [25] investigated the expression of esterase D in retinoblastomas, that is, in the tumor tissue itself. They observed several patients with constitutional heterozygosity for esterase D but with only one type of esterase D expressed in tumor cells despite the presence of two or more normal-appearing chromosomes 13. These findings were consistent with somatic alteration in the 13q14 region being a frequent event in tumor development, even in the presence of two normal-appearing chromosomes.

Gene expression in tumors is often altered, however, and hence, may not be a valid indicator of critical events in tumorigenesis. To define the specific gene alterations in tumor development, Cavenee et al. [9, 10] have utilized a recombinant DNA library enriched for human chromosome 13 to identify polymorphic DNA regions on human chromosome 13. They have demonstrated the regional localization of a series of restriction-enzyme fragment-length polymorphisms (RFLPs) on chromosome 13 that exhibit mendelian, codominant segregation within families, and have used these markers to define the chromosome 13 constitutional and tumor genotypes in retinoblastoma patients. These studies have demonstrated that at least half of retinoblastoma tumors, including bilateral and unilateral sporadic tumors, have selectively lost heterozygosity for the RFLPs on one chromosome 13, presumably leaving the tumor either hemizygous or homozygous for the critical mutant allele at 13q14; loss of heterozygosity was not observed for RFLPs on other chromosomes. These findings suggest that a tumor might achieve hemizygosity or homozygosity for most of one chromosome 13 through any of several mechanisms, including nondisjunction with loss of one chromosome 13 and reduplication of the other chromosome 13, mitotic recombination, or chromosome deletion. Presumably, more local events such as gene conversion, deletion, or point mutation also occur [9, 16, 17].

The loss of heterozygosity for proximal and distal chromosome 13q RFLPs in many tumors suggests that large regions of the chromosome 13 often are lost. In patients with a constitutional chromosome 13q deletion, loss of a major portion of the normal homologue might render the cell "homozygous" for a deletion or without genetic material for a sufficient region of the genome that cell viability could be affected. The number of tumors developing in these patients then may depend on the frequency of the localized events only, which could account for the lower frequency of bilateral tumors in patients with constitutional chromosome 13 deletions [83].

It is presumed that the chromosome 13 retained in the tumor carries the retinoblastoma mutant. To date, this has been confirmed only in one case, one in which the "mutant" chromosome apparently had a submicroscopic deletion involving esterase D, as evidenced by half the normal constitutional level of esterase D activity. In the tumor from that patient, one chromosome 13 was lost, as revealed by cytogenetic and molecular markers, and the esterase D activity was zero, implying loss of the normal chromosome 13 [4, 9]. Confirmation of selective retention of the retinoblastoma mutant may be addressed by parallel family studies in which the retinoblastoma mutant can be identified by linkage to RFLPs segregating with the retinoblastoma phenotype and by tumor-specific studies of the retained chromosome 13 RFLPs.

If it is the loss of normal gene function at homologous loci that leads to tumor, then the question arises as to the normal function of such genes. In 1973 Comings suggested that tumors arise as a result of the activation of certain structural genes called "transforming genes." Presumably, these "transforming genes" had some critical function in normal growth and development, after which time they were normally suppressed. Although these transforming genes could be activated in a number of ways, Comings suggested that autosomal dominant hereditary predisposition to tumor might involve inheritance of a mutation for a regulatory gene. Subsequent somatic mutation for the homologous regulatory gene might lead to loss of suppression of the structural transforming genes and hence tumor development [13]. To date, the normal gene product from the putative retinoblastoma mutation is unknown. Given the cascade of events involving activation of a series of "transforming sequences" by mutation, rearrangement, or amplification evolving in the study of human tumors, however, it seems a likely mechanism [32]. The only evidence at this time of enhanced expression or amplification of a suspected human oncogene in retinoblastoma tumors is that of N-*myc* [48]. The increased N-*myc* expression observed in hereditary and nonhereditary tumors suggests a possible role in tumor development or progression.

Relevance of the retinoblastoma "mutation" to other cancers

It might be easy to dismiss the events occurring in retinoblastoma as unique to an embryonal tumor that occurs at a very early age and hence events that might involve some very specific step of differentiation in retinal development and might not be relevant to carcinogenesis in general. However, there is evidence that the events that predispose to retinoblastoma are common to the development of many other human cancers. Patients with the heritable form of retinoblastoma have an exceptionally high frequency of new malignant tumors. Although the majority of these additional tumors occur in areas exposed to radiation therapy during tumor treatment, a highly significant excess of new tumors in the absence of radiation therapy has been observed as well. The most common additional tumor is osteosarcoma, which arises in irradiated and nonirradiated areas; however, many other tumor types have been observed also [1]. The short time to development of solid tumors in those patients treated with radiation therapy may indicate an effect of radiation on a late stage of carcinogenesis in an already uniquely predisposed cell [78, 79].

The lifetime burden of a retinoblastoma gene is not known. Survivors of bilateral retinoblastoma treated with radiation and followed for up to 30 years may have a 30% probability of dying of another cancer [1]. Given the early age at diagnosis of retinoblastoma, these observations extend only to patients reaching 30 to 35 years of age. The Manchester Tumor Registry reported 27 individuals with familial or hereditary retinoblastoma who survived to adulthood and described their subsequent experience to include nine cancer deaths between the ages of 31 and 62 years from common adult cancers including lung cancer, breast cancer, and bladder cancer [74]. Although the entire cohort is not defined by years at risk, occupational or smoking histories, and so on, the findings suggest an increased risk of common adult onset tumors. It is not clear whether bilateral retinoblastoma patients are unusually susceptible to carcinogens other than radiation or simply prone to develop multiple primary tumors.

In some families, relatives of retinoblastoma patients seem to have a high frequency of other cancers [11, 27, 75, 77]. At least two systematic studies have suggested that close relatives of retinoblastoma patients might be at increased risk of other cancers [5, 21]. A recent systematic study of families of 80 retinoblastoma patients revealed a significant excess of cancer deaths occurring at young ages (before 55 years of age) only in relatives of bilateral or familial retinoblastoma patients [82]. The cancer excess was most striking in those kindreds with multiple siblings with retinoblastoma whose parents and antecedent relatives did not have retinoblastoma. These findings implied that a retinoblastoma gene without ocular expression or some precursor such as a delayed mutation or a submicroscopic balanced chromosomal rearrangement might be associated with the transmission of retinoblastoma and might predispose to other tumors. The excess of other tumors in retinoblastoma survivors and in their close relatives supports the notion that the mechanism that predisposes to retinoblastoma may be relevant to other cancers as well and could affect susceptibility to environmental carcinogens.

Preliminary data from the study of osteosarcomas arising in retinoblastoma survivors suggest that selective loss of heterozygosity for chromosome 13 markers, similar to that observed in retinoblastoma tumors, is common to the osteosarcomas as well [Cavenee et al., personal communication]. Hence, the 13q14 region may harbor genes whose function (or loss of function) affects pathways common to the development of other tumors.

Wilms' tumor

For no other tumors are the inherited or acquired genetic alterations so well defined as for retinoblastoma. Limited observations, however, suggest that a similar set of events may be involved in the development of at least one other embryonal tumor, Wilms' tumor of the kidney. Familial cases of Wilms' tumor are rare, representing no more than 1% of cases, perhaps in part because of the former near-lethality of the tumor and hence a limited number of survivors to transmit a Wilms' tumor predisposition. However, families with Wilms' tumor occurring in direct descent over three generations have been observed, in addition to more common families in which Wilms' tumor has occurred in siblings or cousins. As for retinoblastoma, familial cases tend to have an earlier average age at diagnosis and to be multifocal or bilateral more frequently than unselected cases. Knudson and Strong suggested that for Wilms' tumor, also, there may be a heritable and nonheritable subgroup, the heritable subgroup accounting for some 20% of cases, including all bilateral cases [41, 81]. The penetrance or probability that a "gene carrier" would develop overt Wilms' tumor was estimated to be 63%, so that some 37% of gene carriers might develop no tumors,

48% unilateral tumors, and 15% bilateral tumors. The probability of Wilms' tumor occurring among the offspring of bilateral Wilms' tumor survivors would be roughly 37% and about 5% among survivors of unilateral Wilms' tumor. Data to support these estimates are limited. Although no childhood cancers were observed in 59 liveborn offspring of 35 systematically ascertained Wilms' tumor patients in one study [28], two other childhood cancers were observed in the offspring of 25 Wilms' tumor patients ascertained through the Connecticut Tumor Registry [38]. In the University of Texas M. D. Anderson Hospital experience, bilateral Wilms' tumor developed in 1 of 11 offspring of 5 sporadic, unilateral Wilms' tumor patients who have had children [Strong, unpublished data].

A review of a large series of Wilms' tumor patients by Miller et al. in 1964 revealed an excess of patients with aniridia, genitourinary anomalies, and hemihypertrophy [52]. Subsequent review of the ages at diagnosis and tumor laterality of those subgroups of Wilms' tumor patients revealed that the pattern of Wilms' tumor in patients with aniridia was similar to that of familial Wilms' tumor (early age at diagnosis, high frequency of bilateral tumors), whereas the pattern of Wilms' tumor in patients with genitourinary anomalies and hemihypertrophy was similar to that of the unselected series [41].

Cytogenetic and biochemical marker studies of patients with Wilms' tumor and aniridia have confirmed that the association of Wilms' tumor and aniridia is attributable to a germinal mutation, a constitutional 11p deletion [67]. The critical region has been identified further as the mid-11p13 region [56]. Most patients with Wilms' tumor/aniridia and an 11p13 constitutional chromosome deletion have reduced erythrocyte catalase activity, consistent with hemizygous activity. Other gene dosage and somatic cell hybrid studies support the assignment of catalase activity to the 11p13 region, providing a biochemical marker for the region of the genome that, when deleted, predisposes to Wilms' tumor and aniridia [34]. The gene for catalase appears to be distinct from the Wilms' tumor/aniridia region, localized slightly distal to it [56, 63].

Nearly half of the reported patients with 11p13 deletion and aniridia have developed Wilms' tumor, with 40% of the tumors being bilateral [56]. As aniridia occurs independent of Wilms' tumor and the 11p13 deletion syndrome, chromosomal and biochemical analyses may be useful in screening patients with aniridia to identify those at risk for Wilms' tumor for whom excretory urography, ultrasound, or other surveillance may be indicated. However, rare patients with aniridia and normal chromosomes and normal catalase levels have developed Wilms' tumor [66].

The 11p13 deletion may arise de novo or may be transmitted by a balanced chromosomal translocation [91]. Parents of patients with Wilms' tumor/aniridia and an 11p13 deletion also should be screened for a chromosomal rearrangement. A familial insertional translocation may give rise to a pattern of cancer in cousins and siblings in the absence of a parent affected with Wilms' tumor/aniridia. Genetic counseling and prenatal diagnosis may be warranted.

Most patients with Wilms' tumor, including those with bilateral or familial Wilms' tumor, genitourinary anomalies, or hemihypertrophy, have normal-appearing chromosomes and normal levels of catalase [35]. It has not been possible to test the linkage between a Wilms' tumor-predisposing gene associated with a normal constitutional karyotype and the 11p13 region because human catalase is not sufficiently polymorphic.

As for retinoblastoma, inheritance of a Wilms' tumor predisposing gene or 11p13 deletion alone is not sufficient for tumor development. Cytogenetic changes observed in Wilms' tumors include loss, rearrangement, or deletion of chromosome 11p in some 25% of tumors and in probable precursor nodules [36, 51]. Other nonrandom chromosomal rearrangements, duplications, or deletions are observed frequently [22].

Recently, several investigators have used DNA probes for chromosome 11p RFLPs to investigate the constitutional and tumor DNA from Wilms' tumor patients. In several cases, loss of constitutional heterozygosity specifically for 11p RFLPs has been observed in the Wilms' tumors [20, 43, 60, 65]. The DNA probes used in these studies have been those mapped to the 11p15 subregion, an area quite distal to the 11p13 region presumably significant in Wilms' tumor development. If it is confirmed that there is a "Wilms' tumor locus" in the 11p13 region that may be a gross deletion, a submicroscopic deletion, or a point mutation, and that tumor development involves somatic cell loss of the normal homologue for this region, then the findings of loss of heterozygosity for the 11p15 region suggest that, as for retinoblastomas, large regions of a chromosome arm frequently are lost by mitotic recombination or nondisjunction in tumor development. The findings confirm the notion that the same genes or regions of the genome that, when inherited, predispose to a given tumor also are altered in subsequent somatic genetic changes in tumor development.

Renal cell carcinoma and other cancers

The only other human tumor for which evidence suggests a common chromosomal region may be involved both in a hereditary predisposition to tumor and in tumor-specific acquired "mutation" is renal cell carcinoma, with chromosome 3p [12, 61, 86]. Data suggesting that renal cell carcinoma may be attributable to hereditary factors in general are limited to a small number of familial case reports. The most impressive findings are those of Cohen et al. in which the development of renal cell carcinoma was observed consistently in patients with a constitutional 3;8 balanced translocation transmitted over three generations [12]. The renal cell carcinomas were diagnosed at a median age of 46 years in symptomatic patients and were frequently bilateral or multifocal. Screening of asymptomatic family members revealed renal cell carcinoma in three individuals with the balanced chromosomal rearrangement and a renal cyst in an additional chromosomal rearrangement carrier. The cumulative probability of renal cell carcinoma in a balanced translocation carrier was estimated to be 87% by the age of 59 years.

The chromosomal rearrangement involved the regions 3p14 and 8q24 [86]. The 8q24 region is of special interest as the region in which the c-*myc* oncogene, frequently activated by tumor-specific chromosomal rearrangements in B-cell neoplasia, is located. Preliminary data from the study of constitutional cells from individuals with this 3;8 translocation have not revealed enhanced expression of the c-*myc* gene; however, tumor cells have not been available for examination [F. P. Li, personal communication, 1984].

Most familial renal cell carcinoma is not associated with a constitutional chromosomal rearrangement, although a high frequency of constitutional chromosomal abnormalities has been observed in renal cell carcinoma patients [2, 37, 70].

Limited study of renal cell carcinomas suggests a nonrandom frequency of tumor-specific aberrations of chromosome 3p, including an apparent balanced rearrangement with the breakpoints at 3p13 or 3p14, in the same region affected in the constitutional familial rearrangement, in one tumor from a patient with familial renal cell carcinoma and a normal constitutional karyotype [61, 86]. Polymorphic markers are needed to assess further the role of genes on 3p in renal cell carcinoma development.

Although no other specific hereditary "mutation" can be clearly mapped to a specific chromosome region, the rapid development of RFLPs may soon provide sufficient markers for the human genome that genes critical to the development of other cancers may be identified by cosegregation with markers in families or selective segregation or alteration in tumors, or both [7, 57, 73].

Other data supporting a multistage mutation model are derived from the rare, autosomal recessively inherited syndromes predisposing to human cancer reviewed by German in this volume. We presume that the mechanism by which these conditions predispose to cancer is the greatly increased rate of spontaneous or induced "mutation," including homologous and nonhomologous somatic recombination, translocation, deletion, point mutation, and so on, such that the probability of mutations specific to cancer is very high (reviewed in Strong [80]).

Clinical evidence for a two-mutation model and interaction of the genetic predisposition with environmental carcinogens

Retinoblastoma

In the preceding model we described mutations that are hereditary in an autosomal dominant manner and that predispose to specific cancers. We described the pathway to cancer from the predisposed cells as involving loss of gene function at the homologous locus by somatic recombination, chromosomal deletion, nondisjunction, localized mutation, or gene conversion. The frequency of the latter events may determine the number and distribution of tumors in the individual, although factors such as differentiation, cell death, and growth stimulation may have a role [53]. According to the model, environmental agents that affect the probability of those mutational events should have a dramatic effect on the predisposed cell. Clinical observations of the effects of environmental agents on genetically predisposed individuals support this notion.

The high frequency of spontaneous and radiation-related osteosarcomas and other second malignant neoplasms in survivors of hereditary retinoblastoma has been cited previously. Although the spontaneous or background risk of a new malignant tumor in a survivor of hereditary retinoblastoma is increased significantly over that of the age-matched population, radiation treatment is associated with a significantly greater risk and shorter time to tumor [1, 69, 78, and Abramson, personal communication]. The relative tissue specificity revealed by the excess of sarcomas, both in spontaneous and radiation-exposed sites, and the relatively short time to a radiation-related solid tumor in hereditary retinoblastoma survivors compared with other radiation-treated childhood-cancer patients are consistent with a radiation effect on a late stage in a predisposed cell [78, 79]. Demonstration of an in vitro model for this "susceptibility" to radiation-induced cancer has been inconclusive, with some reports indicating increased sensitivity of fibroblasts from hereditary retinoblastoma patients to radiation-related cell killing and other reports failing to confirm these findings [19, 29, 42, 88]. The clinical observations are consistent with radiation induction of deletions or somatic recombination converting the predisposed cell to a tumor cell and rendering the tumor cell homozygous or hemizygous for a specific mutant.

Nevoid basal cell carcinoma syndrome

The mutation predisposing to retinoblastoma is not unique among autosomal-dominant, cancer-predisposing genes in affecting the risk and specificity of radiation-related tumors. A visually dramatic example is that of the nevoid basal cell carcinoma syndrome (NBCCS), an autosomal dominant disorder characterized by the early onset of multiple basal cell carcinomas (BCC), primarily on sun-exposed areas, beginning in adolescence or early adulthood. This syndrome is also characterized by a number of stigmatizing anomalies including characteristic facies with a large head and widely spaced eyes, multiple bony abnormalities, ectopic calcifications, dentigerous keratocysts of the jaw, and pathognomonic palmar pits that may represent dysplastic basal cells [31]. In addition to the BCCs of the skin, these patients have an increased risk of ovarian fibroma, childhood medulloblastoma, and perhaps other cancers [for reviews, see 29, 64, 78–80].

Clinical observations suggest that the development of BCC in NBCCS patients is affected by the same risk factors as BCC development in the general population. As in the general population, BCCs are rarer in black NBCCS patients than in white NBCCS patients. In white NBCCS patients, hundreds of BCCs may be observed. As in the general population, the BCCs are most frequent in sun-exposed areas, especially the head and neck, although isolated lesions have been observed on almost all parts of the body [reviewed in 80].

Review of patients with NBCCS and childhood medulloblastoma, treated with craniospinal radiation and followed for 1 to 15 years, has revealed that within the first 6 months to 5 years postradiation, "nevi" appear within and around the irradiated area with a remarkably high frequency. Biopsy of these "nevi" reveals typical BCCs, although in young children these lesions may not follow an aggressive or invasive course. Virtual eruptions of "nevi" (BCC) have been observed in other patients with NBCCS following radiation therapy for thymic enlargement, rhabdomyosarcoma, eczema, and other conditions [26, 64, 71, 72], so the response is not related to any particular allele or condition associated with medulloblastoma. The sites of the basal cell carcinomas in the radiation-treated patients clearly are unusual and unlike those observed in NBCCS patients in the absence of radiation therapy, including close relatives of the radiation-exposed patients, so the distinct distribution and early onset is not attributable to a unique or more severe mutant.

Follow-up of the only two reported NBCCS females with medulloblastoma successfully treated by craniospinal radiation reveals that both had developed ovarian fibromas or fibrosarcomas during childhood. Although ovarian fibromas are frequently associated with NBCCS, the typical age at onset is postpuberty [78, 79].

The specific tumor types and the short time to radiation-related tumors in NBCCS patients suggest that the first stage of multistage carcinogenesis involves a tissue-specific genetic predisposition and that the later stage is affected by sun exposure or radiation, affecting the time to tumor, number of tumors, and distribution of tumors, but not the tissue specificity. It is of interest to contrast these findings of tissue-specific genetic susceptibility to cancer affected by different environmental agents with those of xeroderma pigmentosum (XP), an autosomal recessive disorder involving loss of ability to repair sun-induced DNA damage. In XP there is a dramatically increased cancer risk of all sun-exposed cell types, including BCC, squamous cell carcinoma of the skin, melanoma of the skin, and squamous cell carcinoma of the tongue, but no apparent increased risk of cancer of any type including skin by exposure to ionizing radiation [44]. Presumably, XP cells have such a high rate of "mutation" (in the broad sense) on exposure to sun that tissue-specific "cancer" mutations are almost inevitable for any exposed cell type.

Second tumors in childhood-cancer survivors

Limited data from case reports, family reports, and surveys of second tumors among childhood cancer patients suggest that other genetic and host factors affect tissue-specific susceptibility to tumors. Patients with neurofibromatosis, treated by radiation for one tumor, may develop subsequent "radiation-related" neurofibrosarcomas, and patients with the "Li-Fraumeni" syndrome of sarcomas, breast cancer, brain tumors, and other cancers also may have a high frequency of radiation-related sarcomas [81, 84, 90]. These findings, although limited, are consistent with an underlying tissue-specific genetic predisposition to tumor and with the subsequent stage or stages enhanced by radiation exposure.

Conclusions

The relationship between those "mutations" or genes that segregate in families and predispose to human cancer and the "mutations" or cellular genetic changes critical to cancer development has long been debated [29]. In this review I have focused on those autosomal dominant genes that predispose to specific human tumors and that now seem to mark regions of the genome involved in hereditary or nonhereditary tumor development. Sites of the genes for retinoblastoma, Wilms' tumor, and perhaps renal cell carcinoma initially were identified by the occurrence of rare patients with constitutional chromosomal anomalies associated with a specific tumor predisposition. At least for retinoblastoma, it now seems clear, however, that the same region of the genome is involved in hereditary or acquired mutations predisposing to retinoblastoma in the absence of a gross deletion, and that evolution from a predisposed cell to a tumor cell involves alteration of the homologous locus. The evidence reviewed confirms that autosomal dominant mutations involve genes directly on the multistage pathway of cell transformation and neoplasia, that those genes determine the specific tissues at risk, and that tumor development in predisposed tissues may be greatly enhanced by certain environmental "mutagen/carcinogens" (radiation exposure). The findings suggest that agents that cause mitotic recombination, chromosome deletion, nondisjunction, or other mutational events may be very effective carcinogens in the genetically predisposed cells, leading to a high frequency of specific tumors within a short time to tumor.

Evaluation of other autosomal dominant, cancer-predisposing syndromes in light of this model awaits further mapping of cancer predisposing genes. The observation that many retinoblastomas (the tumors themselves) and Wilms' tumors showed loss of heterozygosity for markers covering a large region of a chromosome arm may indicate that an effective strategy for mapping other "cancer genes" would be to examine constitutional and tumor samples from the same individuals with a battery of RFLPs marking each chromosome arm. Identification of a chromosome arm consistently showing tumor-specific loss of heterozygosity might provide a chromosome arm on which to focus a linkage study of mutants segregating in families.

Mapping of other genes would have direct clinical impact with respect to genetic counseling and identification of high-risk individuals for medical surveillance. Already it is feasible to use chromosomal, biochemical, and DNA markers for prenatal diagnosis or detection of high-risk family members in rare, informative familial retinoblastoma or Wilms' tumor/aniridia. Development of additional RFLP markers should permit linkage mapping of many other cancer predisposing genes. Once such genes are mapped, there is the potential for isolating the gene and characterizing the gene structure, function, and regulation. Although the majority of human cancers may be attributable to environmental agents, localization of the rather rare, autosomal dominant genes that predispose to cancer seems likely to identify regions of the genome directly involved in cancer in general. Characterization of these genes affecting primary events in carcinogenesis may yield new strategies for cancer detection and prevention.

Acknowledgments

This investigation was supported by PHS grants CA 27925 and CA 32064, awarded by the National Cancer Institute, DHHS, and by the Retina Research Foundation.

Literature cited

1. Abramson DH, Ellsworth RM, Rosenblatt M, Tretter P, Jereb B, Kitchin FD: Retreatment of retinoblastoma with external beam irradiation. *Arch Ophthalmol* 100:1257–1260, 1982.
2. Asal NR, Muneer RS, Geyer JR, Thompson LM, Riser D, Rennert OM: Cytogenetic studies on renal-cell carcinoma. *Am J Hum Genet* 35:59A, 1983.
3. Balaban G, Gilbert F, Nichols W, Meadows AT, Shields J: Abnormalities of chromosome #13 in retinoblastomas from individuals with normal constitutional karyotypes. *Cancer Genet Cytogenet* 6:213–221, 1982.
4. Benedict, WF, Murphree AL, Banerjee A, Spina CA, Sparkes MC, Sparkes RS: Patient with 13 chromosome deletion: Evidence that the retinoblastoma gene is a recessive cancer gene. *Science* 219:973–975, 1983.
5. Bonaiti-Pellie C, Briard-Guillemot ML: Excess of cancer deaths in grandparents of patients with retinoblastoma. *J Med Genet* 17:95–101, 1980.
6. Bonaiti-Pellie C, Briard-Guillemot ML: Segregation analysis in hereditary retinoblastoma. *Hum Genet* 57:411–419, 1981.
7. Botstein D, White RL, Skolnick M, Davis RW: Construction of a genetic linkage map in man using restriction fragment length polymorphisms. *Am J Hum Genet* 32:314–331, 1980.
8. Carlson EA, Desnick RJ: Mutational mosaicism and genetic counseling in retinoblastoma. *Am J Med Genet* 4:365–381, 1979.
9. Cavenee WK, Dryja TP, Phillips RA, Benedict WF, Godbout R, Gallie BL, Murphree AL, Strong LC, White RL: Expression of recessive alleles by chromosomal mechanisms in retinoblastoma. *Nature* 305:779–784, 1983.
10. Cavenee W, Leach R, Mohandas T, Pearson P, White R: Isolation and regional localization of DNA segments revealing polymorphic loci from human chromosome 13. *Am J Hum Genet* 36:10–24, 1984.
11. Chan H, Pratt CB: A new familial cancer syndrome? A spectrum of malignant and benign tumors including retinoblastoma, carcinoma of the bladder and other genitourinary tumors, thryoid adenoma, and a probable case of multifocal osteosarcoma. *J Natl Cancer Inst* 58:205–207, 1977.
12. Cohen AJ, Li FP, Berg S, Marchetto DJ, Tsai S, Jacobs SC, Brown RS: Hereditary renal-cell carcinoma associated with a chromosomal translocation. *N Engl J Med* 301:592–595, 1979.
13. Comings DE: A general theory of carcinogenesis. *Proc Natl Acad Sci USA* 70:3324–3328, 1973.
14. Connolly MJ, Payne RH, Johnson G, Gallie BL, Allderdice PW, Marshall WH, Lawton RD: Familial, *EsD*-linked retinoblastoma with reduced penetrance and variable expressivity. *Hum Genet* 65:122–124, 1983.
15. Devesa SS: The incidence of retinoblastoma. *Am J Ophthalmol* 80:263–265, 1975.

16. Dryja TP, Cavenee W, White R, Rapaport JM, Petersen R, Albert DM, Bruns GAP: Homozygosity of chromosome 13 in retinoblastoma. *N Engl J Med* 310:550–553, 1984.
17. Dryja TP, Rapaport JM, Weichselbaum R, Bruns GAP: Chromosome 13 restriction fragment length polymorphisms. *Hum Genet* 65:320–324, 1984.
18. Ejima Y, Sasaki MS, Kaneko A, Tanooka H, Hara Y, Hida T, Kinoshita Y: Possible inactivation of part of chromosome 13 due to 13qXp translocation associated with retinoblastoma. *Clin Genet* 21:357–361, 1982.
19. Ejima Y, Sasaki MS, Utsumi H, Kaneko A, Tanooka H: Radiosensitivity of fibroblasts from patients with retinoblastoma and chromosome-13 abnormalities. *Mutat Res* 103:177–184, 1982.
20. Fearon ER, Vogelstein B, Feinberg AP: Somatic deletion and duplication of genes on chromosome 11 in Wilms' tumours. *Nature* 309:176–178, 1984.
21. Fedrick J, Baldwin JA: Incidence of cancer in relatives of children with retinoblastoma. *Br Med J* 1:83–84, 1978.
22. Ferrell RE, Strong LC, Pathak S, Riccardi VM: Wilms tumor (WT) cytogenetics: The variable presence of del(11p) in WT explants. *Am J Hum Genet* 35:63A, 1983.
23. Gallie BL, Ellsworth RM, Abramson DH, Phillips RA: Retinoma: Spontaneous regression of retinoblastoma or benign manifestation of the mutation? *Br J Cancer* 45:513–521, 1982.
24. Gardner HA, Gallie BL, Knight LA, Phillips RA: Multiple karyotypic changes in retinoblastoma tumor cells: Presence of normal chromosome no. 13 in most tumors. *Cancer Genet Cytogenet* 6:201–211, 1982.
25. Godbout R, Dryja TP, Squire J, Gallie BL, Phillips RA: Somatic inactivation of genes on chromosome 13 is a common event in retinoblastoma. *Nature* 304:451–453, 1983.
26. Golitz LE, Norris DA, Luekens CA Jr, Charles DM: Nevoid basal cell carcinoma syndrome: Multiple basal cell carcinomas of the palms after radiation therapy. *Arch Dermatol* 116:1159–1163, 1980.
27. Gordon H: Family studies in retinoblastoma. *Birth Defects* 10:185–190, 1974.
28. Green DM, Fine WE, Li FP: Offspring of patients treated for unilateral Wilms' tumor in childhood. *Cancer* 49:2285–2288, 1982.
29. Harnden D, Morten J, Featherstone T: Dominant susceptibility to cancer in man. *Adv Cancer Res* 41:185–255, 1984.
30. Hashem N, Khalifa S: Retinoblastoma. A model of hereditary fragile chromosomal regions. *Hum Hered* 25:35–49, 1975.
31. Hashimoto K, Howell JB, Yamanishi Y, Holubar K, Bernhard R Jr: Electron microscopic studies of palmar and plantar pits of nevoid basal cell epithelioma. *J Invest Dermatol* 59:380–393, 1972.
32. Heldin CH, Westermark B: Growth factors: Mechanism of action and relation to oncogenes. *Cell* 37:9–20, 1984.
33. Herrmann J: Delayed mutation model: Carotid body tumors and retinoblastoma. *In* Genetics of Human Cancer (Mulvihill JJ, Miller RW, Fraumeni JF Jr, eds) vol 3, New York, Raven Press, 1977, 417–438.
34. Junien C, Turleau C, de Grouchy J, Said R, Rethore MO, Tenconi R, Dufier JL: Regional assignment of catalase (CAT) gene to band 11p13. Association with the aniridia-Wilms' tumor-gonadoblastoma (WAGR) complex. *Ann Genet* 23:165–166, 1980.
35. Junien C, Turleau C, Lenoir GM, Philip T, Said R, Despoisse S, Laurent C, Rethore MO, Kaplan JC, de Grouchy J: Catalase determination in various etiologic forms of Wilms' tumor and gonadoblastoma. *Cancer Genet Cytogenet* 10:51–57, 1983.
36. Kaneko Y, Kondo K, Rowley JD, Moohr JW, Maurer HS: Further chromosome studies on Wilms' tumor cells of patients without aniridia. *Cancer Genet Cytogenet* 10:191–197, 1983.

37. Kantor AF, Blattner WA, Blot WJ, Fraumeni JF Jr, McLaughlin JK, Schuman LM, Lindquist LL, Wang N, Hozier JC: Hereditary renal carcinoma and chromosomal defects. *N Engl J Med* 307:1403–1404, 1982.
38. Kantor AF, Li FP, Fraumeni JF Jr, McCrea Curnen MG, Flannery JT: Childhood cancer in offspring of two Wilms' tumor survivors. *Med Pediatr Oncol* 10:85–89, 1982.
39. Kennedy AR, Cairns J, Little JB: Timing of the steps in transformation of C3H 10T½ cells by X-irradiation. *Nature* 307:85–86, 1984.
40. Knudson AG Jr: Mutation and cancer: A statistical study of retinoblastoma. *Proc Natl Acad Sci USA* 68:820–823, 1971.
41. Knudson AG Jr, Strong, LC: Mutation and cancer: A model for Wilms' tumor of the kidney. *J Natl Cancer Inst* 48:313–324, 1972.
42. Kossakowska AE, Gallie BL, Phillips RA: Fibroblasts from retinoblastoma patients: Enhanced growth in fetal calf serum and a normal response to ionizing radiation. *J Cell Physiol* 111:15–20, 1982.
43. Koufos A, Hansen MF, Lampkin BC, Workman ML, Copeland NG, Jenkins NA, Cavenee, WK: Loss of alleles at loci on human chromosome 11 during genesis of Wilms' tumour. *Nature* 309:170–172, 1984.
44. Kraemer KH, Lee MM, Scotto J: DNA repair protects against cutaneous and internal neoplasia: Evidence from xeroderma pigmentosum. *Carcinogenesis* 5:511–514, 1984.
45. Land H, Parada LF, Weinberg RA: Tumorigenic conversion of primary embryo fibroblasts requires at least two cooperating oncogenes. *Nature* 304:596–602, 1983.
46. Land H, Parada LF, Weinberg RA: Cellular oncogenes and multistep carcinogenesis. *Science* 222:771–778, 1983.
47. Ledbetter DH, Airhart SD, Hittner HM, Caskey CT, Cherry LM, Hsu TC: Extreme chromosome instability in a family with retinoblastoma and Wilms' tumor. *Mammalian Chromosome Newsletter* 25:17, 1984.
48. Lee WH, Murphree AL, Benedict WF: Expression and amplification of the N-*myc* gene in primary retinoblastoma. *Nature* 309:458–460, 1984.
49. Matsunaga E: Hereditary retinoblastoma: Delayed mutation or host resistance? *Am J Hum Genet* 30:406–424, 1978.
50. Matsunaga E: Retinoblastoma: A model for the study of carcinogenesis in humans. *Jpn J Hum Genet* 28:57–71, 1983.
51. McGavran L, Heiderman R, Waldstein G, Berry P: Interstitial deletion of 11p in renal nodule from a child with Wilms' tumor. *Am J Hum Genet* 35:67A, 1983.
52. Miller RW, Fraumeni JF Jr, Manning MD: Association of Wilms' tumor with aniridia, hemihypertrophy and other congenital malformations. *N Engl J Med* 270:922–927, 1964.
53. Moolgavkar SH, Knudson AG Jr: Mutation and cancer: A model for human carcinogenesis. *JNCI* 66:1037–1052, 1981.
54. Motegi T, Minoda K: A decreasing tendency for cytogenetic abnormality in peripheral lymphocytes of retinoblastoma patients with 13q14 deletion mosaicism. *Hum Genet* 66:186–189, 1984.
55. Murphree AL, Benedict WF: Retinoblastoma: Clues to human oncogenesis. *Science* 223:1028–1033, 1984.
56. Narahara K, Kikkawa K, Kimira S, Kimoto H, Ogata M, Kasai R, Hamawaki M, Matsuoka K: Regional mapping of catalase and Wilms' tumor-aniridia, genitourinary abnormalities, and mental retardation triad loci to the chromosome segment 11p1305Tpl306. *Hum Genet* 66:181–185, 1984.
57. Naylor SL, Sakaguchi AY, Barker D, White R, Shows TB: DNA polymorphic loci mapped to human chromosomes 3, 5, 9, 11, 17, 18, and 22. *Proc Natl Acad Sci USA* 81:2447–2451, 1984.

58. Newbold RF, Overell RW: Fibroblast immortality is a prerequisite for transformation by EJ c-Ha-*ras* oncogene. *Nature* 304:648–651, 1983.
59. Nichols WW, Miller RC, Sobel M, Hoffman E, Sparkes RS, Mohandas T, Veomett I, Davis JR: Further observations on a 13qXp translocation associated with retinoblastoma. *Am J Ophthalmol* 89:621–627, 1980.
60. Orkin SH, Goldman DS, Sallan SE: Development of homozygosity for chromosome 11p markers in Wilms' tumour. *Nature* 309:172–174, 1984.
61. Pathak S, Strong LC, Ferrell RE, Trindade A: Familial renal cell carcinoma with a 3;11 chromosome translocation limited to tumor cells. *Science* 217:939–941, 1982.
62. Pereira-Smith OM, Smith JR: Evidence for the recessive nature of cellular immortality. *Science* 221:964–966, 1983.
63. Punnett HH, Marshall LS, Qureshi AR, DiGeorge AM, Kistenmacher ML: Deletion 11p13 with normal catalese activity. *Pediatr Res* 17:217A, 1983.
64. Rayner CRW, Towers JF, Wilson JSP: What is Gorlin's syndrome? The diagnosis and management of the basal cell naevus syndrome, based on a study of thirty-seven patients. *Br J Plast Surg* 30:62–67, 1976.
65. Reeve AE, Housiaux PJ, Gardner RJM, Chewings WE, Grindley RM, Millow LJ: Loss of a Harvey *ras* allele in sporadic Wilms' tumour. *Nature* 309:174–176, 1984.
66. Riccardi VM, Hittner HM, Strong LC, Fernbach DJ, Lebo R, Ferrell RE: Wilms tumor with aniridia/iris dysplasia and apparently normal chromosomes. *J Pediatr* 100:574–577, 1982.
67. Riccardi VM, Sujansky E, Smith AC, Francke U: Chromosomal imbalance in the aniridia-Wilms' tumor association: 11p interstitial deletion. *Pediatrics* 61:604–610, 1978.
68. Ruley HE: Adenovirus early region 1A enables viral and cellular transforming genes to transform primary cells in culture. *Nature* 304:602–606, 1983.
69. Sagerman RH, Cassady JR, Tretter P, Ellsworth RM: Radiation induced neoplasia following external beam therapy for children with retinoblastoma. *Am J Roentgenol* 105:529–535, 1969.
70. Sanchez M, Bahng K, Ibrahim I: Papillary renal cell carcinoma in two brothers associated with translocation of chromosomes 13 and 14. *In* 71st Annual Meeting of the International Academy of Pathology (United States–Canadian Division), March 1–5, 1982. Boston, International Academy of Pathology, 1982, 95.
71. Scharnagel IM, Pack GT: Multiple basal cell epitheliomas in a five year old child. *Am J Dis Child* 77:647–651. 1949.
72. Schweisguth O, Gerard-Marchant R, Lemerle J: Naevomatose baso-cellulaire association a un rhabdomyosarcome congenital. *Arch Franc Ped* 25:1083–1093, 1968.
73. Skolnick MH, Bishop DT, Cannings C, Hasstedt SJ: The impact of RFLPs on human gene mapping. *In* Genetic Epidemiology of Coronary Heart Disease: Past, Present, and Future. Proceedings of a Workshop held in St. Louis, Missouri, August 10–12, 1983. (Rao DC, Elston RC, Kuller LH, Feinleib M, Carter C, Havlik R, eds) New York, Alan R Liss, Inc, 1984, 271–292.
74. Smith JLS, Bedford MA: Retinoblastomas. *In* Tumours in Children 2nd ed (Marsden HB, Steward JK, eds) New York, Springer-Verlag, 1976, 245–281.
75. Sorensen SA, Jensen OA, Klinken L: Familial aggregation of neuroectodermal and gastrointestinal tumors. *Cancer* 52:1977–1980, 1983.
76. Sparkes RS, Murphree AL, Lingua RW, Sparkes MC, Field LL, Funderburk SJ, Benedict WF: Gene for hereditary retinoblastoma assigned to human chromosome 13 by linkage to esterase D. *Science* 219:971–973, 1983.
77. Stefani FH: Maligne Tumoren bei Angehorigen von Retinoblastopatienten. *Klin Mbl Augenheilk* 168:716–718, 1976.
78. Strong LC: Theories of pathogenesis: Mutation and cancer. *In* Genetics of Human Cancer (Mulvihill JJ, Miller RW, Fraumeni JF Jr, eds) New York, Raven Press, 1977, 401–414.

79. Strong LC: Genetic and environmental interactions. *Cancer* 40:1861–1866, 1977.
80. Strong LC: Genetic-environmental interactions. *In* Cancer Epidemiology and Prevention (Schottenfeld D, Fraumeni JF Jr, eds) Philadelphia, WB Saunders Co, 1982, 506–516.
81. Strong LC: Genetics, etiology, and epidemiology of childhood cancer. *In* Clinical Pediatric Oncology, 3rd ed (Sutow WW, Vietti TJ, Fernbach DJ, eds) St. Louis, CV Mosby Co, 1984, 14–41.
82. Strong LC, Herson J, Haas C, Elder K, Chakraborty R, Weiss K, Majumder P: Cancer mortality in relatives of retinoblastoma patients. *JNCI*, 73:303–311, 1984.
83. Strong LC, Riccardi VM, Ferrell RE, Sparkes RS: Familial retinoblastoma and chromosome 13 deletion transmitted via an insertional translocation. *Science* 213:1501–1503, 1981.
84. Tucker MA, Meadows AT, Boice JD Jr, Hoover RN, Fraumeni JF Jr, for the Late Effects Study Group. Cancer risk following treatment of childhood cancer. *In* Radiation Carcinogenesis: Epidemiology and Biological Significance. Progress in Cancer Research and Therapy (Boice JD Jr, Fraumeni JF Jr, eds) New York, Raven Press, 1984, 211–224.
85. Vogel F: Genetics of retinoblastoma. *Hum Genet* 52:1–54, 1979.
86. Wang N, Perkins KL: Involvement of band 3p14 in t(3;8) hereditary renal carcinoma. *Cancer Genet Cytogenet* 11:479–481, 1984.
87. Ward P, Packman S, Loughman W, Sparkes M, Sparkes R, McMahon A, Gregory T, Ablin A: Location of the retinoblastoma susceptibility gene(s) and the human esterase D locus. *J Med Genet* 21:92–95, 1984.
88. Weichselbaum RR, Nove J, Little JB: X-ray sensitivity of diploid fibroblasts from patients with hereditary or sporadic retinoblastoma. *Proc Natl Acad Sci USA* 75:3962–3964, 1978.
89. Weller CV: The inheritance of retinoblastoma and its relationship to practical eugenics. *Cancer Res* 1:517–535, 1941.
90. Williams WR, Strong LC, Norsted T: Identification of a cohort of cancer-prone families ascertained through childhood sarcoma by segregation analysis. *Am J Hum Genet* 36:39S, 1984.
91. Yunis JJ, Ramsay NKC: Familial occurrence of the aniridia-Wilms tumor syndrome with deletion 11p13-14.1. *J Pediatr* 96:1027–1030, 1980.

4.

Chromosomal mechanisms of cancer etiology

R. S. K. CHAGANTI SURESH C. JHANWAR

The significance of mitotic and chromosomal abnormalities seen in cancer cells to the development of tumors began to be recognized during the late nineteenth and early twentieth centuries. Notable among the early investigators was Theodore Boveri, an embryologist who studied sea urchin development. His observations on this species led him to conclude that abnormal embryo development resulted from spindle abnormalities that lead to unbalanced chromosomal complements; from this he made the bold deduction that tumor development may have its basis in mitotic errors that produce abnormal chromosomal complements [5]. That a population of proliferating tumor cells represents an evolutionary cellular genetic system was first recognized by Winge when he proposed the stem-line concept [83]. Subsequently, Foulds extended this concept to tumor development at the pathophysiologic level [20]. Cytogenetic analyses of cultured mammalian cells by Levan [40], Hauschka [26], and others provided evidence for this dynamic view of the genetic system of tumor cells. To paraphrase Hauschka, although the tumor stem-line tends to retain its stability under uniform environmental conditions, it can undergo changes in the direction of greater or lesser malignancy, invasiveness, or drug-resistance and give rise to new stem-lines that invariably differ from the parental ones in their chromosomal constitution [26].

The first major breakthrough relating to chromosome change in primary human tumors was the discovery by Nowell and Hungerford, reported in 1960, that a specific chromosome abnormality (an apparent deletion in one of the G group chromosomes, later designated as the Philadelphia or Ph^1 chromosome) was consistently associated with chronic myelogenous leukemia (CML) [55]. This discovery, made possible by the then recently introduced technical innovations in cytogenetics such as hyposmotic treatment of cells to obtain better spreading of metaphase chromosomes, initiated a new era in tumor cytogenetics. Studies of human tumor chromosomes using these new techniques in a relatively short time enabled the definition of a set of basic properties of neoplastic cells in the primary host, succinctly stated in 1963 by Ford and Clarke [19] in the following way: (1) *a population of neoplastic cells may descend from a single ancestral cell;* (2) *variant, though related, cell types may arise within it;* and (3) *one of such variant cells may replace all others, becoming the dominant component of the population.*

The introduction of the banding techniques in the early 1970s constituted a second major catalyst to the study of tumor cytogenetics. Thus, the Ph^1 chromosome in CML and the $14q^+$ marker chromosome observed to be a consistent feature of Burkitt's lymphoma (BL) were shown to be derived from specific translocations [44, 65]. During the past ten or more years, a vast literature has accumulated that describes in increasing detail the patterns of chromosome changes that occur in various kinds of primary human tumors. The continuous improvements to tumor cell culture and banding methods that have been made during this period also have contributed to the generation of these data. So far, tumors of the lymphohematopoietic system have received the greatest attention from cytogeneticists. Solid tumors have received less attention until recently because, technically, they are much more difficult to handle than leukemias and lymphomas. However, more recent technical innovations are now enabling many investigators to study in detail these tumors as well. The third and by far the most exciting development in the study of tumor chromosomes is represented by the successful integration of classic cytogenetics with molecular biology and virology that has been achieved in the early 1980s.

These considerations apply to neoplastic development that takes place in individuals with presumably normal germ-line–derived chromosomal complements. Certain constitutional chromosome abnormalities that lead to developmental defects also have been shown to be associated with tumor development. Thus, somatic as well as germ-line chromosome changes show an association with carcinogenesis, thus raising a number of important questions; for example, What is the role of somatic chromosome change in the transformation of a cell to the neoplastic state and in the further development of the tumor? Do germ-line chromosome changes predispose to neoplastic transformation? What is the relationship between somatic and germ-line chromosome changes in tumorigenesis? Recent molecular genetic studies of somatic as well as germ-line chromosome changes associated with tumor development are providing new insights into the mechanisms of tumorigenesis. The aim of this chapter is to present models of chromosomal mechanisms currently implicated in tumor development in the light of this information. Descriptive data pertaining to chromosome changes in tumors will not be presented here; this aspect has been treated in detail by other authors in this volume [68, 77].

Chromosome changes in neoplastic cells

The main conclusions drawn from the vast cytogenetic literature relating to human and animal tumors may be stated in the following manner: (1) if studied with the appropriate techniques, virtually all tumors can be seen to exhibit chromosome changes, (2) the changes themselves tend to affect chromosomes nonrandomly within the genome, often showing considerable specificity, which correlates reasonably well with histologic subtypes and sometimes also with etiologic factors, and (3) tumor progression is associated with evolution of increasingly complex karyotypes. The changes encompass the entire gamut of chromosome aberrations resulting from abnormal anaphase segregation as well as breakage and reunion of broken ends of chromosomes. The former results in chromosome numbers that range from monosomy and trisomy to haploidy and polyploidy, and the latter results in such classic lesions as deletion, duplication, inversion, and translocation as well as the more recently recognized lesion, amplification of chromosomal regions. Thus, the genetic system of a tumor cell becomes unstable probably early in its history and remains so throughout its life—the source of variation for the evolution of stem-lines discussed earlier. The significance of this change to the genetic organization of the transformed cell and the establishment, survival, and spread of the tumor that derives from it has been questioned many times in the past. The obvious deduction that specific changes may disrupt the normal dosage or alignment and coordinated function of some important genes (e.g., those that control cell proliferation and cell function), thereby setting the stage for abnormal proliferation or development, or both, has also been stated many times [see References 10, 11, 40, 41, 66 for reviews of these views]. Thus, the probable role of gene dosage and position effect in the development of tumors has been recognized for some time.

Of the two major classes of errors that affect chromosomes in tumor cells, namely, abnormal segregation and chromosome breakage, certain types of the latter class (notably translocation and amplification) have recently been subjected to detailed molecular analyses and yielded new insights into gene action in tumorigenesis. We are far from understanding either the mechanisms of origin of abnormal chromosome segregation or the role of its consequence in the development of tumors.

With regard to chromosome breakage, available data from a variety of tumors indicates that stable rearrangements affect a limited number of sites within the genome that number less than 100 [11, 50]. These sites, which are also called specific breakpoints, are shown in Figure 4–1. The verification of the hypothesis of abnormal gene regulation brought about by position effect of genes states above would require a knowledge of the genes that normally reside at these breakpoints and their fate in rearrangements. The first such verification came recently with the study of BL, a tumor of B cells, which exhibits three specific translocations, one standard and two variant (Figure 4–2). The standard

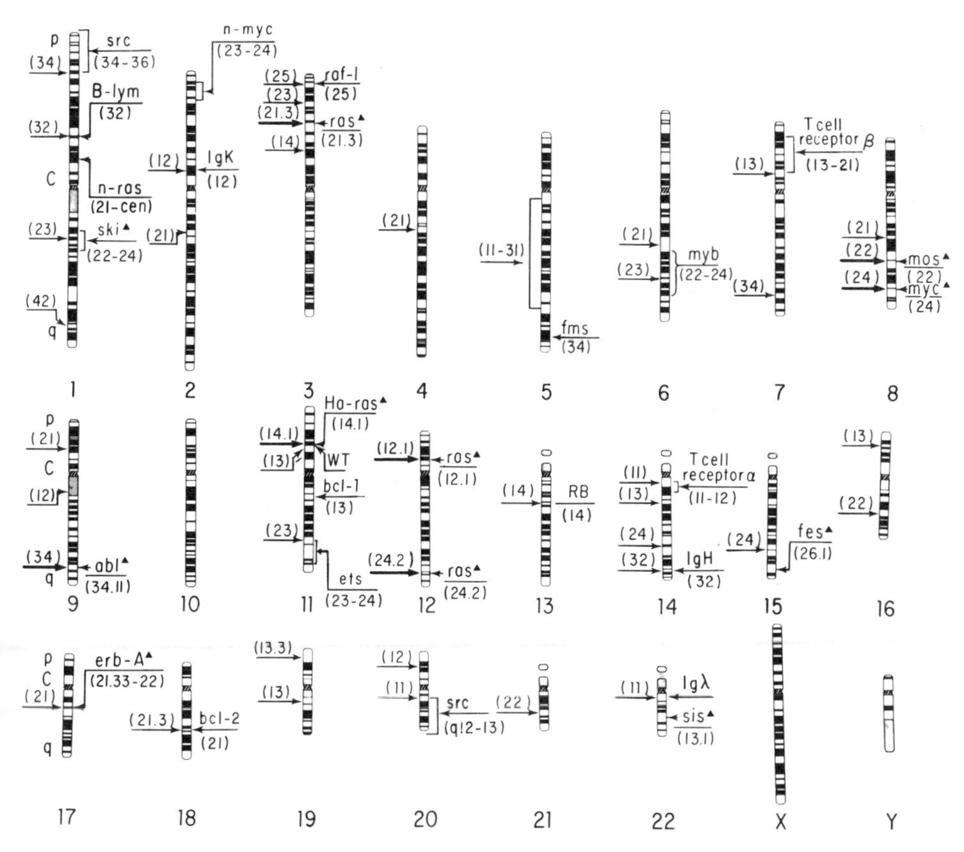

Figure 4–1 Banded idiogram of human chromosomes showing oncogene localizations and nonrandom breakpoints reported in neoplastic cells (c = centromere; p = short arm; q = long arm). The oncogene localizations are indicated by arrows on the right side of each relevant chromosome, with the specific band designation shown in parentheses below the arrow. The data on gene localizations are compiled from the literature [4, 24, 25, 29, 30, 31, 52, 54, 59, 71, 86–90]. Breakpoints are indicated by arrows on the left side of each chromosome, with the specific band designation shown in parentheses above the arrow. The data on breakpoints are compiled from the literature [11, 50, 68, 77].

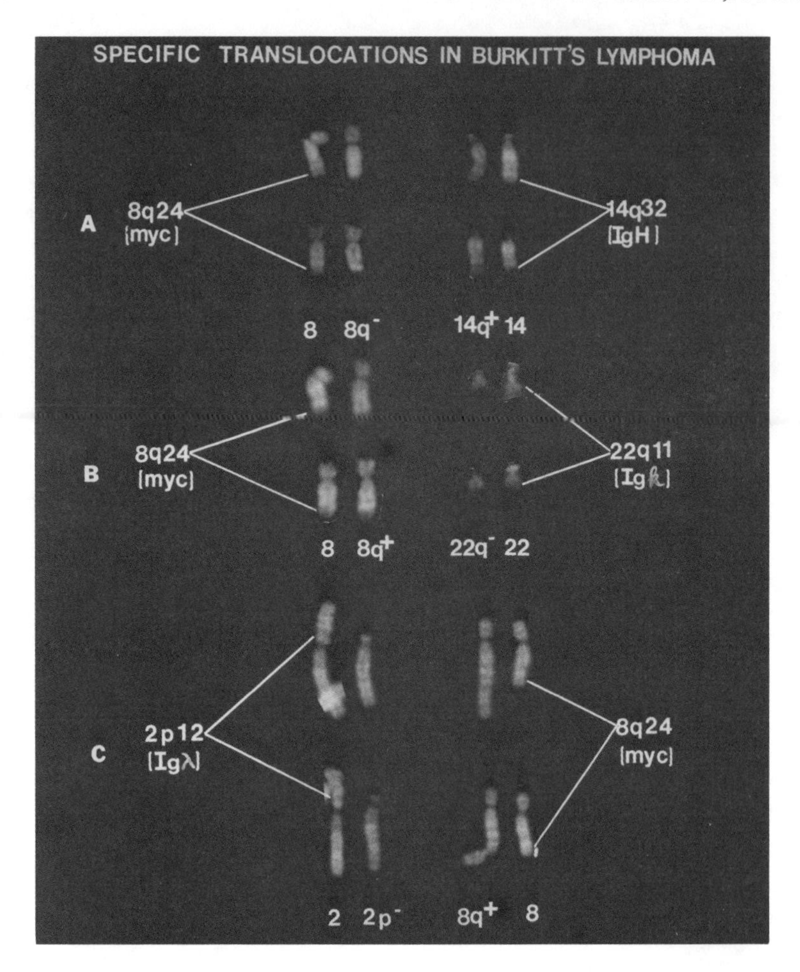

Figure 4–2 Chromosome and gene rearrangements in Burkitt's lymphoma. Partial karyotypes of the standard (*A*) and the variant (*B* and *C*) translocations from two cells each are shown. The specific breakpoints and the genes localized at them are indicated.

translocation engages chromosomes 8 and 14 with breakpoints in bands 8q24 and 14q32, and the variant translocations engage chromosomes 2 and 8 in one and 8 and 22 in the other. The breakpoint in chromosome 8 in all cases is in band 8q24 whereas the breakpoints in chromosomes 2 and 22, respectively, are in bands 2p12 and 22q11 [39]. Gene localization studies have shown that the breakpoints do indeed correspond with the positions of a most interesting set of genes. Thus, chromosomes 14, 2, and 22 carry, at their breakpoints, the determinants for the immunoglobulin heavy chains, kappa light chains, and lambda light chains, respectively [35, 43, 46]. The breakpoint in chromosome 8 carries c-*myc* [54], which belongs to a class of genes known as cellular oncogenes (see below). In these translocations, c-*myc* rearranges with the immunoglobulin genes and, although constitutively inactive in nonembryonal tissues, it undergoes activation in the translocation-carrying cells, presumably under the influence of the immunoglobulin genes that are normally active in these cells [17]. A parallel situation has been observed in mineral oil–induced plasmacytomas in BALBc mice, which are also B-cell tumors. These exhibit translocations involving either chromosomes 6 and 15 or chromosomes 12 and 15 [56]. The murine chromosome 6 carries the kappa light chain determinants (the precise localization on chromosome 6 is not known). The murine heavy chain determinants and the murine c-*myc* are localized on chromosomes 12 and 15, at their respective breakpoints, 12F2 and 15D3 [47]. Just as human c-*myc*, the murine c-*myc* undergoes activation in these tumors [45].

The demonstration that a cellular oncogene undergoes activation associated with a change in position led to an intense effort to identify genes situated at the breakpoints, the prime candidates, based on the human and murine B-cell tumor model just discussed, being oncogenes. The role of oncogenes in transformation and tumor induction has been discussed in detail by Hayward in another article in this volume [27]. Oncogenes will be reviewed here briefly in terms of their relevance to chromosome aberration in tumor cells.

The genomes of RNA tumor viruses (retroviruses) that cause rapid transformation in vertebrate cells include genes that determine their oncogenic property. These are termed the viral oncogenes (or v-*onc* genes). A number of v-*onc* genes have been identified and studied extensively as to their structure and the transforming protein molecules that they encode. The genomes of host vertebrate species contain genes that are homologous to v-*onc* genes which in turn are called cellular oncogenes (or c-*onc*) genes. C-*onc* genes are highly conserved throughout vertebrate evolution and they have sometimes been referred to as proto-oncogenes [2, 3]. Until recently their function was totally unknown, although it has been suggested for some time that they are part of the cell's genetic machinery responsible for the control of proliferation and differentiation [2, 3]. More recent experimental data support such a view. Thus, c-*myc* has been shown to play a role in the control of cell proliferation [34], c-*sis* has been shown to be homologous to the gene that encodes platelet-derived growth factor (PDGF)[81], and c-*erb-B* has been shown to be homologous to the gene that encodes the receptor for epidermal growth factor (EGF)[15]. Finally, various lines of evidence indicate that vertebrate proto-oncogenes are also capable of inducing transformation under appropriate conditions [12].

Although the retrovirally related c-*onc* genes have classically been identified by their homology to v-*onc* genes, more recently, c-*onc* genes with either no known homologies to v-*onc* genes or showing varying degrees of divergence from v-*onc* genes have been isolated from tumor cells in one of three ways: (1) by their ability to transform NIH/3T3 cells following transfection of tumor DNA into them as calcium phosphate precipitate, indicating that the oncogene in question is in an activated state in the given tumor (example: B-*lym-1* [13]), (2) by isolating DNA sequences that occur amplified severalfold in tumor cells and comparing their organization with that of known oncogenes (example: n-*myc* [51]), and (3) by identifying the sequences translocated into genes with known organization such as the immunoglobulin determinants (example: *b cl-1* [79]).

The number of already known c-*onc* genes is impressive, although the total number of such genes present in the human genome is currently unknown. So far, c-*onc* genes have been mapped to 15 sites in the human chromosomal complement. With the exception of two (c-*fes* at 15q26.1 and c-*sis* at 22q13.1), the positions of the rest correspond to those of specific breakpoints reported in rearrangements seen in tumor cells. These data are presented in Figure 4-1. Whether all or most of these chromosome rearrangements also represent gene rearrangements, and if so with what consequence to the development of the tumor, is obviously an issue of considerable fundamental importance. We will return to this topic in a later section of this chapter.

Germ-line chromosome changes that predispose to neoplastic development

The human species is unique among vertebrates in carrying an exceptionally high load of germ-line chromosome mutations. About one in 12 recognizable conceptions carries a chromosome abnormality [78]. The abnormalities include tetraploidy, triploidy, trisomy for virtually each chromosome, monosomy of sex chromosomes, and balanced and unbalanced rearrangements [78]. Although the great majority of chromosomally abnormal concepti are eliminated by way of early spontaneous abortion, sufficient numbers survive to yield an incidence of chromosome abnormalities in newborns of approximately 1 in 165 [78]. Several distinct clinical disorders characterized by specific patterns of abnormal development are now recognized whose basis is chromosome abnormality [78]. Among such disorders, a few have been shown to carry increased predisposition to specific types of leukemia or cancer. These associations, which are listed in the tables in the chapter by German [21] in this volume, present unusual opportunities to examine the role of germ-line chromosome abormalities in cancer predisposition, on the one hand, and the relationship between germ-line and somatic chromosome change in cancer etiology, on the other.

The fact that the neoplasms in each of these disorders are restricted to no more than one or two tissue or organ systems at once indicates that the germ-line chromosome abnormalities per se are not tumorigenic changes. Furthermore, in each of these disorders the target organs present with abnormal development, which indicates that the origin of neoplastic development is rooted in the aberrent development of such organs. Thus, Down's syndrome (DS) children are predisposed to abnormal hematopoiesis and myeloblastic leukemoid reaction, which present during the neonatal period and usually disappear spontaneously over a period of weeks or months [6, 64]. Although it is generally believed that the incidence of leukemia in DS children (but not in DS adults) is increased over that of normal children [48], there is some disagreement as to whether the types of leukemia seen in the former are similar to those seen in the latter [33]. Data on banding analysis of leukemia in DS children are inadequate [reviewed in reference 33] to establish conclusively whether or not the chromosome abnormalities seen in these leukemias are similar to those seen in non-DS children with acute leukemia.

Klinefelter's syndrome patients appear to exhibit an increased predisposition to breast and germ cell cancers [53]. Breast tumors may be related to gynecomastia development, which affects a significant proportion of these patients, and development of germ cell tumors may have its origin in the supression of germ cells, which also characterizes this disorder. The predisposition of patients with mixed gonadal dysgenesis to gonadoblastoma may have a similar basis because these patients invariably develop dysgenetic streak gonads with suppression of germ cell development.

In the D-deletion retinoblastoma (RB) and the aniridia–Wilms' tumor association abnormal development often affects multiple organ systems including those that act as targets for carcinogenesis (eye in retinoblastoma and kidney in aniridia–Wilms' tumor association). The deletion in RB has been shown to vary in size ranging from a submicroscopic gene deletion to loss of a segment of the long arm of chromosome 13, but always including band 13q14, thus localizing the RB gene to this band [84]. The deletion in aniridia–Wilms' tumor association always includes the band 11p13, likewise localizing the Wilms' tumor (WT) gene to this band [63]. Thus, constitutionally these patients are hemizygous for certain key genes. Epidemiologic studies of familial and sporadic RB and WT, especially the former, were the basis for Knudson's genetic model of cancer origin in these disorders according to which two mutations (in the broad sense of the word, including chromosome mutations) are required to precipitate oncogenesis. In the inherited forms of RB and WT, the first is transmitted through the germ-line (e.g., D deletion) and the second affects the homologous locus in the target somatic cells, thus rendering the cells recessive. In the sporadic forms of the disease, both mutations occur in the genetically normal cells of the patient. (See chapter by Strong [74] in this volume for a more complete discussion of the two-hit hypothesis.) Recent molecular data indicate that RB and WT may attain homozygosity for the deletion through somatic mechanisms such as nondisjunction or recombination [18, 38, 58, 61]. However, the key questions to be answered in this regard are whether or not the somatic events that affect these loci represent tumor-initiating changes, and if so, when during development and by what mechanisms are they triggered; and finally, what role does the developmental abnormality play in this process?

These data identify a rare but fascinating pathway to neoplastic development, one in which a germ-line chromosome abnormality ultimately leads to tumor development. Because the neoplastic development in these cases is linked to a derangement of normal development in target organs, a molecular assault on tumorigenesis in these situations eventually can be expected to illuminate mechanisms of normal development.

Genetic changes in neoplastic cells mediated by chromosome changes

The data discussed in the previous section clearly show that more than one cytogenetic mechanism for modulation of gene action is involved during tumorigenesis; these range in their effect from germ-line predisposition to generation of variation for tumor evolution. In this section, the cellular genetic role of each of these cytogenetic mechanisms will be discussed.

Dosage alteration by aneuploidy

Nondisjunction leading to increase and decrease in chromosome number probably was among the first chromosome abnormality types to be recognized in tumor cells. Although rare subsets of tumors present with near-haploid or polyploid chromosome numbers [77], nonrandom loss or gain of specific chromosomes is the more usual observation [77]. By definition, aneuploidy implies change in gene dosage; however, neither its mechanism of origin nor the genes whose dosage alteration plays a role in any form of tumorigenesis are known at present. Trisomy of chromosome 8 is a nonrandom change seen in a number of tumor types, including acute nonlymphocytic leukemia [77]. Constitutional mosaicism for trisomy 8 leads to a syndrome of abnormal development that includes abnormal hematopoiesis [62]. Although the oncogenes c-*mos* and c-*myc* have been localized on this chromosome, their activation status is unknown in any of the above situations. The murine chromosome 15, which also carries c-*myc*, becomes trisomic in the T-cell leukemias that develop in AKR mice [14]. Cytological evidence for a qualitative change (reflected as altered replication pattern) in chromosome 15 associated with nondisjunction in these tumors has been reported recently [73]; however, the level of expression of the *myc* gene in these tumors is not known. Thus, although gene dosage alteration brought about by chromosome nondisjunction is a well-documented genetic change in tumor cells, no examples of its specific role in tumor development have been reported yet.

Hemizygosity by deletion

In addition to the previously discussed constitutional deletions in chromosomes 11 and 13 associated with WT and RB, a number of other cases of specific deletions have been described associated with sporadic neoplastic proliferations. These include the 5q-, 7q-, and 20q- abnormalities reported in a number of myeloid neoplasms [77]; 6q- in lymphomas [77], and 1p- in neuroblastomas [22]. Deletion in one homologue by definition results in hemizygosity for the genes contained in the deleted segment. Besides hemizygosity and consequent alteration in gene dosage, chromosome deletions can affect the function of genes situated at the break sites. Thus, such genes may be activated by removal of contiguous repressor sequences or by bringing them under the influence of active promoters of other genes. Alternatively, genes may be repressed by deletion of their own promoters or by bringing them under the influence of repressors of other genes. However, such mechanisms so far remain theoretical possibilities because most of the cytologically detected deletions did not include molecular identification of deleted genes or their products. Recent evidence suggests that c-*Ha-ras-1* may be deleted in WT [16, 61], and c-*myc* may be deleted in a subset of murine plasmacytomas [82].

Homozygosity by recombination

Some years ago Radman proposed that conversion of a "key" locus from a heterozygous to a homozygous state for a mutant allele by somatic recombination constitutes a promotional event in tumorigenesis [60]. Cytological demonstration of increased somatic crossing over in Bloom's syndrome, an inherited disorder that predisposes to leukemia and cancer [9], provided support for this hypothesis. Recently, genetic evidence for attainment of homozygosity for the chromosomal regions containing the RB and WT genes (at least in a proportion of tumors) has been obtained by comparing the genotypes of constitutional and tumor cells with regard to restriction length fragment polymorphisms at a number of arbitrary loci flanking the genes in question [8, 18, 23, 38, 58, 61]. These data clearly show that regions of chromosomes containing key genes can become homozygous by somatic recombination, one of the conditions predicted by Knudson's hypothesis of cellular recessivity as the genetic mechanisms for induction of RB and WT [36]. However, it is not clear how applicable this mechanism is to tumorigenesis per se, nor has it been shown whether recombination precedes or follows the establishment of the tumor progenitor cells. For example, in the case of an induced mouse T-cell leukemia that is heterozygous at the H-2 locus ($H\text{-}2^b/H\text{-}2^a$) loss of the $H\text{-}2^a$ allele induced by immunoselection leads to homozygosity for the $H\text{-}2^b$ allele, presumably by a somatic mechanism such as recombination, indicating that this may be one way in which tumor cells respond to the selective pressures to which they are subjected [72]; in such cases, somatic recombination and homozygosity may be evolutionary rather than etiologic changes.

Homozygosity by parthenogenesis

Another mechanisms by which tumor cells attain homozygosity is parthenogenesis. This mechanism has been shown to be associated with the origin of two germ cell tumor types.

The first tumor type is benign ovarian teratoma, a tumor derived from unfertilized secondary oocytes. These cells normally have completed first meiosis and are held in the ovary at the dictyate stage until postpubertal maturity and entry into the ovulatory cycle. Comparison of constitutional and teratoma cells at a number of fluorescent chromosomal heteromorphisms and enzyme polymorphisms revealed the following, which indicated their parthenogenetic origin. In the case of chromosomal heteromorphisms, in cases in which the constitutional cells were heterozygous, the tumor cells were always homozygous. In the case of the enzyme polymorphisms, however, proximal loci that were heterozygous in the constitutional cells were homozygous in the tumor cells, although some of the distal loci retained heterozygosity in the tumor cells because of crossing over between the gene locus and the centromere [42].

The other tumor type is the so-called complete hydatidiform mole. By comparison of chromosomal and enzyme markers of tumor cells with those of parental cells, it was shown that the majority of these tumors were homozygous for paternal markers, indicating that they were derived by parthenogenetic development of sperm nuclei that presumably underwent activation in defective or "empty" eggs [28, 57, 80]. Both ovarian teratomas and complete hydatidiform moles are benign proliferations with normal chromosomal complements, although both carry high risk for further malignant transformation and concomitant karyotypic evolution. The molecular mechanisms involved in the development of either of these fascinating tumor are unknown at present.

Activation by rearrangement

As pointed out previously in this chapter, specific rearrangement, especially translocation, is a diagnostic feature of several tumor types [68, 77]. Translocations classically served as the basis for study of position effects in plant, insect, and rodent species. The activation of c-*myc* following translocation into the immunoglobulin genes discussed earlier serves as the paradigm for gene activation by position effect in tumor cells. Other examples of this phenomenon can be expected to be elucidated in the near future.

Amplification by transposition

Amplification of a mammalian chromosomal region was first described by Biedler and Sprengler in Chinese hamster ovary cells (CHO) rendered resistant to methotrexate (MTX) by selection in vitro [1]. In these cells, certain chromosomal regions exhibited unbanded homogeneous staining when banding techniques were applied, earning them the name homogeneously staining regions (HSRs). Subsequently, HSRs in MTX-resistant CHO cells were shown to contain severalfold amplification of the MTX-resistance gene. Studies of different amplified genes showed that in general these present either as chromosomal HSRs or as extrachromosomal elements called double minutes (DMs), which are tiny bipartite, chromosome-like elements capable of autonomous replication. In a given cell, either HSRs or DMs may be present but not both simultaneously [69]. Amplification of c-*onc* genes identified as either HSR-DM structures, increased gene copy numbers, or increased

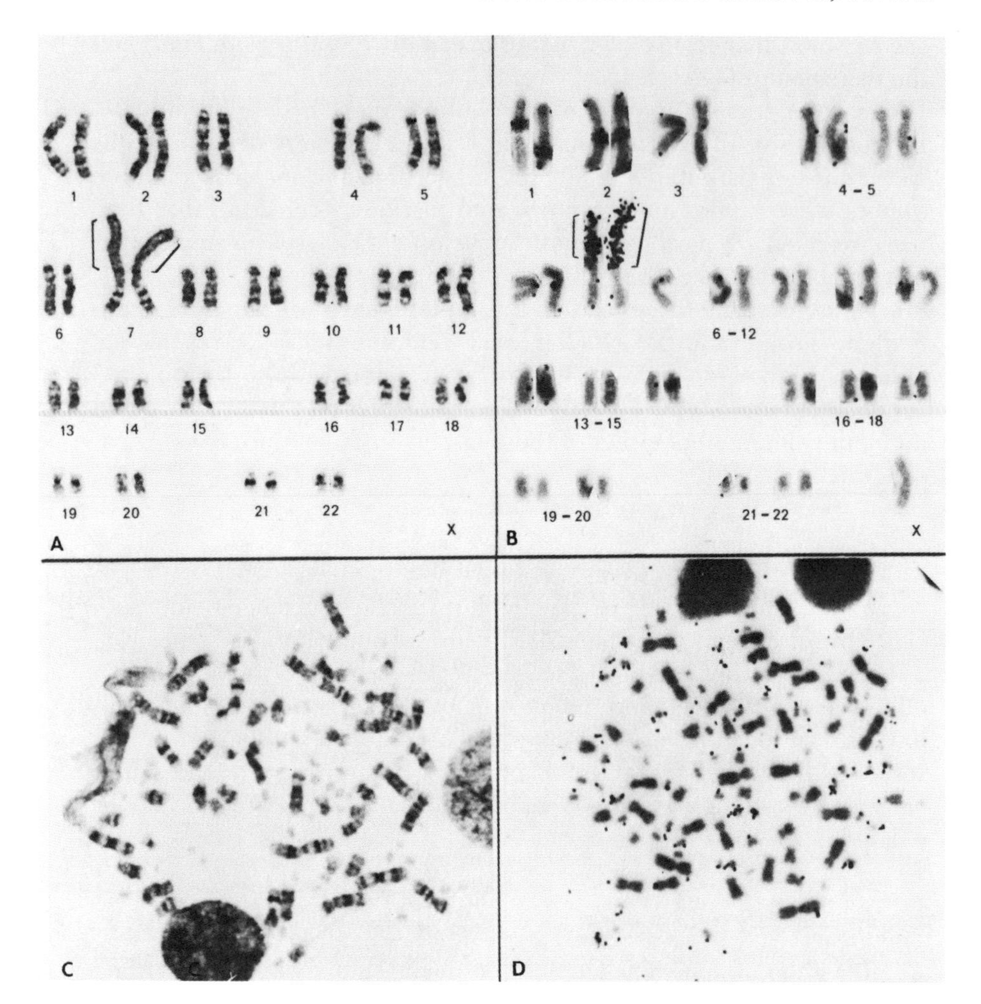

Figure 4–3 N-*myc* amplification in the neuroblastoma cell lines showing HSRs and DMs. *A* and *B* are from the cell line NAP(H). (*A*) A trypsin-Giemsa banded metaphase showing HSRs on both copies of chromosome 7. (*B*) An autoradiogram from an in situ hybridization of a metaphase plate with a DNA probe derived from the amplified sequence in NAP(H) showing hybridization to the HSR regions. *C* and *D* are from the cell line CHIP-234. (*C*) A trypsin-Giemsa banded metaphase showing palely staining DMs. (*D*) An autoradiogram of a metaphase plate hybridized with the NAP(H)DNA probe, showing hybridization to the DMs (Reprinted from Montgomery et al. [51] with permission from the authors and the journal.)

mRNA production has been reported in the case of a number of established tumor cell lines, and more recently in some fresh tumors as well; so far, amplification of oncogenes that belong to the *myc* family (c-*myc* and n-*myc*) has been studied most (Figure 4–3) [7, 32, 37, 51, 67, 70]. An interesting feature of amplification of these genes is that prior to amplification they frequently undergo transposition from their germline positions to other sites in the genome [37, 67].

Oncogene amplification can be viewed as a very powerful mechanism through which tumor cells achieve their evolution. Amplification of proliferation-governing genes such as *myc* will place them at increasing growth advantage whereas amplification of some other genes may aid them in gaining resistance to cellular or iatrogenic factors that attempt to eradicate the tumor.

Origin of chromosome abnormalities in neoplastic cells

So far in this chapter, we have discussed chromosome abnormality in tumor cells from the point of view of its presentation and the possible role that chromosome change–mediated alterations in gene structure and function play in tumorigenesis. The issue that needs to be addressed now relates to their origin and may be stated in the following way: at what point in the history of the neoplasm does its chromosomes become unstable and what factor or factors trigger the instability? Unfortunately, experimental data to address this question are unavailable. However, the available data on tumor cytology and cytogenetics permit at least three mechanisms to be proposed.

First, since the majority of chemical and physical carcinogens also are potent mutagenic and chromosome disrupting agents, the initial carcinogenic insult would also initiate chromosome abnormalities in a population of cells among which cells with certain nonrandom abnormalities gain selective advantage and become established as progenitors of tumor clones. In support of this is evidence from experimental rodent tumors studied by Mitelman and his associates some years ago, which showed a correlation between nonrandom chromosome change and tumor-inducing agents in histologically identical tumors [49]. The types of chromosome changes seen in acute nonlymphocytic leukemia that develops posttreatment in some patients with Hodgkin's disease, ovarian carcinoma, and other neoplasms also appear to correlate with specific components of the treatment regimens (reviewed by Testa and Misawa in this volume [77]).

The second mechanism is based on the possible error-prone nature of certain chromatin regions, for example, transcriptionally active gene sites and fragile sites (chromosomal regions that exhibit spontaneous gaps and breaks at somatic metaphase in cultured blood lymphocytes from clinically normal individuals with normal chromosomal complements [75, 76]). The highly specific translocation break in the immunoglobulin heavy chain region exhibited by the great majority of B-cell neoplasms is an example of the former. With regard to fragile sites, recent studies of Yunis and Soreng showed that 20 of the 51 fragile sites recorded in the human complement maps close to one of two breakpoints found in 26 of 31 specific chromosome abnormalities described in leukemias, lymphomas, and solid tumor; this suggests that errors at these sites would lead to tumor-associated rearrangements [85].

The third mechanism takes into account the well-established cellular genetic properties of cancer: multistep mutational origin and progression based on evolution by selection acting on genetic variation generated within the neoplastic cell population. Thus, following initiation, a variety of intracellular systems of control of cell structure and function may become destabilized either by mutations in the genes that control these systems or by alterations in intracellular environment. Chromosome instability may be generated when genotypic control of cell division and chromosome integrity are disrupted.

Conclusions

The significance of mitotic and chromosome abnormalities encountered in cancer cells has been a subject of continuing interest for cancer cell biologists over the past three quarters of a century. Recognition of certain basic principles of cancer cell genetics such as the clonal nature of tumors and their dynamic genetic systems were early results of these studies. Recent advances in techniques of chromosome analysis and the integration of tumor cytogenetics with immunology and molecular genetics have permitted extraordinary insights into the nature of perturbation gene structure and function undergo in cancer cells. In this chapter, these developments have been reviewed insofar as they address the central issue of cancer cytogenetics, namely, the role of somatic and germ-like chromosome change in the origin and evolution of neoplastic cells.

Acknowledgments

Research supported by the NIH grants CA-34775 and AI-21189, and a grant from the Cancer Research Institute, Inc., New York.

Literature Cited

1. Biedler JL, Spengler BA: Metaphase chromsome anomaly: Association with drug resistance and cell specific products. *Science* 191:185–187, 1976.

2. Bishop JM: Enemies within: The genesis of retrovirus oncogenes. *Cell* 23:5–6, 1981.
3. Bishop, JM: Cellular oncogenes and retroviruses. *Annu Rev Biochem* 52:301–354, 1983.
4. Bonner T, O'Brien SJ, Nash WG, Rapp UR, Morton CC, Leder P: The human homologs of the *raf (mil)* oncogene are located on human chromosomes 3 and 4. *Science* 223:71–74, 1984.
5. Boveri T: Zur Frage der Entstechung maligner Tumoren. Jena, Fischer, 1914.
6. Brodeur GM, Dahl GV, Williams DL, Tipton RE, Kalwinsky DK: Transient leukemoid reaction and trisomy 21 mosaicism in a phenotypically normal newborn. *Blood* 55:691–693, 1980.
7. Brodeur GM, Seeger RC, Schwab M, Varmus HE, Bishop JM: Amplification of N-*myc* in untreated human neuroblastomas correlates with advanced disease stage. *Science* 224:1121–1124, 1984.
8. Cavenee WK, Dryja TP, Phillips RA, Benedict WF, Godbout R, Gallie BL, Murphree AL, Strong LC, White RL: Expression of recessive alleles by chromosomal mechanisms in retinoblastoma. *Nature* 305:779–784, 1983.
9. Chaganti, RSK, Schonberg S, German J: A manyfold increase in sister chromatid exchange in Bloom's syndrome lymphocytes. *Proc Natl Acad Sci USA* 71:4508–4512, 1974.
10. Chaganti RSK: The significance of chromosome change to neoplastic development. *In* Chromosome Mutation and Neoplasia (German J, ed) New York, Alan R Liss, 1983, 359–396.
11. Chaganti RSK: Significance of chromosome change to hematopoietic neoplasms. *Blood* 62:515–524, 1983.
12. Cooper GM: Cellular transforming genes. *Science* 218:801–806, 1982.
13. Diamond A, Cooper CM, Ritz J, Lane MA: Identification and molecular cloning of the human *B-lym* transforming gene activated in Burkitt's lymphoma. *Nature* 305:112–116, 1983.
14. Dofuku R, Biedler JL, Spengler BA, Old LJ: Trisomy of chromosome 15 in spontaneous leukemia of AKR mice. *Proc Natl Acad Sci USA* 72:1515–1517, 1975.
15. Downward J, Yarden Y, Mayes E, Scrace G, Totty N, Stockwell P, Ullrich A, Schlessinger J, Waterfield MD: Close similarity of epidermal growth factor receptor and v-*erb-B* oncogene protein sequences. *Nature* 307:521–527, 1984.
16. Eccles MR, Millow LI, Wilkins RJ, Reeve AE: Harvey-ras allele deletion detected by *in situ* hybridization to Wilms' tumor chromosomes. *Hum Genet* 67:190–192, 1984.
17. Erikson J, Ar-Rushid A, Drwinga HL, Nowell PC, Croce CM: Transcriptional activation of translocated c-*myc* oncogene in Burkitt's lymphoma. *Proc Natl Acad Sci USA* 80:820–824, 1983.
18. Fearon ER, Vogelstein B, Feinberg AP: Somatic deletion and duplication of genes on chromosome11 in Wilms' tumors. *Nature* 309:176–178, 1984.
19. Ford CE, Clarke CM: Cytogenetic evidence of clonal proliferation in primary reticular neoplasms. *Can Cancer Conf* 5:129–146, 1963.
20. Foulds L: The natural history of cancer. *J Chronic Dis* 8:2–37, 1958.
21. German J: Heritable conditions that predispose to cancer. Chapter 5 in this volume.
22. Gilbert F, Balaban G, Moorhead P, Bianchi D, Schlesinger H: Abnormalities of chromosome 1p in human neuroblastoma tumors and cell lines. *Cancer Genet Cytogent* 7:33–42, 1982.
23. Godbout R, Dryja TP, Squire NJ, Gallie BL, Phillips RA: Somatic inactivation of genes on chromosome 13 is a common event in retinoblastoma. *Nature* 304:451–453, 1983.
24. Groffen J, Heisterkamp N, Spurr NK, Dana SL, Wasmuth JJ, Stephenson JR: Regional assignment of the human c-*fms* oncogene to band q34 of chromosome 5. Human gene mapping 7. *Cytogenet Cell Genet* 37:484, 1984.

25. Harper ME, Franchini G, Love J, Simon MI, Gallo RC, Wong-Staal F: Chromosomal localization of human c-*myb* and c-*fes* cellular *onc* genes. *Nature* 304:169–171, 1983.
26. Hauschka TS: The chromosomes in ontogeny and oncogeny. *Cancer Res* 21:957–974, 1961.
27. Hayward WS: Viral and cellular oncogenes in cancer etiology. Chapter 2 in this volume.
28. Jacobs PA, Wilson CM, Sprenkle JA, Rosenshein NB, Migeon BR: Mechanism of origin of complete hydatidiform moles. *Nature* 286:714–716, 1980.
29. Jhanwar SC, Neel BG, Hayward WS, Chaganti RSK: Localization of c-*ras* oncogene family on human germline chromosomes. *Proc Natl Acad Sci USA* 80:4794–7497, 1983.
30. Jhanwar SC, Neel BG, Hayward WS, Chaganti RSK: Localization of the cellular oncogenes ABL, SIS, and FES on human germ-line chromosomes. *Cytogenet Cell Genet* 38:73–75, 1984.
31. Jhanwar SC, Chaganti RSK, Croce CM: Germ-like chromosomal localization of human c-*erb-A* oncogene. *Somatic Cell Mol Genet* 11:99–102,1985.
32. Kanda N, Schreck R, Alt F, Bruns G, Baltimore D, Latt S: Isolation of amplified DNA sequences from IMR-32 human neuroblastoma cells: Facilitation by flourescence activated flow sorting of metaphase chromosomes. *Proc Natl Acad Sci USA* 80:4069–4073, 1983.
33. Kaneko Y, Rowley JD, Variakojis D, Chilcote RR, Moohr JW, Patel D: Chromosome abnormalities in Down's syndrome patients with acute leukemia. *Blood* 58:459–466, 1981.
34. Kelly K, Cochran BH, Stiles CD, Leder P: Cell-specific regulation of the c-*myc* gene by lymphocyte mitogens and platelet-derived growth factor. *Cell* 35:603–610, 1983.
35. Kirsch IR, Morton CC, Nakahara K, Leder P: Human immunoglobulin heavy chain genes map to a region of translocation in malignant B lymphocytes. *Science* 216:301–303, 1982.
36. Kundson AG, Jr: Genetics and etiology of human cancer. *Adv Hum Genet* 8:1–51, 1977.
37. Kohl NE, Kanda N, Schreck RR, Bruns G, Latt SA, Gilbert F, Alt FW: Transposition and amplification of oncogene-related sequences in human neuroblastomas. *Cell* 35:359–367, 1983.
38. Koufos A, Hansen MF, Lampkin BC, Workman ML, Copeland NG, Jenkins NA, Cavenee WK: Loss of alleles at loci on human chromosome 11 during genesis of Wilms' tumor. *Nature* 309:170–172, 1984.
39. Lenoir GM, Preud'homme JL, Bernheim A, Berger R: Correlation between immunoglobulin light chain expression and variant translocation in Burkitt's lymphoma. *Nature* 298:474–476, 1982.
40. Levan A: Relation of chromosome status to the origin and progression of tumors: The evidence of chromosome numbers. *In* Genetics and Cancer. Austin, University of Texas Press, 1959, 151–182.
41. Levan G, Mitelman F: Chromosomes and the etiology of cancer. *In* Chromosomes Today, vol 6 (de la Chapelle A, Sorsa M, eds) Amsterdam, North-Holland Biomedical Press, 1977, 363–371.
42. Linder D, Kaiser-McCaw B, Hecht F: Parthenogenetic origin of benign ovarian teratomas. *N Engl J Med* 292:63–66, 1975.
43. Malcom S, Barton P, Murphy C, Ferguson-Smith MA, Bently DL, Rabbits TH: Localization of human immunoglobulin light chain variable region genes to the sort arm of chromosome 2 by *in situ* hybridization. *Proc Natl Acad Sci USA* 79:4957–4960, 1982.

44. Manolov G, Manolova Y: Marker band in one chromosome 14 from Burkitt lymphomas. *Nature* 237:33–34, 1972.
45. Marcu K, Harris LJ, Stanton LW, Erikson J, Watt R, Corce CM: Transcriptionally-active c-*myc* oncogene is contained within NIARD, a DNA sequence associated with chromosome translocation in B-cell neoplasia. *Proc Natl Acad Sci USA* 80:519–523, 1983.
46. McBride OW, Swan D, Leder P, Hieter P, Hollis G: Chromosomal location of human immunoglobulin light chain constant region genes. *Cytogenet Cell Genet* 32:297–298, 1982.
47. Meo T, Johnson JJ, Beechey CV, Andrews SJ, Peters J, Searle AG: Linkage analysis of murine immunoglobulin heavy chain and serum pre-albumin genes establish their location on chromosome 12 proximal to the *T*(5;12)*31H* breakpoint in band 12F1. *Proc Natl Acad Sci USA* 77:550–553, 1980.
48. Miller RW: Neoplasia and Down's syndrome. *Ann NY Acad Sci* 171:637–644, 1970.
49. Mitelman F: Cytogenetics of experimental neoplasms and non-random chromosome correlations in man. *Clin Hematol* 9:195–219, 1980.
50. Mitelman F: Restricted number of chromosomal regions implicated in aetiology of human cancer and laeukemia. *Nature* 310:325–327, 1984.
51. Montgomery KT, Biedler JL, Spengler BA, Melera PW: Specific DNA sequence amplification in human neuroblastoma cells. *Proc Natl Acad Sci USA* 80:5724–5728, 1983.
52. Morton CC, Taub R, Diamond A, Lane MA, Cooper GM, Leder P: Mapping of human *B-lym-1* transforming gene activated in Burkitt's lymphomas to chromosome 1. *Science* 223:173–175, 1984.
53. Mulvihill JJ: Congenital and genetic diseases. *In* Persons at High Risk of Cancer. An Approach to Cancer Etiology and Control (Fraumeni JF, Jr, ed) New York, Academic Press, 1975, 1–37.
54. Neel BG, Jhanwar SC, Chaganti RSK, Hayward WS: Two human c-*onc* genes are located on the long arm of chromosome 8. *Proc Natl Acad Sci USA* 79:7842–7846, 1982.
55. Nowell PC, Hungerford DA: A minute chromosome in human chronic granulocytic leukemia. *Science* 132:1497, 1960.
56. Ohno S, Babontis M, Wiener F, Spira J, Klein G, Potter M: Non-random chromosome changes involving the distal end of chromosome 15 and the Ig-gene carrying chromosomes (Nos. 12 and 16) in pristane-induced mouse plasmacytomas. *Cell* 18:1001–1007, 1979.
57. Ohama K, Kajii T, Okamoto E, Fukuda Y, Imaizumi K, Tsukahara M, Kobayashi K, Hagiwara K: Dispermic origin of XY hydatidiform moles. *Nature* 292:551–552, 1981.
58. Orkin SH, Goldman DS, Sallan SE: Development of homozygosity for chromosome 11p markers in Wilms' tumor. *Nature* 309:172–174, 1984.
59. Rabin M, Watson M, Barber PE, Ryan J, Berg WR, Ruddle RH: NRAS transforming gene maps to region p11-p13 on chromosome 1 by *in situ* hybridization. *Cytogenet Cell Genet* 38:70–72, 1984.
60. Radman M, Kinsella AR: Chromosomal events in carcinogenic initiation and promotion: Implications for carcinogenicity testing and cancer prevention strategies. *In* Molecular and Cellular Aspects of Carcinogen Screening Tests (Montesano R, Beutsch N, Tomatis L, eds) Lyons, IARC Scientific Publications No. 27, 1980, 75–90.
61. Reeve AE, Housiaux PJ, Gardner RJM, Chewings WE, Grindley RM, Millow LJ: Loss of Harvey *ras* allele in sporadic Wilm's tumour. *Nature* 309:174–176, 1984.
62. Riccardi VM, Forgason J: Chromosome 8 abnormalities as components of neoplastic and hematologic disorders. *Clin Genet* 15:317–327, 1979.

63. Riccardi VM, Sujansky E, Smith AC, Franke U: Chromosomal imbalance in the aniridia-Wilms' tumor association: 11p interstitial deletion. *Pediatrics* 61:604–609, 1978.
64. Rosner F, Lee SL: Down's syndrome and acute leukemia: Myeloblastic or lymphoblastic? *Am J Med* 53:203–218, 1972.
65. Rowley JD: A new and consistent chromosomal abnormality in chronic myelogenous leukemia identified by quinacrine fluorescence and Giemsa staining. *Nature* 243:290–293, 1973.
66. Rowley, JD: Mapping human chromosomal regions related to neoplasia: Evidence from chromosomes 1 and 17. *Proc Natl Acad Sci USA* 74:5729–5733, 1977.
67. Sakai K, Kanda N, Shiloh Y, Donlan T, Schreck R, Shipley J, Dryja T, Phillips R, Chaum E, Chaganti, RSK, Latt S: Molecular and cytological analysis of DNA amplification in retinoblastoma. *Cancer Genet Cytogenet* (in press).
68. Sandberg AA: Chromosome changes in lymphoma and solid tumors. Chapter 12 in this volume.
69. Schimke RT: Gene amplification in cultured animal cells. *Cell* 37:705–713, 1984.
70. Schwab M, Alitalo K, Klempnauer K-H, Varmus HE, Bishop JM, Gilbert F, Brodeur G, Goldstein M, Trent J: Amplified DNA with limited homology to *myc* cellular oncogene is shared by human neuroblastoma cell lines and a neuroblastoma tumor. *Nature* 305:245–248, 1983.
71. Schwab M, Varmus HE, Bishop JM, Grzeschik K-H, Naylor SL, Sakaguchi AY, Brodeur G, Trent J: Chromosome localization in normal human cells and neuroblastoma of a gene related to c-*myc*. *Nature* 308:288–291, 1984.
72. Shen FW, Chaganti RSK, Doucette LA, Litman GW, Steinmetz M, Hood L, Boyse EA: Genomic constitution of an H-2:T1a variant leukemia. *Proc Natl Acad Sci USA* 82:1447–6450, 1984.
73. Spira J: Similar translocation patterns in B cell derived tumors of human, mouse, and rat origin. *In* Chromosomes and Cancer. From Molecules to Man (Rowley JD, Ultmann JE, eds), New York, Academic Press, 1983, 85–97.
74. Strong LC: Mutational models for cancer etiology. Chapter 3 in this volume.
75. Sutherland GR: Heritable fragile sites on human chromosomes I. Factors affecting expression in lymphocyte culture. *Am J Hum Genet* 31:125–135, 1979.
76. Sutherland GR: Heritable fragile sites on human chromosomes II. Distribution, phenotypic effects, and cytogenetics. *Am J Hum Genet* 31:136–148, 1979.
77. Testa JR, Misawa S: Chromosome changes in leukemia. Chapter 11 in this volume.
78. Thompson JS, Thompson MW: Genetics in Medicine. Philadelphia, WB Saunders 1980.
79. Tsujimoto Y, Yunis JJ, Onorato-Showe L, Erikson J, Nowell PC, Croce CM: Molecular cloning of chromosomal breakpoints of B-cell lymphomas and leukemias with the t(11;14) chromosome translocation. *Science* 224:1403–1406, 1984.
80. Wake N, Seki T, Fugita H, Okubo H, Sakai K, Okuyama K, Hayashi H, Shiina Y, Sato H, Kuroda M, Ichinoe K: Malignant potential of homozygous and heterozygous complete moles. *Cancer Res* 44:1226–1230, 1984.
81. Waterfield MD, Scrace, GT, Whittle N, Stroobant P, Johnson A, Wasteson A, Westermark B, Heldin CH, Huang JS, Deuel TF: Platelet-derived growth factor is structurally related to the putative transforming protein p28 sis of simian sarcoma virus. *Nature* 304:35–39, 1983.
82. Wiener F, Ohno S, Babonits M, Sumegi J, Wirschubsky Z, Klein G, Mushinski JF, Potter M: Hemizygous interstitial deletion of chromosome 15 (band D) in three translocation negative murine plasmacytomas. *Proc Natl Acad Sci USA* 81:1159–1163, 1984.
83. Winge O: Zytologische Untersuchungen uber die Natur maligner Tumoren. II. Teerkarzinome bei Mauzen. *Z Zellforsch* 10:683–735, 1930.

84. Yunis JJ, Ramsay N: Retinoblastoma and subband deletion of chromosome 13. *Am J Dis Child* 132:161–163, 1978.
85. Yunis JJ, Soreng L: Constitutional fragile sites and cancer. *Science* 226:1199–1204, 1984.
86. Caccia N, Kronenberg M, Saxe D, Haars R, Bruns GAP, Goverman J, Malissen M, Willard H, Yoshikai Y, Simon M, Hood L, Mak TW: The T-cell receptor β chain genes are located on chromosome 6 in mice and chromosome 7 in humans. *Cell* 37:1091–1099, 1984.*
87. Croce CM, Isobe M, Palumbo A, Duek J, Ming J, Tweardy D, Erikson J, Davis M, Rovera G: Gene for α-chain of human T-cell receptor: Location on chromosome 14 region involved in T-cell neoplasms. *Science* 227:1044–1047, 1985.*
88. De Taisne C, Gegonne A, Stehelin D, Dernheim A, Berger R: Chromosomal localization of the human proto-oncogene c-*ets*. *Nature* 310:581–583, 1984.*
89. Tsujimoto Y, Finger LR, Yunis J, Nowell PC, Croce CM: Cloning of the chromosome breakpoint of neoplastic β cell with the t(14;18) chromosome translocation. *Science* 26:1097–1099, 1984.*
90. Unpublished data from the author's laboratory.*

*References 86 through 90 are published and unpublished works describing gene localizations (Figure 4–1) that became available while this book was in production.

B. THE ROLE OF HEREDITY IN CLINICAL CANCER

5.

Heritable conditions that predispose to cancer

JAMES GERMAN

In roughly similar environments, some people develop a clinical cancer and others do not. Why should this be? Thinking about cancer etiology is different from thinking about, say, causative agents in infectious disease. If a *Treponema pallidum* enters the tissues and proliferates, one can expect to get syphilis; if a critical number of *Salmonella typhi* enter the stomach, one can expect to get salmonellosis. With cancer, however, although predisposing environmental agents such as sunlight, ionizing radiation, and certain chemicals have long been recognized, chance plays an enormous role, for the following reason: a population of cells that has the clinically recognized characteristics of cancer is the derivative of a single cell that underwent neoplastic transformation. A normal somatic cell can become neoplastic only through certain mutational events, and very possibly these events must occur in a specific sequence. Although the events may be of different types and may affect any of a number of genetic loci, a degree of specificity is known to exist, at least for certain cell types (e.g., of one particular breakpoint in a chromosome, or of activation of one particular oncogene). Therefore, it will be in only an occasional person that the necessary, rare stochastic events will occur in the proper sequence and in just the right cell type. Nevertheless, it is natural to ask whether significant differences exist from one person to the next in cancer susceptibility. Are the rare mutational events more likely to occur in cells of certain people? Or, a separate but related question, is a neoplastic cell or clone more likely to progress to overt (clinical) cancer in some persons than in others?

Evidence for the existence of a strong genetic component in the generality of human cancer has never been obtained. (However, the difficulties and pitfalls inherent in studies aimed at estimating the genetic contribution to the occurrence of common cancers in the general population have been discussed appositely by Peto [18].) Once a pedigree analysis of any particular patient convinces the physician that the family is not one of the exceptional ones in which one of the recognized cancer-predisposing genetic determinants is segregating (see tables later in this chapter), in the vast majority of cases the patient can be told—more exactly, reassured—that members of the family are not genetically predisposed to cancer. The physician, of course, must be on the alert for the exceptional family that actually is genetically predisposed, in which case appropriate counseling and risk figures can be provided. In some cases the pedigree analysis will drastically alter the physician's handling of the family. Whether genetic predisposition to cancer does or does not exist, offering to provide the affected patients and their families accurate information about the disease affecting them—and then providing it if the offer is accepted, as it usually is—is an integral part of good medicine today.

When is genetics important?

Pedigrees of so-called cancer families have appeared in the medical literature for over a century [22]. Some families exhibit a remarkable clustering of cancers. The familial occurrence of a trait certainly can be the first indication that it is genetically determined; however, all that's familial is not genetic. Clustering of cancer in families can be explained in one of two ways. In *Explanation a,* susceptibility and resistance to clinical cancer, although ultimately genetically determined, are quite constant from person to person throughout the general population, and families with an exceptionally great or small number of affected members represent nothing more than the inevitable extremes of the normal distribution of a random event. According to *Explanation b,* certain families really are different, having many affected members either because of excessive exposure of the family to some carcinogenic environment or because of the transmission of some genetic determinant or determinants that strongly predispose to cancer. Clearly, *Explanation b* removes certain families completely from the normal distribution postulated in *Explanation a,* and puts them in a different category with respect to counseling.

It is safe to conclude that in many cases familial clustering of cancer results simply from the chance occurrence of a common disorder, and there, genetics need not be invoked as an explanation. However, as is shown in the chapters on recessive and dominant inheritance in human cancer [8, 20], in rare families, clustering is caused by discrete genetic factors (referred to hereafter as genes) that undergo straightforward mendelian segregation; these genes are recognizable because they are responsible for diagnosable phenotypes, often distinct clinical syndromes. A family in which such a single, cancer-predisposing gene is segregating can be usually identified readily if a history is taken to determine the distribution in the family of cases of disease (a so-called genetic or family history) and if the person with cancer who brings the family to the physician's attention receives an adequate clinical evaluation. In addition to these already recognized genes, however, pedigree evidence is slowly accumulating that points to the existence of others that produce no yet recognizable trait other than cancer proneness itself but that can be transmitted for many generations and be responsible for a dominant inheritance pattern of site-specific cancer. (The two best examples are (1) the impressive "Family G," the subject of medical reports since 1913, in which an exceptionally large number of adenocarcinomas of the colon and uterus occur [14, 29], and (2) certain families in which considerably more childhood sarcomas, early-onset breast cancers, and leukemias than expected occur [12]. These families are believed to be transmitting genes for one of the "cancer-family syndromes.") Finally, so-called multifactorial inheritance is believed to be responsible for increased cancer incidence in some kindreds [8]. The taking of a careful family history concerning cancer (type, location, age of onset, multiplicity) will in many cases identify these families [commented on further in 8, 11, 22, 29].

In summary, the risk of cancer for an individual or a family can be increased substantially over that in the general population by the inheritance of the recognized genetic constitutions just discussed. The only other basis known for increased cancer risk, and one which also can affect either single individuals or multiple members of a family, is excessive exposure to a carcinogenic environment. Although the subject of environmental carcinogenesis is outside the scope of this volume, it and genetics are best considered simultaneously when explanations are sought for any of the following: familial clustering of cancer, onset of cancer in an exceptionally young person, or the occurrence of multiple neoplasms in one person. In other words, the features generally associated with genetically determined cancer also can be those of cancer caused by recognized environmental carcinogens.

Tabulations of cancer-predisposing genetic conditions

In recent years, listings of environmental carcinogenic agents and situations have been published frequently (e.g., [7]). As companions to those lists, the tables in this chapter list genetically determined conditions that predispose significantly to benign or malignant neoplasia. I prepared these tables primarily for the use of physicians dealing with cancer, and, therefore, they contain far fewer entries than the extensive list published earlier by Mulvihill [16, 17].

Tables 5–1 and 5–2 list cancer-predisposing clinical entities that are inherited in a simple mendelian fashion. Each entity in the two lists is rare; collectively they make only a small impact on clinical oncology. In some of the entities, neoplasia itself is the major clinical manifestation, or at least a major complication, the risk of which is great enough to be of significance in the overall management of affected persons. In other entities, only a few cancers have ever been reported; these are less important in clinical oncology, because cancer is not a major consideration in their medical management. In 1975 [16] and again in 1977 [17] Mulvihill examined the then-current editions of McKusick's cata-

Table 5–1 Recessively transmitted clinical entities that predispose to neoplasia.

Clinical entity	Reference*
Immunodeficiency-associated disorders	
Ataxia-telangiectasia†	20890
Bloom's syndrome†	21090
Wiskott-Aldrich syndrome‡	30100
Other genetically determined immunodeficiencies [23]	
Defects of DNA repair	
Xeroderma pigmentosum (XP), excision-repair defective†	27870-4, 27876, 27878
XP, "variant"	27875
Disturbances of sexual development	
Testicular feminization‡	31370
XY gonadal dysgenesis‡	30610
Miscellaneous entities	
Albinism	20310, 20320
Chediak-Higashi syndrome	21450
Dyskeratosis congenita‡	30500
Epidermolysis bullosa dystrophica, Hallopeau-Siemens type	22660
Fanconi's anemia†	22765, 22766
Glycogen-storage disease (von Gierke's disease)	23220
Hemochromatosis	23520
Retinoblastoma/osteogenic sarcoma§	18020
Turcot's syndrome	27630
Werner's syndrome†	27770

*Number of entry in V.A. McKusick's *Mendelian Inheritance in Man* [15].

†Also classifiable as a chromosome-breakage syndrome [8].

‡X-linked transmission.

§For an argument for recessive inheritance, see References 8 and 24.

Table 5–2 Dominantly transmitted clinical entities that predispose to neoplasia.

Clinical entity	Reference*
Gastrointestinal tract tumors predominant	
Polyposis, intestinal, adenomatous, type I (familial polyposis of the colon)	17510
Polyposis, intestinal, hamartomatous, type II (Peutz-Jeghers syndrome)	17520
Polyposis, intestinal, adenomatous, type III (Gardner's syndrome)	17530
Tylosis with esophageal cancer	14850
Integumentary tumors predominant	
Nevoid basal cell carcinoma syndrome	10940
Epithelioma, hereditary multiple benign cystic	13270
Epithelioma, self-healing squamous	13280
Focal dermal hypoplasia (Goltz's syndrome)	30560
Dysplastic nevus syndrome (melanoma)	15560
Nevi (pigmented moles)	16290
Scleroatrophic and keratotic dermatosis of limbs	18160
Nervous-system tumors predominant	
Carotid-body tumor (chemodectoma)	16800
Neurinoma, bilateral acoustic	10100
Neurofibromatosis	16220, 16222
Pheochromocytoma	17130
Other entities	
Acoustic neurinoma, bilateral	10100
Cylindromatosis	12385
Exostosis, hereditary multiple	13370
Fibrocystic pulmonary dysplasia	13500
Glomus tumors, multiple	13800
Leiomyomas, hereditary multiple, of skin	15080
Lipomas, multiple	15190
Multiple endocrine neoplasia syndrome (MEN), type I	13110
MEN, type II	17140
MEN, type II	16230
Multiple hamartomas syndrome	15835
Von Hippel-Lindau syndrome	19330
Tuberous sclerosis	19110

*Number of entry in V.A. McKusick's *Mendelian Inheritance in Man* [15].

log of inherited human phenotypes *Mendelian Inheritance in Man* and reported that in 8 to 9% of the hundreds of known single-gene or possibly single-gene traits, neoplasia was "a sole feature, a frequent concomitant, or just a rare complication." He tabulated more than 200 such conditions [17] in order to emphasize "the large number of genes that might be involved in cancer susceptibility," which (as stated) is different from my purpose here. Consequently, entries in Tables 5–1

Table 5–3 Familial clustering of neoplasia possibly due to dominant inheritance.

Clinical entity	Reference*
Breast cancer [1, 2, 11, 30]	21200
Cancer-family syndrome I (hereditary adenocarcinomatosis [colon, endometrium]) [13, 29]	11440
Cancer-family syndrome II (breast, soft tissue) [12]	
Colon cancer [2, 29]	11450
Muir-Torre syndrome†	15832
Ovarian tumor	16700
Wilms' tumor (associated with aniridia, gonadoblastoma, genital ambiguity, mental deficiency)	19407

*Number of entry in V.A. McKusick's *Mendelian Inheritance in Man* [15].

†Multiple cutaneous sebaceous neoplasms and keratoacanthomas with gastrointestinal and other cancers.

and 5–2 represent only 3 to 4% of the more than 1600 recognized, genetically determined traits listed in the current catalog [15].

Table 5–3 lists some prominent examples of familial clustering of cancer. Table 5–4 lists chromosomal disorders that can be associated with cancer.

Table 5-1: A gene that determines a recessively transmitted disorder usually affects either the structure or the amount of synthesis of an enzyme. A single normal allele at the locus is sufficient for normal functioning of the metabolic pathway concerned, so that only persons homozygous for the undesirable allele present an abnormal phenotype, that is, the clinical disorder. Persons who are heterozygous for rare genes that, when homozygous, are responsible for a disorder that pre-

Table 5–4 Constitutional chromosomal abnormalities that predispose to neoplasia.

Abnormality	Neoplasm
Inherited	
Translocation affecting Nos. 3 & 8 [5]*	Hypernephroma
Translocation affecting 11p [10, 19]	Wilms's tumor, with aniridia
Translocation-insertion affecting 13q [25, 28]	Retinoblastoma; osteosarcoma
De novo	
Deletion affecting 11p13 [10]	Wilms's tumor, with aniridia
Deletion affecting 13q14 [10]	Retinoblastoma
Down's syndrome (trisomy 21)†	Acute leukemia in infancy
Klinefelter's syndrome (47,XXY)	Germ-line-cell tumor [21]
Gonadal dysgenesis due to mosaicism 45,X/46,XY	Gonadoblastoma

*A single family.

†Trisomy either from nondisjunction (47 chromosomes) or by translocation [9].

disposes to cancer do not appear to be significantly cancer-prone. (Published reports that heterozygotes develop cancer at excessive rates have been made in the case of two of the disorders, xeroderma pigmentosum [26] and ataxia-telangiectasia [27], but these reports as yet are unconfirmed.)

For the vast majority of recessive cancer-predisposing conditions the enzymes affected remain unidentified. Eventually, identification of the specific defects in several of the conditions should provide a better understanding of neoplastic transformation and tumor progression. On the other hand, in a few of the recessive conditions listed in Table 5–1 the emergence of a neoplasm may be so far removed from the main function of the mutant gene's product that elucidation of the defect will have little significance with respect to a basic understanding of neoplasia itself, for example, skin cancer in albinism or hepatoma in the extensively damaged livers of hemochromatosis and von Gierke's disease.

Table 5–2: In a large proportion of the dominantly transmitted entities, neoplasia of some distinctive type is itself the major or sole clinical feature (e.g., the intestinal polyposes, the dysplastic nevus syndrome).

In contrast to the recessively transmitted entities, enzyme defects usually are not demonstrable in the dominants. In the dominantly transmitted disorders, the eventual demonstration of a variety of interesting defects can be anticipated, including regulatory defects, cell-surface-receptor defects, and abnormal structural proteins. In the dominants as in the recessives, new insight into neoplasia itself will emerge as the basic defects are elucidated.

Table 5–3: A listing of familial, but not necessarily genetic, neoplasia is included here for completeness. The genes possibly responsible for the familial aggregations of neoplasia are largely unidentified, and in at least some cases both genetic and environmental factors may be acting together in systems of multifactorial inheritance. An argument can be made for moving at least two of the entities in this table to Table 5–2 (the "Dominant" table), namely, the so-called cancer-family syndromes [12, 13, 14, 29]. Dominant inheritance seems also to account for some familial clustering of both colon and breast cancer. Given the lack of understanding of the significance of the familial occurrence of certain of the common cancers in the entities represented in Table 5–3, it seems prudent for counseling purposes to assume that unitary genetic determinants that greatly increase the risk of clinical cancer are segregating in at least some affected families. We may expect clarification in the next few years of the role of genetics in the common cancers that occasionally show familial aggregation. This will come about in part through application of the remarkably sophisticated methods of genetic epidemiology recently developed, or under development now, to very carefully collected sets of human family and human population data.

Such methods can probably eventually be used by a physician to decide, first, whether a significant excess of cancer exists in a given family under consideration [3], and, if so, the probable pattern of transmission [2, 11, 30].

Table 5–4: A few inherited chromosome rearrangements that predispose to some specific cancer are known, and each is mentioned or discussed elsewhere in this book. Analysis of the "dominant" cancer-associated conditions listed in Table 5–2 probably will reveal additional examples of inherited chromosome abnormality, in some cases affecting very short chromosome segments, especially when molecular cytogenetic techniques are employed along with conventional banding procedures, as has been done recently in retinoblastoma [4, 24].

Also, certain noninherited constitutional chromosome imbalances that predispose to neoplasia occur de novo in the affected individual. Table 5–4 indicates that imbalance affecting 11p and 13q can predispose to Wilms's tumor and retinoblastoma, respectively. The table lists two trisomic states. Persons born with one of these, trisomy 21, manifest a poorly understood dysregulation of hematopoiesis that occasionally eventuates in acute leukemia in infancy. The leukemia—sometimes just a leukemoid picture—may be present at birth, or it may appear during infancy or early childhood. (Trisomy 21 has not been shown to predispose to cancer of other types.) In Klinefelter's syndrome, the phenotype produced by an extra X chromosome in males, several examples of a rare type of tumor derived from germ-line cells have been reported, enough to permit the conclusion that the chromosome complement 47,XXY predisposes to that tumor. Breast cancer does occur in persons with Klinefelter's syndrome, but whether it is more common there than in other males is unknown. Finally, in the gonads or gonadal remnants of persons with chromosomal mosaicism in which cells bearing a Y chromosome (46,XY) or a structurally abnormal Y derivative are mixed with cells totally lacking the Y chromosome (45,X), gonadoblastoma and certain other more malignant tumors of gonadal origin frequently occur.

Also for completeness, mention must be made of certain known associations of genetically determined traits and a cancer predisposition. With respect to the ABO blood group locus, persons of erythrocyte Type A are at a somewhat greater risk of stomach cancer and several other cancers than persons of Type O. With respect to the major histocompatibility locus, population studies have revealed a few significant associations between certain cancers and a specific HLA type [6], for example, nasopharyngeal carcinoma, certain leukemias, and Hodgkin's disease. Such associations, although of theoretical interest, as yet have little significance for the physician caring for a family affected with cancer.

Finally, in the listing of inherited, genetic factors that are important in neoplasia, cellular oncogenes must be mentioned. Oncogenes are discussed elsewhere in this book, but they are not tabulated or emphasized in this particular chapter because they currently are understood to play roles in normal development and the control of normal cellular proliferation. Their abnormal activation by any of several types of mutational events is believed to divert a relatively nonproliferative stem cell into irreversible, unscheduled cell cycling. They do not predispose to cancer per se.

Literature Cited

1. Anderson DE: Genetic study of breast cancer: Identification of a high risk group. *Cancer* 34:1090–1097, 1974.
2. Anderson DE, Williams WR: Familial cancer: Implications for healthy relatives. Chapter 15 in this volume.
3. Bale SJ, Chakravarti A, Strong LC: Aggregation of colon cancer in family data. *Genet Epidemiol* 1:53–61, 1984.
4. Benedict WF, Murphree AL, Banerjee A, Spina CA, Sparkes MC, Sparkes RS: Patient with 13 chromosome deletion: Evidence that the retinoblastoma gene is a recessive cancer gene. *Science* 219:973–975, 1983.
5. Cohen AJ, Li FP, Berg S, Marchetto DJ, Tsai S, Jacobs SC, Brown RS: Hereditary renal cell carcinoma associated with chromosomal translocation. *N Engl J Med* 301:592–595, 1979.
6. Dausset J, Colombani J, Hors J: Major histocompatibility complex and cancer, with special reference to human familial tumours (Hodgkin's disease and other malignancies). *In* Inheritance of Susceptibility to Cancer in Man (Bodmer WF, ed) Oxford, Oxford University Press, 1983, 119–147.
7. Doll R, Peto R: The causes of cancer. *J Natl Cancer Inst* 66:1203, 1981.
8. German J: Recessive inheritance in human cancer. Chapter 6 in this volume.
9. German J, DeMayo AP, Bearn AG: Inheritance of an abnormal chromosome in Down's syndrome (mongolism) with leukemia. *Am J Hum Genet* 14:31–43, 1962.
10. Harnden DG, Herbert A: Association of constitutional chromosome rearrangements with neoplasia. *In* Inheritance of Susceptibility to Cancer in Man (Bodmer WF, ed) Oxford, Oxford University Press, 1983, 150–173.
11. King MC: Genetics and epidemiology of susceptibility to human breast cancer. Chapter 8 in this volume.
12. Li FP, Fraumeni JF, Jr: Prospective study of a family cancer syndrome. *JAMA* 247:2692–2694, 1982.
13. Lynch H, Krush AJ: Differential diagnosis of the cancer family syndrome. *Surg Gynecol Obstet* 136:221–224, 1973.
14. Lynch HT, Krush AJ, Thomas RJ, Lynch J: Cancer family syndrome. *In* Cancer Genetics (Lynch HT, ed) Springfield, IL, Charles C Thomas, 1977, 355–388.
15. McKusick VA: Mendelian Inheritance in Man. Catalogs of Autosomal Dominant, Autosomal Recessive and X-Linked Phenotypes, 6th Ed. Baltimore, Johns Hopkins University Press, 1983.
16. Mulvihill JJ: Congenital and genetic diseases. *In* Persons at High Risk of Cancer (Fraumeni JF Jr, ed) New York, Academic Press, 1975, 3–37.
17. Mulvihill JJ: Genetic repertory of human neoplasia. *In* Genetics of Human Cancer (Mulvihill JJ, Miller RW, Fraumeni JF Jr, eds), Raven Press, New York, 1977, 137–143.

18. Peto J: Genetic predisposition to cancer. *In* Banbury Report 4—Cancer Incidence in Defined Populations. Cold Spring Harbor, NY, Cold Spring Harbor Laboratory, 1980, 203–213.
19. Riccardi VM, Hittner HM, Francke U, Yunis JJ, Ledbetter D, Borges W: The aniridia-Wilms' tumor association: The critical role of chromosome band 11p13. *Cancer Genet Cytogenet* 2:131–137, 1980.
20. Schimke RN: Dominant inheritance in human cancer. Chapter 7 in this volume.
21. Schimke RN, Madigan CM, Silver BJ, Fabian CJ, Stephens RL: Choriocarcinoma, thyrotoxicosis, and the Klinefelter syndrome. *Cancer Genet Cytogenet* 9:1–8, 1983.
22. Schneider NR: Familial aggregation of cancer. Chapter 9 in this volume.
23. Spector BD, Perry GS, III, Kersey JH: Genetically determined immunodeficiency disease (GDID) and malignancy: Report from the Immunodeficiency-Cancer Registry. *Clin Immunol Immunopathol* 11:12–29, 1978.
24. Strong LC: Mutational models for cancer etiology. Chapter 3 in this volume.
25. Strong LC, Riccardi VM, Ferrell RE, Sparkes RS: Familial retinoblastoma and chromosome 13 deletion transmitted via an insertional translocation. *Science* 213:1501–1503, 1981.
26. Swift M, Chase C: Cancer in families with xeroderma pigmentosum. *J Natl Cancer Inst* 62:1415–1421, 1979.
27. Swift M, Sholman L, Perry M, Chase C: Malignant neoplasms in the families of patients with ataxia-telangiectasia. *Cancer Res* 36:209–215, 1976.
28. Turleau C, Grouchy Jde, Chavin-Colin F, Despoisses S, Leblanc A: Two cases of del(13q) retinoblastoma and two cases of partial trisomy due to a familial insertion. *Ann Genet* 26:158–160, 1983.
29. Williams CJ: Managing families genetically predisposed to cancer: The "cancer-family syndrome" as a model. Chapter 14 in this volume.
30. Williams WR, Anderson DE: Genetic epidemiology of breast cancer: Segregation analysis of 200 Danish pedigrees. *Genet Epidemiol* 1:7–20, 1984.

6.

Recessive inheritance in human cancer

JAMES GERMAN

The role of dominant inheritance in human cancer predisposition is considered elsewhere in this volume [23], as are the familial aspects of breast and colon cancer [1, 13]. This chapter addresses recessive inheritance, including so-called multifactorial inheritance. Although much has been written about certain rare, recessively transmitted disorders that predispose to cancer, little consideration has been given to the role of recessive inheritance in general in the causation of cancer in the population at large. After some general statements about recessive and multifactorial inheritance themselves, this chapter will examine and discuss the evidence for such inheritance in the determination of human cancer.

Recessive inheritance

An autosomal recessive trait (the "trait" may be a clinical disorder) appears in a person only if he or she is homozygous for the particular gene determining the trait. The same gene is inherited from each of that person's parents, neither of whom manifests the trait; in fact, carriers of the gene, heterozygotes, usually are identified as such only at the time of the birth of a homozygous-affected offspring. Phenotypically normal persons are presumed to carry several rare, potentially disease-producing genes, and probably will transmit each of them to half of their descendants with impunity. (Examples of potentially detrimental genes we might unknowingly carry are those for phenylketonuria, the Tay-Sachs disease, and some form of deafness.) The possibility of genetic disease arises only if one heterozygous individual marries a person heterozygous for the same gene. In the case of such rare genes, if several members of a family manifest the clinical disorder, those affected as a rule will be limited to a single sibship, the genetic history otherwise being negative. That most recessive genes of clinical significance are rare is reflected by an increased consanguinity rate among parents of affected homozygotes. Some known recessive genes are not rare (e.g., the gene[s] for cystic fibrosis, and the Tay-Sachs gene in Ashkenazi Jews with Polish ancestry); an increased consanguinity rate is not expected nor is it found among the parents of persons homozygous for common recessive genes. Several hundred recessively transmitted traits that qualify as disease are known in man [17], indicating that a large number of potentially detrimental genes are transmitted in the healthy human population.

Multifactorial inheritance

Because members of a family share not only their genetic determinants but, frequently, their environment and behavior patterns as well, and because of the biological interaction that takes place between these several factors, the model of multifactorial inheritance usually is invoked to explain the etiology of many common disorders. By this concept, if aggregation of an unusually great number of genes that predispose individually to a given trait occurs by chance in a single kindred, an exceptionally large number of family members will manifest the trait, given the appropriate environment. Multifactorial inheritance is believed to be important in the etiology of those congenital malformations that occur with a relatively great frequency in man ($>1/1000$ births) (e.g., defects of the neural tube, heart, and palate) as well as of many of the common diseases of adulthood (e.g., hypertension and various acquired heart diseases). Because in multifactorial inheritance the various genes that determine the expression of some trait nearly always go unidentified and because the genes are postulated to act as a consortium, not being associated individually with any particular phenotype, it is appropriate to mention this type of inheritance in a consideration of the role of recessive inheritance in cancer.

Rare, recessively transmitted disorders that predispose the affected homozygote to cancer

The most important recessively transmitted, cancer-predisposing disorders fall into three groups (Tables 6–1 through 6–3), although con-

Table 6–1 Recessively inherited cancer predisposition: Immunodeficiencies.

Ataxia-telangiectasia*†
Bloom's syndrome*
Severe combined immunodeficiency
Wiskott-Aldrich syndrome‡
Other types [26]

*Also classified as a chromosome-breakage syndrome in Table 6–2.

†Genetic heterogeneity reported.

‡X-linked inheritance.

Table 6–2 Recessively inherited cancer predisposition: Chromosome-breakage syndromes.

Ataxia-telangiectasia*†
Bloom's syndrome*
Fanconi's anemia†
Werner's syndrome
Xeroderma pigmentosum, excision-repair deficient†‡

*Also classified as a genetically determined immunodeficiency in Table 6–1.

†Genetic heterogeneity reported.

‡Also classifed as a DNA-repair deficiency in Table 6–3.

siderable overlap among the groups exists (indicated in the table footnotes). Table 6–4 lists several additional recessively transmitted entities in which neoplasia occurs often enough to be of clinical significance. Recessive disorders in which cancer of some type has been reported other than those in these tables are known [9, 21], but many of them are themselves poorly understood entities or are accompanied only occasionally by neoplasia, or both.

Considerable scientific and medical attention has been directed recently to several of these disorders that predispose to clinical neoplasia and that exhibit a classic mendelian pattern of recessive transmission. (The conditions, all very rare, serve as naturally occurring experimental models for use in investigations in a number of basic areas of science.) For those not working primarily in the field of human cancer genetics, this attention may have suggested incorrectly that recessive inheritance is of considerable importance in cancer occurrence in the general population. The rarity of these disorders both individually and collectively, and the fact that they do not make a significant numerical impact on cancer incidence in the general population, may have been

Table 6–3 Recessively inherited cancer predisposition: DNA-repair deficiencies.

Entity	Complementation group
Xeroderma pigmentosum, excision-repair deficient type ("classic" XP)*	A B C D E F G H I
Xeroderma pigmentosum, daughter-strand-gap repair deficient type ("variant" XP)	

*Also classified as a chromosome-breakage syndrome in Table 6–2.

Table 6–4 Recessively inherited cancer predisposition: Miscellaneous entities.

Albinism
Chediak-Higashi syndrome
Dyskeratosis congenita*
Genetic syndromes that feature chronic liver disease†
Retinoblastoma/osteogenic sarcoma
Testicular feminization*
XY gonadal dysgenesis*

*X-linked inheritance.

†For example, glycogen-storage disease due to glucose-6-phosphatase deficiency; hemochromatosis.

underemphasized. Furthermore, confirmed studies that would indicate that the carriers of any of these rare genes are cancer-prone are unavailable.

Genetically determined immunodeficiencies

The first of several genetically determined immunodeficiencies to be recognized was described a little more than 30 years ago. By 1971 it was apparent that cancer occurred with increased frequency in the majority of these rare conditions (Table 6–1); a registry of cancer in affected persons has documented this dramatically [26]. Later, an increased frequency of cancer was recognized to exist in immunodeficiencies induced for medical reasons, mainly in association with organ transplantation; more recently, cancer-proneness has been recognized to be a feature of the acquired immunodeficiency syndrome (AIDS). Together, these observations suggest that the immune system plays some role in neoplasia and raise the possibility that depressed immune function, whatever its cause, may be a factor in the emergence of clinical cancer.

Chromosome-breakage syndromes

Several disorders feature increased genomic instability that can be recognized microscopically—"chromosome breakage" (Table 6–2). Bloom's syndrome is the prototype of this group; in this condition, as in the case of the other genetically determined immunodeficiencies, an international registry of affected persons documents their striking cancer predisposition [8]. Ataxia-telangiectasia, Fanconi's anemia, Werner's syndrome, and sometimes xeroderma pigmentosum are also considered chromosome-breakage syndromes. (Brief descriptions of four of these entities appear elsewhere in this book [10].) Each of these rare, cancer-predisposing conditions has both a distinctive pattern of chromosome instability that affects certain cell types preferentially [22] and a distinctive distribution of cancer types [7]. It seems reasonable to hypothesize that the genetically determined genomic instability in each of these conditions increases the chance that the mutation(s) required for neoplastic transformation [28] will occur—thus their cancer-proneness. It is worth noting that two of the chromosome-breakage syndromes listed in Table 6–2, namely, Bloom's syndrome and ataxia-telangiectasia, also are genetically determined immunodeficiency diseases and, therefore, also listed in Table 6–1; this may be significant because these two conditions stand alone with respect to their enormously increased occurrence of clinical cancer, of many types.

DNA-repair defects

The prototype of the DNA-repair defects is the "classic" form of xeroderma pigmentosum, that is, the excision-repair-defective type (Table 6–3). In fact, xeroderma remains the only bona fide repair-defective disorder, although evidence of defective repair of certain types of DNA damage has been reported for both Fanconi's anemia and ataxia-telangiectasia. In xeroderma, the defective or retarded repair of lesions in DNA produced either by the interaction of DNA with noxious environmental agents or by intracellular factors is considered responsible for the mutations that lead to neoplastic transformation.

Other rare, recessively transmitted disorders

Several other rare disorders occasionally are associated with cancer. In the entities listed in Tables 6–1 through 6–3, the cancer propensity is a major consideration. For those in Table 4, with the exception of retinoblastoma, cancer, although distinctly more common than in the general population, usually does not assume major significance in the program of clinical management, or else can be easily avoided or controlled. Some idea exists of the basis for the increased cancer occurrence for several of the entities in Table 6–4. Skin cancer in albinism, for example, may be supposed to result from excessive sunlight damage to dermal stem cells inadequately shielded by melanin; and in glycogen-storage disease and hemochromatosis, cancer of the liver is presumably secondary to chronic liver disease, which itself is a complicating feature of the disease. As yet, only one of the conditions listed in Table 6–4 has emerged as a useful model for the experimental study of cancer—hereditary retinoblastoma (see following section)—in contrast to the situation in the immunodeficiency diseases, the chromosome-breakage syndromes, and the DNA-repair defects.

Recessiveness at the cellular level—a different aspect of recessive inheritance

The several chromosome-breakage syndromes and DNA-repair deficiencies mentioned above demonstrate that classic recessive inheritance sometimes can be important in determining the probability of cancer occurring. There, homozygosity in a somatic cell for any of several mutant genes apparently increases the cell's risk of undergoing neoplastic transformation. As was mentioned, the heterozygote for these several genes does not appear to be unusually cancer-prone. However, in the somatic tissues of a heterozygote for, say, the Bloom's syndrome or the ataxia-telangiectasia gene, crossing-over theoretically can give rise to a clone of cells that would be homozygous at the mutant locus. Cells of that clone should then have chromosome instability, and they presumably would be at increased risk of neoplastic transformation. Whether or not this occurs in carriers of these genes is unknown.

The retinoblastoma model

In the case of a mutation at another locus, however—namely, the so-called retinoblastoma (RB) locus mapped to chromosome no. 13—somatic crossing-over and other genetic mechanisms that can give rise to homozygosity of the mutant locus have been shown to occur and to be crucial in determining neoplasia. In hereditary RB, the mutant locus itself is inherited through the germ line in a straightforward mendelian manner, that is, from a heterozygote in the previous generation. The somatic cells of the person inheriting the mutation are heterozygous, but phenotypically they are normal; specifically, as far as is known they proliferate in a normally controlled way and form normally functioning tissues, such as of the retina or the bone. Then, through some mutational event that affects the nonmutant RB locus on the homologous no. 13 chromosome—for example, point mutation, deletion, crossing-over at a point between the locus and the centromere, or possibly a mitotic error involving nondisjunction—homozygosity, or possibly hemizygosity, for the RB mutation is brought about in a cell of an otherwise heterozygous person [3]. If the cell in which the mutation occurs is a stem cell of the retina or of osteogenic tissue, neoplastic transformation can be the consequence. The homozygous (or hemizygous) cell proliferates relatively autonomously, in due time giving rise to a clone recognizable as a tumor, of the retina or bone.

That necessary second mutational event occurs with such great frequency that most persons who inherit the crucial allele (or deletion) at the RB locus can expect to develop not just one but several neoplasms in their retinal tissue and, if they survive those, in osteogenic tissue.

Because of the great frequency of cancer in carriers of this mutation, pedigree analysis usually has been interpreted to indicate dominant transmission of cancer itself. However, hereditary RB is properly said to be transmitted recessively because, as first was pointed out by DeMars [5]:

> many of the pedigrees that are labelled as autosomal dominants . . . could actually be interpreted as autosomal recessives. We must relate the terms dominant and recessive not only to the level of the individual as a whole . . . but also to the level of the individual cell, or cells, which are involved . . . I think many pedigrees are consistent with the notion that one of the parents in the families might be heterozygous for a recessive and that the neoplasms appear as a result of subsequent somatic mutations in which individual cells become homozygous for a recessive neoplasm-causing gene.

(RB sometimes is described as recessive at the cellular level and dominant at the pedigree level. The provisional nature of the use of the terms *recessive* and *dominant* in relation to genes and the traits they determine is emphasized by the observations in RB.)

Knudson's two-mutation hypothesis for cancer, developed first to explain both familial and nonfamilial RB [14] and subsequently extended to other neoplasms [28], proposes that two or more mutations are required for neoplastic transformation of a somatic cell and that in a proportion of many types of cancers the first of these can be inherited through the germ line. In the case of such inheritance, although all cells in the body would be heterozygous for one of the mutant genes, they would function normally. Thus, in persons heterozygous for the RB mutation, retinal and osteogenic cells, although predisposed to retinoblastoma and osteosarcoma by virtue of being heterozygous for the crucial mutation, appear phenotypically normal—recessive transmission.

Lack of evidence for recessive inheritance in the generality of human cancer

If recessive genes exist that predispose significantly to cancer as encountered in the general population, indications of their existence should be detectable by the appropriate analyses of affected families. What has been published pertaining to this?

Pedigree analysis of persons with cancer

Recessive inheritance would be indicated to be significant if sibs were found to have cancer more often than other first-degree relatives, as well as more often than would be predicted from the overall cancer incidence in the general population. For multifactorial inheritance, both sibs and other first-degree relatives (i.e., parents and children) of the index case would be affected more often than more distant relatives. In a study of the adult patient population of Memorial Hospital of New York City, Schneider et al. [24] found, in agreement with earlier investigators elsewhere, that, as a group, first-degree relatives of cancer patients have somewhat more cancer than would be expected in the general population. However, the cancer distribution in the families studied did not suggest any simple genetic basis. In a genetic epidemiological study of 18 families in which more than one case of breast cancer had occurred, King [13] obtained results "consistent with the hypothesis that breast and some associated cancers have a genetic etiology." She then asked whether the distributions of affected persons in the families were compatible with any of several different genetic hypotheses. Although the distributions seemed most compatible with autosomal dominant inheritance, the hypothesis of autosomal recessive "could not be rejected." King emphasized that the analysis of large samples of such families will be required to discriminate between the various genetic hypotheses that can be made.

Another relevant observation is that neoplasia affects more than one child in a family slightly more often than would be expected by chance [2, 6, 19]. Also, isolated reports have appeared of sibs with childhood leukemia; Steinberg [27] cited three such families, in two of which parental consanguinity was recorded, and correctly observed that in very rare families, recessive inheritance may be of importance in the causation of leukemia. In families not known to be transmitting a gene that predisposes strongly to cancer (such as a retinoblastoma mutation, or the Bloom's syndrome gene in both parents), the risk to a child who is the sib of a child with cancer has been estimated on the basis of genetic epidemiologic studies to be approximately 1 in 300, or about double that in the general population of 1 in 600 [6]. Although compatible with the possibility that recessive or multifactorial inheritance plays a role in human cancer, such observations certainly may have another basis, notably, shared carcinogenic environments. In any case, the evidence derived from pedigree analyses of populations of cancer patients at large suggests that recessive inheritance plays no important causative role in most human cancer.

Twin studies

If recessive or multifactorial inheritance is important in determining human cancer, monozygous (MZ) twins should exhibit a greater concordance for the disease than dizygous (DZ) twins; concordance with respect to location and type might even be expected. Yet, according to Holm et al. [12], who analyzed the extensive data in the Danish Twin Register, MZ and DZ twins fail to differ with respect to the occurrence of common malignant tumors, so that "no significant genetic determination of these cancers could be demonstrated."

Concordance has been reported to be greater for acute leukemia in childhood in MZ than in DZ twins [18–20]. However, a nongenetic explanation is available in at least some instances for the appearance of cancer in both members of a pair of MZ twins: transformation of a normal cell into a neoplastic one may occur in one twin before birth, giving rise to a clone of leukemic cells that disperses itself to both twins by way of their interconnected fetal circulations [4]. Thus, concordance for leukemia during infancy in MZ twins may not have its usual significance and is not necessarily indicative of recessive or multifactorial inheritance.

Inbreeding

A high rate of inbreeding within a population increases the frequency of homozygosity of rare genes and, therefore, the frequency with which rare autosomal recessive disorders occur. Traits determined by multifactorial inheritance might also be expected to be increased by inbreeding. If either recessive or multifactorial inheritance is important in the generality of human cancer, the existence of a high consanguinity rate in a population would be expected to increase the overall occurrence of cancer. Marriage between close relatives is common in Mohammedan countries, for example, 32% in Egypt [11]; therefore, if there are multiple recessive genes that are important in cancer determination, an exceptionally high cancer incidence would be expected in those populations. This has not been documented, however; unfortunately, reliable cancer-incidence figures from such countries usually are not available because of the medical standards as well as underreporting during family interviews because cancer is sometimes considered a stigma in such societies. A Japanese study did show excessive parental consanguinity for leukemic sibs compared with nonfamilial leukemics, with first-cousin parental relationships being recorded in 6 of 20 versus 9 of 200, respectively [15]. However, a careful genetic study of 249 American children with acute leukemia disclosed only a single instance of parental consanguinity [27], which argues against recessive inheritance being of etiologic importance in most cases of that type of neoplastic disease in the West.

Genetic isolates (e.g., inbred populations dwelling for many generations on small islands or in remote mountain valleys, or isolated by religion) characteristically have increased numbers of persons with recessively transmitted disease. This is a consequence of the so-called founder effect, combined with the high consanguinity rate. Any one such population will have just a few "bad" genes, for example, genes that when homozygous determine some form of disease such as a type of blindness, deafness, or skeletal dysplasia. The undesirable genes carried will depend on which genes had by chance been carried by the founders of the population. The next island, valley, or religious group probably will have a different set of undesirable genes and, therefore, different genetic diseases. Two brief reports of the cancer incidence in inbred populations have been published [16, 29]; however, I have been unable to learn of any genetic isolate that exhibits an impressively increased incidence of cancer. Should one be located, and if an environmental basis for the excessive cancer can be ruled out, the population will deserve intensive study by cancer epidemiologists and tumor biologists in an effort to identify and characterize recessive genes that are of general importance in human cancer. (One genetic isolate of interest in relation to cancer is a subpopulation of Moroccan Jews now in Israel but that had long dwelt in the Atlas Mountains. Ataxia-telangiectasia, a condition that in turn predisposes to cancer [Tables 6–1 and 6–2], occurs much more frequently in those Jews than in either the general population or other Jewish groups, presumably through the founder effect and inbreeding.) (For the rare, autosomal-recessively transmitted, cancer-predisposing disorders mentioned earlier [Tables 6–1 through 6–4], parental consanguinity rates are high, as expected.)

To summarize this section, it may be stated that if recessive genetic factors are important in determining a significant proportion of human cancer, not only do they remain unidentified but firm evidence for their existence has yet to be obtained. With respect to multifactorial inheritance, it may explain some familial clustering of cancer, but the vagueness of our understanding of clustering and of the concept of multifactorial inheritance itself makes this of little practical value in clinical medicine. For families that have several members with cancer but in which the distribution and type of cancer do not indicate dominant transmission of a site- or type-specific tumor (as is the case, for example, in familial polyposis of the colon and in the so-called cancer-family syndromes), and in which no rare genetically determined, cancer-predisposing syndrome or entity is segregating, it almost never can be determined whether the family represents one extreme of a normal distribution of cancer or that multifactorial inheritance is responsible.

Conclusions

Evidence has never been obtained for the segregation of recessive genes that predispose a large segment of the human population to cancer. In fact, in the generality of human cancer, heredity appears to be of little importance in determining who will and who will not be affected [9].

A small number of genes are known that do predispose significantly to cancer (listed in Tables 6–1 through 6–4 and in References 9 and 23). They are responsible for clinical disorders, some of which follow recessive and others dominant patterns of transmission. Even collectively, their numerical impact on the bulk of human cancer is insignificant. Nevertheless, the detection of the segregation of any one of those genes in a given family has great importance for that particular family. Upon encountering such a family, the clinician knowledgeable in the principles of genetics can provide accurate counseling, with respect to both the recurrence risk in the family of persons with the disorder and the risk of cancer in those affected, which unfortunately for several well-characterized disorders are great.

Familial clustering of cancer not associated with one of the simply inherited clinical disorders sometimes does occur. This is inevitable because of the frequency of cancer itself, and ordinarily chance alone appears to be an adequate explanation; that is, genetics need not be invoked to explain the clustering. Nevertheless, no reason exists a priori for rejecting so-called multifactorial inheritance as important in the predisposition to cancer, and in some families it is entirely possible that it is responsible for excessive instances of cancer. In that type of inheritance, however, the hypothetical genes involved go unrecognized, as do the environmental agents with which they are supposed to interact.

Rarely, a dominant gene that does not produce a recognizable clinical phenotype appears to be segregating in a family as the explanation for the clustering of cancer, as is discussed elsewhere in this volume [1, 9, 13, 23]. Unfortunately, only an analysis of a carefully constructed pedigree (see reference 30 for an example) can help clinicians decide whether they are dealing with one of these rare families; the study of additional genetic markers in such cases is not yet of any practical value.

That persons with cancer in general do have slightly but significantly more relatives with cancer than would be predicted from the general population-incidence figures is compatible with the hypothesis that recessive, cancer-determining genes really do exist, including those that play a role in multifactorial inheritance. However, even if true, such a conclusion is of minimal practical value in clinical oncology, that is, for counseling regarding recurrence risk in relatives of persons with cancer [25]. (On the other hand, fortunately, empiric risk figures are available for certain common cancers, notably breast and colon [1], and they are of considerable value in counseling affected families.)

Literature cited

1. Anderson DE, Williams WR: Familial cancer: Implications for healthy relatives. Chapter 15 in this volume.
2. Barber R, Spiers P: Oxford Survey of Childhood Cancers; Progress Report II Monthly Bulletin, Ministry of Health and the Public Health Laboratory Service 23:46–52, 1964.
3. Cavanee WK, Dryja TP, Phillips RA, Benedict WF, Godbout R, Gallie BL, Murphree AL, Strong LC, White RL: Expression of recessive alleles by chromosomal mechanisms in retinoblastoma. *Nature* 305:779–784, 1983.
4. Chaganti RSK, Miller DR, Meyers PA, German J: Cytogenetic evidence of the intrauterine origin of acute leukemia in monozygotic twins. *N Engl J Med* 300:1032–1034, 1979.
5. DeMars R: Discussion of paper by D. E. Anderson. *In* Genetic Concepts and Neoplasia, 23rd Annual Symposium on Fundamental Cancer Research (1969). Baltimore, Williams & Wilkins, 1970, 105–106.
6. Draper GJ, Heaf MM, Kinnier Wilson LM: Occurrence of childhood cancers among sibs and estimation of familial risks. *J Med Genet* 14:81–90, 1977.
7. German J: Patterns of neoplasia associated with the chromosome-breakage syndrome. *In* Chromosome Mutation and Neoplasia (German J, ed) New York, Alan R Liss, 1983, 97–134.
8. German J, Bloom D, Passarge E: Bloom's syndrome. XI. Progress report for 1983. *Clin Genet* 26:166–174, 1984.
9. German J: Heritable conditions that predispose to cancer. Chapter 5 in this volume.
10. German J: The significance of identifying a cancer-predisposed person: Lessons from the chromosome-breakage syndromes. Chapter 13 in this volume.
11. Hashem N: Consanguinity patterns among Egyptians with genetic or congenital anomalies. *Ain-Shams Med J* 19:261–270, 1968.
12. Holm NV, Hauge M, Jensen OM: Studies of cancer aetiology in a complete twin population: Breast cancer, colorectal cancer and leukaemia. *Cancer Surveys* 1:17–32, 1982.
13. King MC: Genetic epidemiology of human cancer: Application to familial breast cancer. Chapter 8 in this volume.
14. Knudson AG, Jr: Mutation and cancer: A statistical study of retinoblastoma. *Proc Natl Acad Med* 68:820–823, 1971.
15. Kurita S, Kamei Y, Ota K: Genetic studies on familial leukemia. *Cancer* 34:1098–1101, 1974.
16. Martin AO, Dunn JK, Rosenthal F, Kemel S, Grace M, Elias S, Sarto GE, Steinberg AG, Simpson JL: Relationship of inbreeding to cancer in an isolate. *Am J Hum Genet* 30:120A, 1978.
17. McKusick VA: Mendelian Inheritance in Man, 6th Ed. Baltimore, Johns Hopkins University Press, 1983.
18. MacMahon B, Levy M: Prenatal origin of childhood leukemia: Evidence from twins. *N Engl J Med* 270:1082–1085, 1964.

19. Miller RW: Down's syndrome (mongolism), other congenital malformations and cancers among the sibs of leukemic children. *N Engl J Med* 268:393–401, 1963.
20. Miller RW: Deaths from childhood leukemia and solid tumors among twins and other sibs in the United States, 1960–67. *J Natl Cancer Inst* 46:203–209, 1971.
21. Mulvihill JJ: Genetic repertory of human neoplasia. *In* Genetics of Human Cancer (Mulvihill JJ, Miller RW, Fraumeni JF, Jr, eds) New York, Raven Press, 1977, 137–143.
22. Ray JH, German J: The cytogenetics of the "chromosome-breakage syndromes." *In* Chromosome Mutation and Neoplasia (German J, ed) New York, Alan R Liss, 1983, 135–167.
23. Schimke RN: Dominant inheritance in human cancer. Chapter 7 in this volume.
24. Schneider NR, Chaganti SR, German J, Chaganti RSK: Familial predisposition to cancer and age at onset of disease in randomly selected cancer patients. *Am J Hum Genet* 35:454–467, 1983.
25. Schneider NR: Familial aggregation of cancer. Chapter 9 in this volume.
26. Spector BD, Perry GS III, Kersey JH: Genetically determined immunodeficiency disease (GDID) and malignancy: Report from the Immunodeficiency-Cancer Registry. *Clin Immunol Immunopathol* 11:12–29, 1978.
27. Steinberg AG: The genetics of acute leukemia in children. *Cancer* 13:985–999, 1960.
28. Strong LC: Mutational models for cancer etiology. Chapter 3 in this volume.
29. Thurmon TF, Robertson KP: Genetic considerations in human cancer incidence. *Public Health Rep* 94:471–476, 1979.
30. Williams CJ: Managing families genetically predisposed to cancer: The "cancer-family syndrome" as a model. Chapter 14 in this volume.

7.

Dominant inheritance in human cancer

R. NEIL SCHIMKE

The title of this chapter is somewhat misleading. When reference is made to cancer genes and their mode of inheritance, what is implied is an inherited predisposition to malignancy transmitted in families in a pattern consistent with a mendelian trait. In some disorders, such as familial polyposis coli, penetrance of the predisposing gene is virtually complete, in terms of not only intestinal polyps but also eventual colonic neoplasia. In other conditions, like neurofibromatosis, penetrance of the gene is high but the incidence of malignancy is low. Both of these observations—the delayed onset but inevitable development of malignancy in polyposis coli and the varying incidence of neoplasia in neurofibromatosis—are consistent with the idea that some environmental insult must be superimposed on the germinal mutation to trigger malignant degeneration. These somatic triggers quite likely are diverse and to some extent may be tissue specific.

There are a host of possible mutations that could provide the germinal substrate for later somatic induction of cancer. These mutations might be cancer-promoting in an active sense by allowing for unrestrained cell growth (or failing to actively suppress it) or cancer-facilitating by interfering with normal developmental processes, thereby inappropriately exposing a given tissue to an unfavorable postnatal and potentially carcinogenic milieu. Dominantly inherited mutations could be the usual single gene alterations or they could be cytogenetic abnormalities such as transmissible chromosome deletions. Chromosome rearrangements could lead to loss of sequential integrity of a series of related genes by separation or by position effect. Moreover, not all site-specific familial tumors need to be caused by the same genetic event; for example, colon tumors develop not only in polyposis coli but also in the Gardner and Torre-Muir syndromes. The genes responsible for these disorders are distinct; yet they lead to a common neoplasm. It is likely that all heritable tumors or tumor syndromes are genetically, and thus etiologically, heterogeneous. The entire topic has been reviewed recently [83].

Embryonal tumors

The prototype of a heritable embryonal tumor is *retinoblastoma*. It is clear that a sizable proportion, perhaps as many as 25 to 30%, of retinoblastomas are genetic [48]. The features of the heritable tumor, other than family history, that distinguish them from nongenetic tumors are more frequent bilaterality and earlier age of onset. The retinoblastoma gene is 80 to 90% penetrant in terms of retinoblastoma per se, but there is evidence that the gene has pleiotropic effects outside the retina, manifested as pineal tumors (ectopic retinoblastomas ?) [5], sarcomas [82], and perhaps even bladder carcinoma [42]. An additional 5% of retinoblastoma patients have been found to have a deletion of the chromosome region 13q14 [106]. This deletion may be sporadic or may be transmitted by means of such mechanisms as parental chromosome inversions [96] or translocations [76, 80, 98]. A very instructive study of a family illustrating a chromosomal basis for incomplete penetrance of familial retinoblastoma has been published by Strong et al. [98]. The tumors in this family were all unilateral, as were the majority of tumors in the sporadic deletion cases. In some sporadic cases, only the tumor has had the deletion, whereas the constitutional karyotype is normal [7]. However, Motegi [69] found mosaicism in 2 of 42 retinoblastoma patients with 51 and 9%, respectively, of peripheral lymphocytes carrying the requisite deleted segment of chromosome 13. Both had bilateral tumors, and each was a sporadic case.

The 13q14 deletion also includes a polymorphic enzyme locus, esterase D [97]. Qualitative and quantitative study of esterase D as well as restriction enzyme polymorphisms of chromosome 13 have provided strong evidence that an alteration (mutations, deletion, mitotic recombination) of both number 13 chromosomes is necessary before a tumor occurs [71]. Whether or not all familial retinoblastomas appear because of changes in chromosome 13 remains unproved [65].

Early estimates of the heritable fraction of *Wilms' tumor* have recently been revised downward by the results of a recent report of a national study group that noted only about 1% of cases to be familial [11]. An additional 5% had bilateral lesions and probably could be assumed to have genetic disease, perhaps on a mutational basis, although it will be years before this question can be answered with certainty. The bilateral cases may be heterogeneous, because there was an increased incidence of hypospadias and other anomalies in this group, implying perhaps that a postzygotic event in the urogenital ridge could account for both bilateral tumors and developmental anomalies in some instances. The data for survivors of unilateral Wilms' tumor support the idea that few of these tumors are heritable [33]. When definitely familial, Wilms' tumor is dominantly inherited, with penetrance of about 60 to 70% [66]. A chromosome alteration, specifically a deletion of 11p13, has been found to be associated with a syndrome comprising aniridia, mental retardation, genital ambiguity in males, renal anomalies, and Wilms' tumor [77]. The tumor is not inevitable in the deletion-based syndrome, as evidenced by the published reports including that of discordant monozygotic twins [67, 100]. Alternatively, not all children with Wilms' tumor–aniridia have an obvious deletion by constitutional chromosome analysis [78], although mosaicism is difficult to exclude, as some tumors have been shown to carry the deletion in the absence of a constitutional karyotypic abnormality [93]. There is evidence that, at least in some Wilms' tumors, the same sort of mechanism is operational as in retinoblastoma, but in this case it is both number 11 chromosomes that are altered [51]. A familial instance of the 11p- syndrome has resulted from an insertional translocation [107]. Patients with deletions, familial or sporadic, constitute only a minor component of the entire Wilms' tumor spectrum. An additional but not recognizably heritable Wilms' tumor subgroup would seem to be composed of patients who also have an XY karyotype and gonadal dysgenesis [74]. This group may not always be clearly demarcated from the Wilms'–aniridia patients who have been reported to have gonadoblastoma [100], a tumor also seen in patients with dysgenetic testes. It is conceivable that aniridia, when not complete, could be missed, or in any case could be an inconsistent marker when the typical chromosome abnormality is not present. Patients with congenital hemihypertrophy or with the Beckwith-Wiedemann syndrome are yet another minor subgroup at risk for Wilms' tumor. Both of these conditions may be dominantly inherited, and both seem to confer the increased risk for other embryonal tumors such as hepatoblastoma [94].

The other major solid neoplasm of childhood is *neuroblastoma.* There are relatively few reported affected families, perhaps in part because survival is generally so poor. The available pedigrees are consistent with autosomal dominant inheritance with penetrance figures similar to those of Wilms' tumor [49]. The incidence of neuroblastoma in situ is quite high as ascertained from routine autopsy series, but few of these lesions progress to frank tumors [8]. Then, too, spontaneous maturation of obvious tumors to more benign variants such as ganglioneuromas is not uncommon. Both heterochromia irides and aganglionic megacolon have been touted as alternative nontumorous expressions of the neuroblastoma gene [86]. One special class of neuroblastoma patients, class IV-S, is especially interesting. Despite having what appears to be disseminated disease in adrenals, liver, skin, and bone marrow, these patients do surprisingly well and often survive without treatment [28].

This phenomenon has led to the postulate that class IV-S patients represent a postnatal manifestation of the germinal mutation that predisposes to hereditary neuroblastoma—that is, perhaps an even further prolongation of the usual delay in regression of in situ neuroblastoma [50]. Intriguing as this formulation is, it remains to be proved because there are no data on offspring of IV-S survivors. It is sometimes difficult to separate these patients from those with truly metastatic disease. The recent discovery that serum ferritin levels are normal and E-rosette inhibitory factor is absent in most class IV-S patients may be useful because serum ferritin levels are elevated and E-rosette factor is present in patients with true stage IV, that is, metastatic disease [39].

Constitutional cytogenetic studies in neuroblastoma patients have not been overly rewarding. Direct preparations of chromosomes from the tumors themselves and from derived cell lines have revealed deletions of the short arm of chromosome 1, larger marker chromosomes containing long unbanded segments, termed homogeneously staining regions (HSRs), and double minutes, the latter perhaps derived from the HSR segments [6, 12]. Unfortunately, these alterations are not specific for neuroblastomas and they have no current diagnostic or therapeutic significance.

One recognized teratogen that could provoke neuroblastoma in a genetically predisposed host is hydantoin, since the tumor has been described in a number of children with the fetal hydantoin syndrome [27].

There is little evidence for a hereditary basis for other embryonal tumors. Environmental factors seem to be paramount for soft tissue *sarcomas* such as rhabdomyosarcomas [36]. However, both rhabdomyosarcomas and other embryonal neoplasms such as Wilms' tumor occur in one form of the cancer-family syndrome (see later section), thereby indicating further heterogeneity in the genetic etiology of this unusual group of neoplasms. Presacral teratomas, often accompanied by sacral deformity, have been reported by Ashcraft et al. [4] in multiple generations. In some first-degree relatives in these families, sacral defects and no tumors were present. The whole complex segregates as an autosomal dominant trait.

Hamartoma syndromes

A hamartoma is a solid or cystic tumor formed by aggregation of excessive amounts of either single or multiple tissue elements, generally in a normal anatomic location [101]. In a hamartoma syndrome there is a more generalized distribution of these tissue aggregates. The classic malignant hamartoma syndromes (some prefer to use the less descriptive term phacomatoses) include neurofibromatosis, the von Hippel-Lindau syndrome, and tuberous sclerosis.

Neurofibromatosis is one of the most common genetic diseases of humans with an incidence of about 1 per 3000 births. There is some feeling that the disorder is heterogeneous, comprising at least central, peripheral, and mixed forms, but except for dominantly inherited acoustic neuroma, which could be construed as a pure central form [105], the evidence is not conclusive. A number of benign and malignant tumors have been recorded in neurofibromatosis (Table 7–1), but the overall frequently of malignancy probably does not exceed 5% [86]. Most of the tumors can be explained on the basis of disordered development of neuroectoderm. Some, such as Wilms' tumor and childhood-onset chronic myelogenous leukemia, cannot unless unusual growth factors elaborated by the disordered neural elements are in some fashion responsible.

By definition, cancer is a prominent part of the *von Hippel-Lindau syndrome.* In addition to the characteristically located hemangioblastomas, perhaps 20% or more of affected individuals develop hypernephromas, which may be bilateral, and an additional 5% or so have pheochromocytoma. A peculiar nonsecretory islet-cell tumor has also been described in a few instances [44]. This syndrome is frequently underdiagnosed even in known families, as is shown by some recent experience with computed tomographic (CT) scanning in purportedly normal relatives [53].

The central and cutaneous features of *tuberous sclerosis* are well known, but outside the central nervous system (CNS) the hamartomas are rarely malignant [68]. The brain tumors may be asymptomatic and discovered on brain CT scans or may be responsible for a host of CNS symptoms. The ocular gliomas, if present early, may be confused with retinoblastoma. Cardiac rhabdomyomas are rare, but renal angiomyoli-

Table 7–1 Tumors reported with neurofibromatosis.

Neurinoma	Pheochromocytoma
Glioma	Paraganglioma
Ganglioneuroma	Carcinoid
Neuroblastoma	Leukemia
Schwannoma	Wilms' tumor
Meningioma	Chronic leukemia
Fibroma	Rhabdomyosarcoma
Sarcoma	

pomas are reasonably common. The former can precipitate heart failure, and the latter is often complicated by recurrent nephrolithiasis.

Other dominantly inherited disorders that could be classified as hamartoma syndromes include the *basal cell nevus syndrome* and the *Cowden syndrome.* The first condition features adult-onset skin tumors, but medulloblastomas and other brain tumors may be seen in children, and sarcomas, ovarian fibromas and carcinomas, and gastrointestinal hamartomas have been recorded [95]. Interestingly, irradiation therapy given for the brain tumor often leads to premature development of basal cell carcinomas in the posterior neck and shoulder area. The Cowden syndrome comprises a variety of fibromas, papillomas, angiomas, lipomas, hamartomas, and cysts throughout the body, but breast and papillary thyroid carcinomas have also been seen [13, 102].

Endocrine system

There are three well-recognized multiple endocrine neoplasia (MEN) syndromes [84], all dominantly inherited. *MEN type I* comprises tumors or hyperplasia of parathyroid, pancreatic islet cells, and pituitary. The adrenal cortex and thyroid follicular cells are less consistently involved. In roughly 60% of patients, two glands are involved, and 20% eventually have three or more. Symptoms are thus quite variable depending on the secretory status of the various affected glands. The most consistent true malignancy occurs in the pancreas with the potential presentation being that of insulinoma, glucagonoma, gastrinoma, or the pancreatic cholera syndrome. Pancreatic tumors may also secrete ectopic hormones such as adrenocorticotropic hormone (ACTH) or parathyroid hormone (PTH). Other tumors occasionally described include foregut carcinoids, lipomas, schwannomas, and cutaneous leiomyomas.

MEN II features medullary thyroid carcinoma (MTC), pheochromocytoma, and parathyroid adenomas/hyperplasia. As is generally true with heritable tumor syndromes, the lesions are commonly multifocal within the gland or, in the case of paired organs, bilateral. The pheochromocytomas may be extraadrenal as well. The precancerous form of MTC is C-cell hyperplasia, a stage that can be diagnosed by provocative testing with calcium/pentagastrin infusions, which stimulate the release of abnormal quantities of calcitonin from the hyperplastic cells. If the fasting serum calcitonin level is elevated, frank MTC is present and is likely already in adjacent nodes.

Although both of these syndromes may not be fully penetrant in affected families until the fifth or sixth decade, *MEN III* is usually diagnosable in the first few years of life because of the characteristic facies and body habitus that have given rise to the alternative designation of this condition as the mucosal neuroma syndrome. As in MEN II, MTC and pheochromocytoma are the cancers involved, but hyperparathyroidism has not been established. The MTC behaves in a much more malignant fashion in MEN III, with the average age at death being about 27 in MEN III versus nearly 60 in MEN II [103].

Another combination of tumors, which may be presumptively considered *MEN IV*, is that of pheochromocytoma and islet-cell carcinomas [14]. Generally, the pheochromocytomas appear earlier than with either MEN II or III, and the islet cell tumors are nonfunctional unlike those in MEN I, which are secretory most of the time either clinically or by immunohistochemical study. However, whether or not this is truly a separate syndrome is problematic, because some patients and their relatives have evidence of the von Hippel-Lindau syndrome [44]. It is conceivable that the latter condition is heterogeneous, comprising individuals with hemangioblastomas and no endocrinopathy and those with endocrine tumors.

Many of the glands affected by the MEN syndromes may be affected by separate mutations. Thus, dominantly inherited but independent hyperparathyroidism, MTC, and pheochromocytoma have all been separately recorded in the literature [88]. A related but overlapping condition is familial paragangliomas, in which pheochromocytomas may develop, but the presenting tumors are more likely to be in the chemoreceptor elements such as the carotid bodies [72].

Reproductive system

Testicular cancer is much less common than ovarian cancer, and the tumors are almost always of germ-cell origin. In children of both sexes, teratoma and stromal tumors are relatively more common. Familial examples of testicular tumors have been reported infrequently, and as a whole cannot account for more than 1% of the total [18]. Sibs, twins, and two generations have been seen, but the age distribution is similar to the sporadic tumors, and bilateral involvement, although it has been described, is uncommon [45, 90]. The tumor types may be histologically variable even in the same family. Perhaps a reasonable genetic interpretation would be that a very small proportion of testicular cancers is inherited as a sex-limited autosomal trait. It is important to exclude dominantly inherited genetic conditions that independently predispose to cryptorchidism because this abnormal physiologic state is known to increase the risk of malignancy substantially [92]. There is also an increased risk of cancer in dysgenetic testes, but most syndromes with such maldevelopment are either recessive or of chromosomal origin.

Ovarian cancers tend to affect older women, involve the germinal epithelium, and are virtually never bilateral [91]. Although this bespeaks the absence of important genetic factors, multigenerational pedigrees have been seen for arrhenoblastomas [91], dysgerminomas [88], fibromas [26], and adenocarcinomas [29]. In some families there is a strong association with other tumors, particularly of the breast but of the colon as well [29, 60]. Perhaps these families constitute examples of the cancer-family syndrome (see below). Overall, the percentage of ovarian tumors that could be considered heritable is no more than 5%. Autosomal, sex limited, dominant inheritance could be inferred from a compilation of the available pedigrees. It is important to note that there are distinct genetic conditions in which ovarian cancer has been seen, such as the Peutz-Jeghers, Gardner, and basal cell nevus syndromes, all of which are dominantly inherited [88].

Endometrial carcinoma may occur in families; one series demonstrated that 16 of 154 patients had an affected first-degree relative [57]. Evaluation of the pedigrees revealed that about 10% of the multiply affected families had relatives with other adenocarcinomas, mostly breast and ovary. Again, these findings are compatible with the cancer-family syndrome. Uterine carcinoma has been seen in the *Torre-Muir syndrome*, a dominantly inherited condition with broad malignant potential (Table 7–2) [3]. Independent of these entities there is no good evidence that, even when present in families, endometrial carcinoma has any firm mendelian basis.

There is also no good evidence for any simply inherited form of *cervical* or *prostatic cancer*, although a statistical relationship between prostate cancer in men and breast, endometrial, and ovarian cancer was established in one study [99].

Although the *breast* is not part of the reproductive apparatus per se, it is so strongly influenced by the same hormones that modulate fertility that it can usefully be discussed here. The relationship between breast, uterine, and ovarian carcinoma has already been mentioned. Whether or not this association is really distinct from that between breast and colon carcinoma is open to question. This series of lesions is consistent

Table 7–2 Location of tumors in the Torre-Muir syndrome.

Skin	Stomach	Vulva
Esophagus	Small bowel	Uterus
Larynx	Colon	Bladder
Bronchus	Breast	Ureter

Table 7–3 Type of tumors seen in cancer-family syndromes.

Lesion Site	
Type I	Type II
Breast	Breast
Endometrium	Sarcomas
Ovary	Embryonal neoplasms
Prostate	Acute leukemia
Colon	Adrenal cortex
Stomach	Hodgkin's disease
Skin	Thyroid
Melanoma	Bladder
Pancreas	Pancreas

with one form of the *cancer-family syndrome* (type I), which features adenocarcinomas at a variety of sites (Table 7–3) [58]. Another well-defined cancer-family syndrome (type II) also contains an excess of breast cancer, but other nonadenomatous tumors prevail in these pedigrees [54]. Together, these two syndromes may account for about 10% of the families with breast carcinoma [2]. In these two syndromes the predisposition for cancer, usually in more than one tissue and often at more than one site in any tissue, behaves as an autosomal dominant trait with penetrance of 60% or more. Familial breast carcinoma outside the confines of these syndromes undoubtedly also exists. Characteristically, the familial form shows premenopausal onset and a high frequency of bilaterality [2]. Again, this predisposition seems to segregate as an autosomal dominant trait with high penetrance in women. In favor of the dominant hypothesis is a report of probable linkage of such a susceptibility allele to the glutamate-pyruvate transaminase locus [46]. Interestingly, in some families the susceptibility appeared to be transmitted through an unaffected male. Only about 1% of breast cancer appears in males, and this figure includes patients with Klinefelter's syndrome who are known to be at increased risk. Familial male breast cancer has been reported over two generations on two occasions [89]. What environmental factors might interact with the proposed susceptibility genes in either sex are not known despite intense epidemiologic investigation.

Skin

Squamous cell carcinoma may complicate certain dominantly inherited skin dystrophies such as epidermolysis bullosa and ectodermal dysplasia [85]. Other than these conditions and the previously mentioned basal cell nevus syndrome, familial aggregation of nonmelanotic skin cancer is not impressive and seems more likely to be environmental.

Although this may also be true for the bulk of melanoma patients, it is noteworthy that kindreds containing multiple affected individuals have been reported with sufficient frequency to suggest that an autosomal dominant form exists [61]. However, because the melanomas in families tend to affect light-skinned individuals, and because of the complexities of neural crest migration and differentiation, more than one locus may be involved [1]. It is also conceivable that the bulk of, if not all, such pedigrees, belong to a category called the atypical mole/melanoma or the dysplastic nevus syndrome [61]. In this entity, the melanocytes show cellular atypia accompanied by lymphocyte infiltrates. Melanomas also have been seen in the cancer-family syndromes [59]. Familial intraocular melanoma may also occasionally be dominantly inherited [34].

Respiratory tract

There is no convincing evidence for any simply inherited predisposition to cancer of the lung or upper airway. In view of the fact that neoplasms in this anatomic location are among the most common in the civilized world, it is perhaps surprising that more families with two or more affected individuals have not been reported. Goffman and Mulvihill [31] recorded two such families, both of which had multiple affected members. All those affected had significant exposure to tobacco.

Central nervous system

Other than acoustic neuroma and the hamartoma syndromes, familial aggregation of brain tumors is not common. The risk to relatives may be increased somewhat but not to such an extent that a postulate of mendelian inheritance can be supported [41, 70]. This does not exclude the possibility of genetic heterogeneity. In one report, five members of a family spanning two generations had meningiomas, but one of the individuals also had acoustic neuromas and still another unaffected individual had cafe-au-lait spots [24]. This family may have a central form of neurofibromatosis. Chromosome 22 is commonly absent from meningioma tumor tissue, and it would be of interest to see if the same cytogenetic alteration occurred in tumors from different members in the family [108].

Leukemia/lymphoma

For the most part, genetic studies in the leukemia/lymphoma patients have centered on cytogenetic alterations, a topic reviewed in detail elsewhere in this volume. The proportion of leukemias that are derived from heritable factors is not known, as estimates range from 0 to 25%. A number of impressive pedigrees have been published for acute leukemia, including some multigenerational ones [23, 37, 56]. A reasonable interpretation would be that a small percentage of cases of acute leukemia may result from an autosomal dominant predisposition. However, neither heterogeneity nor polygenic inheritance can be excluded.

Family aggregation of chronic myelocytic leukemia has not shown any significant departure from chance expectations. The data relating to chronic lymphocytic leukemia (CLL) are more impressive. For the most part, affected family members have been siblings, but two generations have been described [9, 10, 20]. In some families, the affected members have an identical clinical course with similar immune defects, whereas in others the immune deficit is discordant and the disease may take a different form, such as lymphosarcoma or acute leukemia [17]. Interestingly, the CLL families often contained members with nonleukemic malignancies as well [64]. There is probably no good reason at present to separate CLL from non-Hodgkin's lymphoma from a genetic perspective.

The data regarding Hodgkin's disease are also quite suggestive of a genetic predisposition, but no convincing evidence has emerged for either dominant or recessive single gene inheritance [35]. Leukemia and lymphoma regularly complicate the simply inherited immune deficiency diseases, the chromosome breakage syndromes, and some other hematologic disorders, but these are virtually all recessively inherited.

Urinary tract

Hypernephroma in families is rare, only about 25 such instances having been documented [32, 62]. In some of these families, the affected individuals may well have had the von Hippel-Lindau syndrome, a condition known to predispose to such malignancy and one in which the malignancy may be the only overt manifestations of the gene [79]. One family is of interest in that it contained ten individuals with hypernephroma, most of whom had bilateral disease over three generations [16]. All the available individuals studied had a constitutional balanced translocation between chromosomes 3 and 8. Other sporadic and familial cases studied by the same authors showed no cytogenetic abnormalities. In another family three individuals were affected with unilateral renal cell carcinoma [73]. In one patient the tumor, not the constitutional karyotype, showed a 3/11 translocation with specific loss of the proximal end of 3p. However, the breakpoint in chromosome 3 in the first family (3p21) is different from that of the latter family (3p13 or 14). Renal cell carcinoma reported in other pedigrees seems to occur in a pattern consistent with an autosomal dominant trait.

Familial examples of cancer of the remainder of the urinary tract are even more scarce. A mother and son with ureteral cancer have been seen. Bladder cancer, generally transitional cell, has been recorded in five families in a pattern consistent with dominant inheritance [63]. Environmental factors are generally considered to play a prime role in the etiology of bladder cancer, but on the average the patients in these families were substantially younger and no common environmental agent could be identified. The risk to the relatives of the usual 60 to 70-year-old patient with this neoplasm would have to be considered negligible.

Gastrointestinal tract

The best known forms of hereditary gastrointestinal cancer are the various polyposis syndromes; however, these account for probably no more than 1% of all large bowel cancers, whereas nonpolyposis syndromes constitute between 10 and 25% [55]. Table 7–4 enumerates the conditions that feature colorectal cancer and are dominantly inherited [85]. This list can be considered only tentative because further heter-

Table 7–4 Dominant colorectal cancer.

With complex syndromes
Gardner syndrome
Peutz-Jeghers syndrome
Torre-Muir syndrome
With polyps alone
Familial polyposis coli
Multiple discrete polyps
Diffuse gastrointestinal polyposis
Gastrocolonic polyposis
Juvenile polyposis
Without polyps
Gastrocolonic carcinoma
Colon carcinoma
Colon/breast cancer
Cancer-family syndrome

Table 7–5 Some cancers reported in the Gardner syndrome

Colon polyps and carcinoma	Thyroid carcinoma
Carcinoma of duodenum	Adrenal carcinoma
Periampullary carcinoma	Bladder carcinoma
Osteosarcoma	Fibrosarcoma

ogeneity may well exist, and conversely, some of the conditions may be identical, that is, colon and breast cancer and the cancer-family syndrome. Probably the most interesting disorder listed in Table 7–4 is the *Gardner syndrome,* a condition with extreme malignant potential both within and outside the gastrointestinal tract (Table 7–5). Cultured cells from a variety of different sites in patients with the syndrome have been shown to have increased rates of viral transformation [75], numerical and structural chromosome aberrations [30] including tetraploidy [21], and increased sensitivity to both X-ray and ultraviolet (UV) light [47]. To some extent, cells from *familial polyposis coli* patients show similar alterations. How these in vitro observations relate to the basic genetic defects are unknown.

Intestinal tumors in the Peutz-Jeghers syndrome may arise either in hamartomatous polyps [15] or in independent sites [43]. Of the nonpolyposis conditions, the Torre-Muir syndrome has the greatest potential for cancers at a variety of locations including a number outside the gastrointestinal tract [3].

Other families with site-specific cancers exclusive of the colorectal area have been described, for example, *pancreas* [22], *gallbladder* [25], *stomach* [104], and *liver* [38], the latter only in sibs. *Bile duct carcinoma* has been recorded in familial polyposis coli [52]. Gastric polyposis without colonic polyposis has been described over three generations in a Portuguese kindred [81]. Although esophageal cancer is felt to be largely environmental in etiology, it is a consistent feature of dominantly inherited tylosis [40], and it complicates such diverse entities as the recessive Bloom and Fanconi syndromes and epidermolysis bullosa dystrophica [87]. In toto, these various familial carcinomas support the idea that a substantial number of genes predispose to gastrointestinal malignancy.

Skeletal system

Familial occurrence of tumors of the skeletal system is quite uncommon. *Osteogenic sarcoma* is a pleiotropic manifestation of hereditary retinoblastoma. It is also included in the spectrum of the type II cancer-family syndrome and may complicate Paget's disease, neurofibromatosis, and multiple exostosis on rare occasions. It is important to exclude these conditions before concluding that a simply inherited form of sarcoma exists. In 1979 Colyer [19] reported affected sibs and reviewed 13 previously recorded families, some of whom showed two-generation involvement. The tumors usually were not multiple, and they tended to develop at adolescence in the ends of growing long bones. In general, the recurrence risk to the relatives of the usual patient with osteogenic sarcoma is small.

Conclusions

Genetic factors clearly play a significant role in causation of some cancers. The heterogeneity within dominantly inherited cancer and cancer syndromes attests to the multiplicity of action of these various genes. How these mutations predispose their hosts to malignancy remains uncertain. Study of mutant gene action in specific families may provide some insight into the mechanisms of carcinogenesis in general and thereby provide for the development of specific therapy tailored to the precise molecular lesion.

Literature cited

1. Anderson DE: Clinical characteristics of the genetic variety of cutaneous melanoma in man. *Cancer* 28:720–725, 1971.
2. Anderson DE: Genetic study of breast cancer: Identification of a high risk group. *Cancer* 34:1090–1097, 1974.
3. Anderson DE: An inherited form of large bowel cancer. *Cancer* 45:1103–1107, 1980.
4. Ashcraft KW, Holder TM, Harris DJ: Familial presacral teratomas. *Birth Defects* 11:143–146, 1975.
5. Bader JL, Meadows AT, Zimmerman LE, Rorke LB, Voute PA, Champion LAA, Muller RW: Bilateral retinoblastoma with ectopic intracranial retinoblastoma: Trilateral retinoblastoma. *Cancer Genet Cytogenet* 5:203–213, 1982.
6. Balaban-Malenbaum G, Gilbert F: The proposed origin of double minutes from homogeneously staining region (HSR)-marker chromosomes in human neuroblastoma hybrid cell lines. *Cancer Genet Cytogenet* 2:339–348, 1980.
7. Balaban-Malenbaum G, Gilbert F, Nichols WW, Hill R, Shields J, Meadows AT: A deleted chromosome No. 13 in human retinoblastoma cells: Relevance to tumorigenesis. *Cancer Genet Cytogenet* 3:243–250, 1981.
8. Beckwith JB, Perrin EV: In situ neuroblastomas: A contribution to the natural history of neural crest tumors. *Am J Pathol* 43:1089–1104, 1963.
9. Blattner WA, Dean JH, Fraumeni JF Jr: Familial lymphoproliferative malignancy: Clinical and laboratory follow-up. *Ann Intern Med* 90:943–944, 1979.

10. Branda RF, Ackerman SK, Handwerger BS, Howe RB, Douglas SD: Lymphocyte studies in familial chronic lymphocytic leukemia. *Am J Med* 64:508–514, 1978.
11. Breslow NE, Beckwith JB: Epidemiologic features of Wilm's tumor: Results of the national Wilm's tumor study. *J Natl Cancer Inst* 68:429–436, 1982.
12. Brodeur GM, Green AA, Hayes FA: Cytogenetic studies of primary human neuroblastomas. *In* Advances in Neuroblastoma Research (Evans AF, ed), New York, Raven Press, 1980, 73–80.
13. Brownstein, MH, Wolf M, Bikowski JB: Cowden's disease. *Cancer* 41:2393–2398, 1978.
14. Carney JA, Go VL, Gordon H, Northcutt RC, Pearse AG, Sheps SG: Familial pheochromocytoma and islet cell tumor of the pancreas. *Am J Med* 68:515–521, 1980.
15. Cochet B, Carrel J, Desbaillets L, Widgren S: Peutz-Jeghers syndrome associated with gastrointestinal carcinoma. *Gut* 20:169–175, 1979.
16. Cohen AJ, Li FP, Berg S, Marchetto DJ, Tsai S, Jacobs SC, Brown RS: Hereditary renal-cell carcinoma associated with a chromosomal translocation. *N Engl J Med* 301:592–595, 1979.
17. Cohen HG, Shimm D, Paris SA, Buckley CE III, Kremer WB: Hairy cell leukemia-associated familial lymphoproliferative disorder: Immunologic abnormalities in unaffected family members. *Ann Intern Med* 90:174–179, 1979.
18. Collins DH, Pugh RC: Classification and frequency of testicular tumors. *Br J Urol* 36:1–11, 1964.
19. Colyer RA: Osteogenic sarcoma in siblings. *Johns Hopkins Med J* 145:131–135, 1979.
20. Conley CL, Misiti J, Laster AJ: Genetic factors predisposing to chronic lymphocytic leukemia and to autoimmune disease. *Medicine* 59:323–334, 1980.
21. Danes BS: Increased *in vitro* tetraploidy: Tissue specific within the heritable colorectal cancer syndromes with polyposis coli. *Cancer* 41:2330–2334, 1978.
22. Danes BS, Lynch HT: A familial aggregation of pancreatic cancer. *JAMA* 247:2798–2802, 1982.
23. Davidson RJ, Walker W, Watt JL, Page BM: Familial erythroleukemia: A cytogenetic and haematologic study. *Scand J Haematol* 20:351–359, 1978.
24. Delleman JW, DeJong JG, Bleeker GM: Meningiomas in five members of a family over two generations, in one member simultaneously with acoustic neurinomas. *Neurology* 28:567–570, 1978.
25. Devor EJ, Buechley RW: Gallbladder cancer in Hispanic New Mexicans. *Cancer Genet Cytogenet* 1:139–145, 1979.
26. Dumont-Herskowitz RA, Safari HS, Senior B: Ovarian fibromata in four successive generations. *J Pediatr* 93:621–624, 1978.
27. Ehrenhard LT, Chaganti RS: Cancer in the fetal hydantoin syndrome. *Lancet* 2:97, 1981.
28. Evans AE, Chatten J, D'Angio GJ, Gerson JM, Robinson J, Schnaufer L: A review of 17 IV-S neuroblastoma patients at the Children's Hospital of Philadelphia. *Cancer* 45:833–839, 1980.
29. Fraumeni JF Jr, Grundy GW, Creagan ET: Six families prone to ovarian cancer. *Cancer* 36:364–369, 1975.
30. Gardner EJ, Rogers SW, Woodward S: Numerical and structural chromosome aberrations in cultured lymphocytes and cutaneous fibroblasts of patients with multiple adenomas of the colorectum. *Cancer* 49:1413–1419, 1982.
31. Goffman TE, Mulvihill JJ: Familial respiratory tract cancer. *JAMA* 247:1020–1023, 1982.
32. Goldman SM, Fishman EK, Abeshouse G, Cohen JH: Renal cell carcinoma diagnosed in three generations of a single family. *South Med J* 72:1457–1459, 1979.
33. Green DM, Fine WE, Li FP: Offspring of patients treated for unilateral Wilm's tumor in childhood. *Cancer* 49:2285–2288, 1982.

34. Green GJ, Hong WK, Everett JR, Bhutani R, Amick RM: Familial intraocular malignant melanoma: A case report. *Cancer* 41:2481–2483, 1978.
35. Grufferman S, Cole P, Smith PG, Lukes RJ: Hodgkin's disease in siblings. *N Engl J Med* 296:248–250, 1977.
36. Grufferman S, Wang HH, DeLong ER, Kimm SY, Delzell ES, Falleta JM: Environmental factors in the etiology of rhabdomyosarcoma in childhood. *J Natl Cancer Inst* 68:107–113, 1982.
37. Gunz FW, Gunz JP, Vincent PC, Bergin M, Johnson FL, Bashir H, Kirk RL: Thirteen cases of leukemia in a family. *J Natl Cancer Inst* 60:1243–1250, 1978.
38. Hagstrom RM, Baker TD: Primary hepatocellular carcinoma in three male siblings. *Cancer* 22:142–150, 1968.
39. Hann HW, Evans AE, Cohen IJ, Leitmeyer JE: Biologic differences between neuroblastoma stages IV-S and IV. *N Engl J Med* 305:425–429, 1981.
40. Harper PS, Harper RM, Howel-Evans A: Carcinoma of the oesphagus with tylosis: *Q J Med* 39:317–333, 1970.
41. Horton WA: Genetics of central nervous system tumors. *Birth Defects* 12:91–98, 1976.
42. Howe JW, Manson N: Familial retinoblastoma—A cautionary tale. *J Pediatr Ophthal* 13:278–282, 1976.
43. Hsu SD, Zaharopoulos P, May JT, Costanza JJ: Peutz-Jeghers syndrome with intestinal carcinoma. *Cancer* 44:1529–1532, 1979.
44. Hull MT, Warfel KA, Muller J, Higgins JT: Familial islet cell tumors in von Hippel-Lindau's disease. *Cancer* 44:1523–1526, 1979.
45. Kademian MT, Caldwell WL: Testicular seminoma: A case report of 2 brothers with seminoma and a review of the literature of testicular malignancies occurring in closely related family members. *J Urol* 116:380–381, 1976.
46. King MC, Go RC, Elston RC, Lynch HT, Petrakis NL: Allele increasing susceptibility to human breast cancer may be linked to the glutamate-pyruvate transaminase locus. *Science* 208:406–408, 1980.
47. Kinsella TJ, Little JB, Nove J, Weichselbaum RR, Li FP, Meyer RJ, Marcetto DJ, Patterson WB: Heterogeneous response to x-ray and ultraviolet light irradiations of cultured skin fibroblasts in two families with Gardner's syndrome. *J Natl Cancer Inst* 68:697–701, 1982.
48. Knudson AG Jr: Mutation and cancer: Statistical study of retinoblastoma. *Proc Natl Acad Sci USA* 68:820–823, 1971.
49. Knudson AG, Strong LC: Mutation and cancer: Neuroblastoma and pheochromocytoma. *Am J Hum Genet* 24:514–532, 1972.
50. Knudson AG Jr, Meadows AT: Regression of neuroblastoma IV-S: A genetic hypothesis. *N Engl J Med* 302:1254–1256, 1980.
51. Koufos A, Hansen MF, Lampkin BC, Workman ML, Copeland NG, Jenkins NA, Cavenee WK: Loss of alleles at loci on human chromosome 11 during genesis of Wilms' tumor. *Nature* 309:170–172, 1984.
52. Lees CD, Herman RE: Familial polyposis coli associated with bile duct cancer. *Am J Surg* 141:378–380, 1981.
53. Levine E, Collins DL, Horton WA, Schimke RN: CT screening of the abdomen in von Hippel-Lindau disease. *AJR* 139:505–510, 1982.
54. Li FP, Fraumeni JF Jr: Prospective study of a family cancer syndrome. *JAMA* 247:2692–2694, 1982.
55. Lovette E: Family studies in cancer of the colon and rectum. *Br J Surg* 63:13–18, 1976.
56. Luddy RE, Champion LA, Schwartz AD: A fatal myeloproliferative syndrome in a family with thrombocytopenia and platelet dysfunction. *Cancer* 41:1959–1963, 1978.

57. Lynch HT, Krush AF, Larsen AL: Heredity and endometrial carcinoma. *South Med J* 60:231–235, 1967.
58. Lynch H, Krush AJ: Differential diagnosis of the cancer family syndrome. *Surg Gynecol Obstet* 136:221–224, 1973.
59. Lynch HT, Frichot BC, Lynch P, Lynch J, Guirgis HA: Family studies of malignant melanoma and associated cancer. *Surg Gynecol Obstet* 141:517–522, 1975.
60. Lynch HT, Harris RE, Guirgis HA, Maloney K. Carmody LL, Lynch JF: Familial association of breast/ovarian cancer. *Cancer* 41:1543–1545, 1978.
61. Lynch HT, Fusaro RM, Pester J, Lynch JF: Familial atypical multiple mole melanoma (FAMMM) syndrome: Genetic heterogeneity and malignant melanoma. *Br J Cancer* 42:58–70, 1980.
62. Lyons AR, Logan H, Johnston GW: Hypernephroma in two brothers. *Br Med J* 1:816–817, 1977.
63. Mahboubi AO, Ahlvin RC, Mahboubi EO: Familial aggregation of urothelial carcinoma. *J Urol* 126:691–692, 1981.
64. Mamison D, Weinerman BH: Subsequent neoplasia in chronic lymphocytic leukemia. *JAMA* 232:267–269, 1975.
65. Matsunaga E: Retinoblastoma: Host resistance and 13q- chromosome deletion. *Hum Genet* 56:53–58, 1980.
66. Matsunaga E: Genetics of Wilm's tumor. *Hum Genet* 57:231–246, 1981.
67. Mauer HS, Pendergrass TW, Borges W, Honig GR: The role of genetic factors in the etiology of Wilm's tumor. *Cancer* 43:205–208, 1979.
68. Monaghan HP, Krafchik BR, MacGregor DL, Fitz CR: Tuberous sclerosis complex in children. *Am J Dis Child* 135:912–917, 1981.
69. Motegi T: Lymphocyte chromosome survey in 42 patients with retinoblastoma: Effort to detect 13q14 deletion mosaicism. *Hum Genet* 58:168–173, 1981.
70. Mulcahy GM, Harlan WL: Occurrence of central nervous system tumors, with special reference to genetic factors. *In* Cancer Genetics (Lynch HT, ed), Springfield, IL, Charles C Thomas, 1980, 263–325.
71. Murphree AL, Benedict WF: Retinoblastoma: Clues to human oncogenesis. *Science* 223:1028–1033, 1984.
72. Parry DM, Li FP, Strong LC, Carney JA, Schottenfeld D, Reimer RR, Grufferman S: Carotid body tumors in humans: Genetics and epidemiology. *J Natl Cancer Inst* 68:573–578, 1982.
73. Pathak S, Strong LC, Ferrell RE, Trindade A: Familial renal cell carcinoma with a 3;11 chromosome translocation limited to tumor cells. *Science* 217:939–941, 1982.
74. Rajfer J: Association between Wilm's tumor and gonadal dysgenesis. *J Urol* 125:388–390, 1981.
75. Rasheed S, Gardner MB: Growth prospectives and susceptibility to viral transformation of skin fibroblasts from individuals at high genetic risk for colorectal cancer. *J Natl Cancer Inst* 66:43–49, 1981.
76. Riccardi VM, Hittner HM, Francke V, Pippin S, Holmquist GP, Kretzer FL, Fovall R: Partial triplication and deletion of 13q: Study of a family presenting with bilateral retinoblastomas. *Clin Genet* 15:332–345, 1979.
77. Riccardi VM, Hittner HM, Francke U, Yunis JJ, Ledbetter D, Borges W: The aniridia-Wilms' tumor association: The critical role of chromosome band 11p13. *Cancer Genet Cytogenet* 2:131–137, 1980.
78. Riccardi VM, Hittner HM, Strong LC, Fernbach DJ, Lebo R, Ferrell RE: Wilms' tumor with aniridia/iris dysplasia and apparently normal chromosomes. *J Pediatr* 100:574–577, 1982.
79. Richards RD, Mebust WK, Schimke RN: A prospective study in von Hippel-Lindau disease. *J Urol* 110:27–30, 1973.

80. Rivera H, Turleau C, deGrouchy J, Junieu C, Despoisse S, Zucker JM: Retinoblastoma-del (13q14): Report of two patients, one with a trisomic sib due to maternal insertion. *Hum Genet* 59:211–214, 1981.
81. Santos JG, Magalhaes J: Familial gastric polyposis. *J Genet Hum* 28:293–297, 1980.
82. Schimke RN, Lowman JT, Cowan GA: Retinoblastoma and osteogenic sarcoma in siblings. *Cancer* 34:2077–2079, 1974.
83. Schimke RN: Genetics and Cancer in Man. Edinburgh, Churchill Livingstone, 1978.
84. Schimke RN: Syndromes with multiple endocrine gland involvement. *Prog Med Genet* 3:143–176, 1979.
85. Schimke RN: Genetic syndromes with GI cancer. *In* The Genetics and Heterogeneity of Common Gastrointestinal Disorders. (Rotter JI, Sandoff IM, Rimoin DL, eds) New York, Academic Press, 1980, 377–389.
86. Schimke RN: The neurocristopathy concept: Fact or fiction. *In* Advances in Neuroblastoma Research (Evans AE, ed) New York, Academic Press, 1980, 13–24.
87. Schimke RN: Genetics and cancer in children. *In* Genetic Issues in Pediatric and Obstetrical Practice (Kaback M, ed) Chicago, Year Book Medical Publishers, 1981, 413–441.
88. Schimke RN: Cancer genetics. *In* Principles and Practice of Medical Genetics (Emery AE, and Rimoin DH, eds), Edinburgh, Churchill Livingstone, 1983, Vol 2, pp. 1401–1426.
89. Schwartz RM, Newell RB Jr, Hauch JF, Fairweather WH: A study of familial male breast carcinoma and a second report. *Cancer* 46:2697–2701, 1980.
90. Shinohara M, Komatsu H, Kawamura T: Familial testicular teratoma in 2 children: Familial report and review of the literature. *J Urol* 123:552–555, 1980.
91. Simpson JL, Photopoulos G: Hereditary aspects of ovarian and testicular neoplasia. *Birth Defects* 12:51–60, 1976.
92. Simpson JL, Photopoulos G: The relationship of neoplasia to disorders of abnormal sexual differentiation. *Birth Defects* 12:15–50, 1976.
93. Slater RM, deKraker J: Chromosome number 11 and Wilms' tumor. *Cancer Genet Cytogenet* 5:237–242, 1982.
94. Sotelo-Avila C, Gonzalez-Crussi F, Fowler JW: Complete and incomplete forms of Beckwith-Wiedemann syndrome: Their oncogenic potential. *J Pediatr* 96:49–50, 1980.
95. Southwick GJ, Schwartz RA: The basal cell nevus syndrome. *Cancer* 44:2294–2305, 1979.
96. Sparkes RS, Muller H, Klisak I, Abram JA: Retinoblastoma with 13q-chromosome deletion associated with maternal paracentric inversion of 13q. *Science* 203:1027–1029, 1979.
97. Sparkes RS, Sparkes MC, Wilson MG, Towner JW, Benedict W, Murphree AL, Yunis JJ: Regional assignment of genes for human esterase D and retinoblastoma to chromosome 13q14. *Science* 208:1042–1044, 1980.
98. Strong LC, Riccardi VM, Ferrell RE, Sparkes RS: Familial retinoblastoma and chromosome 13 deletion transmitted via an insertional translocation. *Science* 213:1501–1503, 1981.
99. Thiessen EJ: Concerning a familial association between breast cancer and both prostatic and uterine malignancies. *Cancer* 34:1102–1107, 1974.
100. Turleau C, deGrouchy J, Dufier JL, Phue LH, Schmelck PH, Rappaport R, Nihoul-Fekete C, Diebold N: Aniridia, pseudohermaphroiditism, gonadoblastoma, mental retardation and del 11p13. *Hum Genet* 57:300–306, 1981.
101. Warkany J: Congenital malformations. Chicago, Year Book Medical Publishers, 1971, 1252–1269.
102. Weinstock JV, Kawanishi H: Gastrointestinal polyposis with orocutaneous hamartomas (Cowden's disease). *Gastroenterology* 74:890–895, 1978.

103. Williams ED: Thyroidectomy for genetically determined medullary thyroid carcinoma. *Lancet* 1:1309–1310, 1977.
104. Wolff CM, Isaacson EA: An analysis of 5 "stomach cancer families" in the state of Utah. *Cancer* 14:1005–1016, 1961.
105. Young DF, Eldridge R, Gardner WJ: Bilateral acoustic neuromas in a large kindred. *JAMA* 214:347–353, 1970.
106. Yunis JJ, Ramsay N: Retinoblastoma and subband deletion of chromosome 13. *Am J Dis Child* 132:161–163, 1978.
107. Yunis JJ, Ramsay NK: Familial occurrence of the aniridia-Wilm's tumor syndrome with deletion 11p13-14.1. *J Pediatr* 96:1027–1030, 1980.
108. Zankl H, Zang KD: Correlations between clinical and cytogenetical data in 180 human meningiomas. *Cancer Genetic Cytogenet* 1:351–356, 1980.

8.

Genetic epidemiology of human cancer: Application to familial breast cancer

MARY-CLAIRE KING

Investigation of the genetics and epidemiology of cancer in families can elucidate how genetic and environmental, cultural, and behavioral factors influence cancer risk. Three samples of particular interest for such studies are extended families at increased risk of disease, series of nuclear families sampled according to a known method of ascertainment, and twins with the disease. Our logic is to address three questions: Is cancer of one or more sites familial? Is familial susceptibility inherited? Are the inherited risk factors genetic? If susceptibility genes for various cancers can be identified, then their mode of inheritance, means of expression, modification by environmental factors, and impact on the total public health burden of cancer can be investigated [15].

Genetic and epidemiologic perspectives

Human cancers constitute a highly heterogeneous group of diseases. Nearly any organ of the body may be affected, and causes appear even more varied than clinical symptoms. Furthermore, many if not most cases of cancer may result from the interaction of biologic and environmental influences [16]. Therefore, we may best be able to understand the multifactorial causes of human neoplastic diseases by using analytic tools that incorporate both genetic susceptibility and environmental exposure as risk factors for human cancer. Historically, genetic analysis has explored the mechanism of mendelian inheritance of susceptibility to disease, whereas epidemiologic investigations have concentrated on those demographic, behavioral, cultural, and environmental factors that influence distribution of diseases in human populations. Integrating these two approaches as *genetic epidemiology* offers a powerful means of studying how genetic factors interact with cultural and environmental factors in influencing the distribution of neoplastic diseases in human populations.

Certain clusters of relatives may be especially informative for studies of genetic and environmental determinants of cancer incidence. First, the study of possible genetic susceptibility to human cancer in families is most frequently stimulated by the description of extended families at high risk of cancer. There is an extensive clinical literature describing families with many cases of breast cancer, colon cancer, and stomach cancer, and several syndromes involving multiple neoplastic diseases have also been reported [24]. Second, the incidence of cancer in a series of families of cancer patients can be investigated by constructing pedigrees for every incident case of the neoplastic disease of interest in a specific population and time. For example, for each case of cancer of a specified site or histologic type in a tumor registry, one might obtain the age and sex of each parent, sibling, and child of the proband, whether each person is alive or dead, and the person's age at diagnosis of any cancer. Analogous data can be collected for the families of healthy persons of the same age, race, and sex as each cancer patient. Such samples have been investigated for many of the most common cancers [1]. Third, studies of cancer in twins can be informative, both for comparing concordance rates for specific cancers in monozygotic versus dyzygotic twin pairs and for comparing cultural and environmental risk factors among twin pairs discordant for disease [19]. Principal limitations to the study of cancer in twins include the rarity of twin pairs with any one neoplastic disease and the assumptions about twins' environments implicit in heritability studies [15].

Investigation of familial breast cancer

The study of breast cancer in families illustrates a pathway of genetic and epidemiologic investigation appropriate to the study of familial cancer of any site.

Does breast cancer cluster in families?

Virtually since breast cancer has been recognized, physicians have observed, treated, and tried to counsel families in which the disease appears much more frequently than one would expect by chance. Such reports led to the hypothesis that close female relatives of breast cancer patients are at increased risk of breast cancer compared to women of the same age and race whose relatives are not affected. This epidemiologic hypothesis has been tested by case-control studies of breast cancer incidence (or mortality) in nuclear families of breast cancer patients compared to breast cancer incidence (or mortality) in families of unaffected individuals. Consistent epidemiologic evidence now indicates that the single factor most dramatically increasing a woman's risk of breast cancer is probably the presence of the disease in her immediate family, especially if more than one relative has been affected, or if the relative was affected bilaterally or at a young age. Thus, anecdotal reports of high-risk families frequently represent real clusters of increased risk, rather than several cases of a common disease occurring in the same family by chance [22].

Is familial clustering of breast cancer due to inherited factors?

Given that breast cancer clusters in some families, is this familial clustering a result of inherited factors or common exposure of relatives to environmental carcinogens? We can approach this question through segregation, or pedigree, analysis [8]. Pedigree analysis allows us to determine whether cancer in a family or families occurs in a pattern consistent with mendelian inheritance of a gene that greatly increases susceptibility to the disease. More generally, such models test for transmission of cancer susceptibility from parents to children in families.

So far, our most complete analysis is of 18 families at high risk of breast cancer [11]. Each family in this sample included at least three first-degree relatives with breast cancer. Because breast cancer may occur in families also at high risk of various other cancers, we classified families into three groups based on associated tumors and ages of breast cancer diagnosis. Group I, consisting of 11 families, includes pedigrees with an intermediate age of breast cancer diagnosis (modal age 48.2 years). The 5 pedigrees with both breast and ovarian cancer are included in this group. Group II consists of 4 families, of which 2 have both breast and endometrial cancer, and is characterized by diagnosis at much later ages (64 years modal). In group III are 2 pedigrees with an excess of childhood cancers including brain tumors, leukemia, adrenal-cortical carcinoma, and very early-onset breast cancer. Five hypotheses were tested for each family:

1. An autosomal dominant allele increases susceptibility to breast, ovarian, or endometrial cancer in women.
2. An autosomal dominant allele increases susceptibility to breast, ovarian, or endometrial cancer in women and to other cancers in men and women.
3. An autosomal recessive allele increases susceptibility to breast, ovarian, or endometrial cancer in women.
4. An X-linked dominant allele increases susceptibility to breast, ovarian, or endometrial cancer in women.

5. An environmental hypothesis, which states that all women of the same age in a family are at equally increased risk of breast cancer.

With the exception of two families, our results are consistent with the hypothesis that breast and some associated cancers in these families have a genetic etiology. However, traits such as breast cancer that have variable age of onset and hence incomplete penetrance require very large samples to discriminate among the various genetic hypotheses. Although autosomal dominant inheritance of breast cancer susceptibility is most likely, neither the recessive nor the X-linked hypothesis could be rejected. The most likely mode of inheritance of breast and ovarian cancer susceptibility in the group I families is through an autosomal dominant allele, by which the risk of breast cancer to a genetically susceptible woman is about 12% by age 35, about 50% by age 50, and about 87% by age 80. However, by this hypothesis, women in these families who do not carry a susceptibility allele are at no increased risk of breast or ovarian cancer. Under the best autosomal dominant hypothesis for the group II families, the risk of breast or endometrial cancer to a genetically susceptible woman is about 1% by age 35, about 17% by age 50, and about 80% by age 80. Women in these families who do not carry a susceptibility allele still have a roughly 16% risk of developing breast or endometrial cancer by age 80, but this risk is not significantly greater than for Caucasian women in the United States as a whole. In group III the most likely mode of inheritance involves both an autosomal dominant susceptibility allele and a factor, possibly environmental but possibly an artifact of selection, leading to diagnoses at younger ages in the more recent generations of these families.

Are the inherited risk factors genetic?

Our statistical analyses of large families with many cases of breast cancer indicate that in some such families, susceptibility to breast cancer appears to be inherited, and that genetic models for inheritance of susceptibility can explain the observed patterns of breast cancer occurrence. However, the consistency of a genetic model for inheritance of breast cancer susceptibility does not preclude cultural or environmental mechanisms for the inheritance of cancer susceptibility. Behavioral factors associated with risk to specific cancers may be culturally transmitted from parents to their children. Cultural inheritance can and frequently does mimic genetic inheritance [6]. For example, families with many cases of lung cancer may occur not only because of genetic susceptibility but also because the smoking habit is passed from parents to children. In other words, smoking and, therefore, susceptibility to lung cancer are in large part inherited culturally [18, 25]. Although we know of no cultural influence of comparable importance for breast cancer, it nevertheless is necessary to test the genetic model in those families whose pattern of breast cancer occurrence is consistent with mendelian inheritance.

If a single gene exists that has a strong influence on breast cancer susceptibility in some families, then an effective means of verifying a model of genetic transmission would be through linkage analysis. The rationale for applying linkage analysis to disease susceptibility in families is that if a hypothetical gene influencing disease does exist, it must be a length of DNA on one of the 23 pairs of human chromosomes. If the chromosomal location of the postulated susceptibility gene can be determined, then it is reasonable to conclude that it does exist, even if its protein product remains unknown.

For each family in groups I and II in which breast cancer occurrence was consistent with a genetic model, 20 polymorphic genetic markers were analyzed for linkage to postulated dominant and recessive alleles, increasing susceptibility to breast cancer. The results were combined for families in group I and group II. (Too few relatives were available for analysis in the group III families.) For the group I families the strongest indication of linkage was for a dominant susceptibility allele closely linked to the gene coding for glutamate-pyruvate transaminase (GPT) [13, 14]. The odds in favor of linkage of breast cancer susceptibility to GPT in the group I families are about 90 to 1, and the hypothetical susceptibility allele appears to be quite close to the GPT locus. (GPT itself, however, has no causal role in breast cancer. The GPT gene simply serves as a flag, or marker, for a chromosome that may carry a susceptibility allele, whose direct function is unknown.) No markers other than GPT appear to be linked to a dominant susceptibility allele in the group I families. Tests for linkage of a recessive susceptibility allele to each of the genetic markers yielded only one slightly positive result: roughly six to one odds in favor of linkage to the acid phosphatase (ACP) locus. Only one family carried any information for ACP, and this result easily could be due to chance.

For the families in group II there is virtually no evidence for linkage of either a dominant or recessive susceptibility allele to any of the markers analyzed. Because group II includes only four families and because many relatives were no longer living, these families are less informative than group I families. In particular, there is no evidence for linkage of GPT to a dominant susceptibility allele. However, only a few possible linkages can be clearly excluded for these families.

Confirmation of linkage between a postulated disease-susceptibility allele and a genetic marker generally requires the combining of results from several families, often from more than one study. For the linkage analysis of breast cancer in families, it has been all the more true that a single family is unlikely to provide adequate information to establish a linkage; the disease is diagnosed relatively late in life and at varying ages, and men are almost never affected. Therefore, a young woman or any man could be genetically susceptible to the disease without expressing any clinical symptoms at the time of examination. In the segregation and linkage analysis of breast cancer in extended families, we took into account variable age at diagnosis and sex-limited expression by using mendelian segregation assumptions, together with each woman's number of cancer-free years, or her age at cancer diagnosis, in the estimation of age- and sex-specific susceptibility. However, the nature of breast cancer leads to linkage results that depend heavily on the marker genotypes of breast (and ovarian and endometrial) cancer patients and of elderly, unaffected women. Unless many such women in a single family are available for analysis, individual families will provide little information. Thus, it is not surprising that no individual families will yield statistically significant results and that several families contributed to the positive result for GPT-linkage analysis.

The use of sequential tests for the detection of linkage is theoretically sound, but application to the analysis of breast and associated cancers poses a practical problem. Breast cancer is almost certainly heterogeneous etiologically. Therefore, combining linkage results from a series of high-risk families, even though the same genetic hypothesis is consistent with the segregation of breast cancer susceptibility in each, may make the detection of a true linkage more difficult if the families represent breast cancers of different etiologies and, therefore, different susceptibility genes.

Furthermore, the usefulness of linkage analysis depends not only on the power and efficiency of the statistical techniques employed but also on the number of polymorphic markers available for analysis. That is, even if a breast cancer susceptibility allele were segregating in a family, it is possible to demonstrate linkage to a marker locus only if such a polymorphic marker locus happens to be on the same chromosome as the susceptibility gene and to be reasonably close to it. Furthermore, because not all genetic markers are segregating in informative patterns in all families, a great many marker systems would be required to "cover" the 23 chromosomes. Although the human genome is by no means completely covered by markers at present, a great deal of progress has been made in the last decade in the development of techniques for revealing polymorphisms in DNA sequences [4]. Many markers now are available, and techniques for preserving cell lines from lymphocytes are sufficiently advanced that samples can be saved indefinitely and new markers tested as they are developed. Among the most intriguing DNA polymorphisms for linkage analysis in families at high risk of cancer will be those in human oncogenes [7]. Polymorphisms in these DNA sequences are more likely to be close to biologically crucial genes, or even to be involved in transformation themselves.

Perhaps the most difficult problem in the mathematical theory of pedigree analysis, as well as in the applications of this theory to the analysis of diseases in families, is correctly distinguishing polygenic inheritance from cultural transmission of nongenetic risk factors and correctly identifying combinations of polygenic inheritance, effects of major genes, and cultural inheritance. Segregation analysis can identify a major gene strongly influencing disease susceptibility in families, even if environmental and polygenic effects also occur [8]. However, applications of mathematical models only now are being developed for the delineation of the effects of major genes, polygenic influences, and culturally transmitted risk factors in the same family [17].

Epidemiologic questions posed by genetic susceptibility to certain cancers

If a gene that greatly increases cancer susceptibility in some families can be identified, it will have become possible to distinguish relatives in such families at high risk from those at low risk. As indicated previously, in the families in which linkage of breast cancer susceptibility to GPT appears likely, the risk of breast cancer to a genetically susceptible woman reaches 70% by age 60, whereas women in the same families who do not carry a susceptibility allele are at no increased risk of breast cancer at all. The identification of a gene increasing susceptibility to certain cancers raises a number of other epidemiologic questions as well.

Why do some genetically susceptible people remain free of disease?

This important clinical question not only is an epidemiologic puzzle but it also addresses the genetic concept of incomplete penetrance. An epidemiologic case-control study may offer the best opportunity to answer it: genetically susceptible relatives affected with cancer are compared with genetically susceptible but unaffected relatives in the same family. The goal is to identify environmental or cultural factors influencing cancer risk among people with the same genetic susceptibility. For example, smoking histories of lung cancer patients in high-risk families could be compared with smoking histories of their relatives without lung cancer. A more subtle example would be to compare exposure to Epstein-Barr virus among Hodgkin's disease patients to the Epstein-Barr virus exposure of their HLA-identical and healthy siblings [3]. In families at high risk of breast cancer we have been interested in exposure to oral contraceptives and menopausal estrogens among affected women compared to their genetically susceptible female relatives without breast cancer. Does exposure to these exogenous estrogens exacerbate genetic susceptibility to breast cancer in high-risk women? Or, more generally, is exposure to exogenous estrogens riskier for women with a family history of breast cancer than for other women?

How is the susceptibility gene expressed prior to onset of disease?

The goal in answering this question is to identify preclinical biologic markers of disease. For example, in families at high risk of breast cancer, healthy young women whose risk of breast cancer, based on genetic analysis, is very high can be compared with appropriately matched healthy young women at low risk of the disease. Using this approach, young women in high-risk families have been found to have reduced excretion of estrogen glucuronides [9], although their circulating estrogen levels were no different from those of low-risk women of the same age [10]. This reduced excretion could result from increased uptake and retention of estrogens in target tissues, including the breast [21]. Furthermore, young women with a family history of breast cancer who later developed the disease themselves differed in thyroid function both from young women from breast-cancer–free families who subsequently developed the disease and from women with and without a family history of breast cancer who remained healthy [5]. In addition, young women at high risk of breast cancer because of family history of the disease differed from other young women of the same age and race in their plasma concentrations of estrone plus estradiol and prolactin [12]. Finally, clinically asymptomatic women with a significant family history of breast cancer are more likely to have cytologically abnormal breast epithelial cells than asymptomatic women without such a family history [23]; among women with both benign breast disease and significant family history of breast cancer the likelihood of abnormal breast epithelial cells is even greater. Each of these results may help explain the etiology of familial breast cancer, that is, to understand how the susceptibility gene is expressed at the physiologic level.

How is the susceptibility gene expressed in disease?

An epidemiologic approach can address this question effectively. In order to determine how familial and nonfamilial disease differ, familial and nonfamilial breast cancer patients can be compared. It is well established that familial breast cancer is diagnosed at younger ages than breast cancer among women in the population as a whole and that familial breast cancer is more likely to be bilateral [1]. In general, familial cancer of all types occurs at younger ages and is more likely to be bilateral at sites where this is possible [22]. Familial breast cancer has as good or better survival than nonfamilial breast cancer in women of the same age [11].

Recently, characteristics of the estrogen receptors in breast tumors of familial patients have been compared to estrogen-receptor characteristics in tumors of nonfamilial breast cancer patients. Estrogen receptor concentration did not differ significantly between familial and nonfamilial patients. However, the dissociation constants of the estrogen receptors from familial patients' tumors were significantly higher than those from tumors of nonfamilial patients. The higher dissociation constants in familial breast cancer patients could be due either to a decreased binding affinity of the estrogen receptor molecule in some familial patients' tumors or to higher tissue-estrogen concentrations in familial breast cancer patients. That is, elevated dissociation constants of some familial breast cancer patients may reflect a subtle underlying endocrine anomaly that increases susceptibility to breast cancer. Specifically, inherited susceptibility to breast cancer could result from increased uptake and retention of estrogen in breast tissue. If this condition persisted after development of breast cancer, tumor samples from some familial patients would contain elevated quantities of estrogen, leading to increased dissociation constants in estrogen receptor assays [21]. Thus, comparison of familial and nonfamilial breast cancer patients may help to elucidate both the epidemiology and etiology of familial breast cancer.

What proportion of cancer is due to genetically inherited factors?

The question of what proportion of cancer is due to genetically inherited factors is important for medical counseling and for offering screening programs to the highest-risk women. The frequency of family history of breast cancer among patients compared with controls indicates that approximately 15% of breast cancer among white women in the United States is associated with a positive family history [2, 20]. However, only a part of this familial effect is genetic. From analysis of individual families in whom breast cancer is very common, it is possible to estimate the proportion of familial risk that is due to inheritance of genetic susceptibility to disease. Preliminary results indicate that perhaps 5 to 10% of breast cancer may be due to inherited susceptibility [14]. As this chapter has indicated, study of these unusual inherited cancers is especially informative because they can reveal clues about the genetic bases of breast cancer as a whole. That is, the same genes may be involved in breast cell transformation, whether these genes have inherited alterations or are changed as the result of somatic mutation [15].

Conclusions

The genetics of cancer susceptibility probably can best be understood and interpreted by employing the tools of both genetic analysis and epidemiology. In the study of breast cancer in families, this approach has permitted the elucidation of at least one possible mechanism for the transmission of genetic susceptibility to the disease and also makes it possible to study the means of expression of genetic susceptibility throughout the natural history of the disease. Finally, it is possible to estimate the magnitude of the clinical and public health burden of genetic susceptibility to breast cancer and to understand how environmental and cultural characteristics may modify the expression of genetic susceptibility.

Acknowledgment

This work was supported in part by NIH grant CA 27632.

Literature cited

1. Anderson DE: Familial predisposition. *In* Cancer Epidemiology and Prevention (Schottenfeld D, Fraumeni, Jr, eds) Philadelphia, Saunders, 1982, 483–493.
2. Anderson DE: Breast cancer in families. *Cancer* 40:1855–1860, 1977.
3. Berberich FR, Berberich MS, King MC, Engleman EG, Grumet FC: Hodgkin's disease susceptibility linkage to the HLA locus demonstrated by a new concordance method. *Hum Immunol* 6:207–217, 1983.
4. Botstein D, White RL, Skolnick M, Davis R: Construction of a genetic linkage map in man using restriction fragment length polymorphisms. *Am J Hum Genet* 3:314–331, 1980.
5. Bulbrook RD, Thomas BS, Fantl VE, Hayward JL: A prospective study of the relation between thyroid function and subsequent breast cancer. *In* Hormones and Breast

Cancer, Banbury Report (Pike MC, Siiteri PK, Welsch CW, eds) vol 8, New York, Cold Spring Harbor Laboratory, 1981, 131–142.

6. Cavalli-Sforza LL, Feldman MW, Chen KH, Dornbusch SM: Theory and observation in cultural transmission. *Science* 218:19–27, 1982.
7. Cooper GM: Cellular transforming genes. *Science* 218:801–806, 1982.
8. Elston RC: Segregation analysis. *In* Advances in Human Genetics (Harris H, Hirschhorn K, eds) vol 11, New York, Plenum, 1980, 61–120.
9. Fishman J, Fukushima D, O'Connor J, Lynch HT: Low urinary estrogen glucuronides in women at risk for familial breast cancer. *Science* 204:1089, 1978.
10. Fishman J, Fukushima D, O'Connor J, Rosenfeld RS, Lynch HT, Lynch JF, Guirgis H, Maloney K: Plasma hormone profiles of young women at high risk for familial breast cancer. *Cancer Res* 38:4006, 1978.
11. Go RCP, King MC, Elston RC, Lynch HT: Genetic epidemiology of breast and associated cancers in high risk families. I. Segregation analysis. *J Natl Cancer Inst* 71:455–461, 1983.
12. Henderson BE, Pike MC: Prolactin—an important hormone in breast neoplasia? *In* Hormones and Breast Cancer, Banbury Report (Pike MC, Siiteri PK, Welsch CW, eds) vol 8, New York, Cold Spring Harbor Laboratory, 1981, 115–130.
13. King MC, Go RCP, Elston RC, Lynch HT, Petrakis NL: Allele increasing susceptibility to human breast cancer may be linked to the glutamate-pyruvate transaminase locus. *Science* 208:406–408, 1980.
14. King MC, Go RCP, Lynch HT, Elston RC, Terasaki PI, Petrakis NL, Rodgers GC, Lattanzio D, Bailey-Wilson J: Genetic epidemiology of breast and associated cancers in high-risk families. II. Linkage analysis. *J Natl Cancer Inst* 71:463–467, 1983.
15. King MC, Lee GT, Spinner NR, Thomson G, Wrensch MR: Genetic epidemiology. *Ann Rev Pub Health* 5:1–52, 1984.
16. Knudsen AG: Genetics and cancer. *In* Cancer (Burchenal JH, Oettgen HF, eds) vol 1, New York, Grune & Stratton, 1981, 381–396.
17. Morton NE: Outline of Genetic Epidemiology. New York, Karger, 1982.
18. Mulvihill JJ: Host factors in human lung tumors: An example of ecogenetics in oncology. *J Natl Cancer Inst* 57:3–7, 1976.
19. Nance WE: The value of population-based twin registries for genetic and epidemiologic research. *In* Cancer Incidence in Defined Populations, Banbury Report (Cairns J, Lyon JL, Skolnick M, eds) vol 4, New York, Cold Spring Harbor Laboratory, 1980, 215–234.
20. Ottman R: Endocrinology and epidemiology of familial breast cancer, PhD dissertation, University of California, Berkeley, 1980.
21. Ottman R, Hoffman PG, Siiteri PK: Estrogen receptor assays in familial and nonfamilial breast cancer. *In* Hormones and Breast Cancer, Banbury Report (Pike MC, Siiteri PK, Welsch CW, eds) vol 8, New York, Cold Spring Harbor Laboratory, 1981, 197–211.
22. Petrakis NL, Bernster VL, King MC: Epidemiology of breast cancer. *In* Cancer Epidemiology and Prevention (Schottenfeld D, Fraumeni JF, eds) Philadelphia, Saunders, 1982, 855–870.
23. Petrakis NL, Ernster VL, Sacks ST, King EB, Schweitzer FJ, Hunt TK, King MC: Epithelial dysplasia in nipple aspirates of breast fluid: Association with family history and other breast cancer risk factors. *J Natl Cancer Inst* 67:227–284, 1981.
24. Schimke RN: Genetics and Cancer in Man. New York, Churchill-Livingstone, 1978.
25. Tokuhata GK, Lilienfeld AM: Familial aggregation of lung cancer in humans. *J Natl Cancer Inst* 30:289–312, 1963.

9.

Familial aggregation of cancer

NANCY R. SCHNEIDER

Epidemiology is concerned with delineating patterns of disease in populations as those patterns reveal clues pertaining to etiology of disease and attributes of populations associated with risk or susceptibility. Descriptions of disease patterns have almost always preceded and lent direction to the eventual elucidation of pathogenesis. Historically, infectious diseases are the most familiar examples; in the more recent past, epidemiologic research has revealed the etiologic association of cigarette smoking and lung cancer. Genetic epidemiology includes the study of familial traits in populations, usually for the purpose of assessing genetic versus nongenetic contributions to the phenotype, often a disease, under study. Genetic epidemiologic investigations of cancer began more than 100 years ago and continue in the present.

Aggregations of cancer in families have attracted the attention of clinicians for at least 200 years. A striking example of such an aggregation is Napoleon Bonaparte's family [42]. Napoleon himself, one of his sisters, and his father were known to have died of gastric carcinoma; in addition, stomach cancer was suspected to have caused the deaths of two more of his sisters, a brother, and a grandfather. Probably the most widely known familial aggregation of cancer, frequently cited in the literature, was that reported by Paul Broca in France in 1866 [9]. In this remarkable family, generally believed to be that of Broca's wife, many of the female members in three generations died of breast cancer or "liver" cancer. The majority of the affected individuals in both Napoleon's family and that reported by Broca died of cancer when they were younger than 50 years of age. Such familial aggregations convinced nineteenth century physicians that cancer, like tuberculosis, was an hereditary disease.

Reasons for familial aggregation of cancer

Most clinicians know of families in which cancer has occurred with unusual frequency. Reasons for familial aggregation of cancer can be

Table 9–1 Reasons for familial aggregation of cancer.

Genetic
Mendelian (single-gene mutation)
Nonmendelian (polygenic or multifactorial)
Nongenetic
Common exposure to carcinogenic agent or life-style
Coincidence

genetic or nongenetic (Table 9–1). Cancers with a major genetic component in their etiology may show either a mendelian pattern of inheritance within a kindred or, more commonly, a nonmendelian pattern. Most single-gene mutations that cause a pronounced predisposition to malignancy and show mendelian patterns of inheritance are individually uncommon; these are discussed elsewhere in this volume [20, 37]. Prominent nonmendelian familial predisposition to cancer probably has a polygenic or multifactorial basis. Nongenetic familial aggregations of cancer occur when family members share not genes, but exposures to carcinogenic agents or life-styles; examples include smoking, diet, and radiation. It is important, however, to remember that not all those exposed to these agents develop cancer; for example, still unidentified host factors undoubtedly distinguish the minority of cigarette smokers who develop lung cancer from those smokers who do not. A synergistic effect of smoking and familial predisposition in some cases of lung cancer has been well documented [45]. In this classic study, mortality from lung cancer was four times greater among nonsmoking relatives (parents and siblings) of lung cancer patients than among nonsmoking relatives of healthy matched controls, evidence of familial predisposition to lung cancer. However, lung cancer mortality among smoking relatives of lung cancer patients was more than twice that of smoking relatives of controls, and nearly 14 times greater than lung cancer mortality of nonsmoking control relatives.

Finally, the most important nongenetic reason for aggregation of cancer in families is coincidence. Data from the American Cancer Society [12] show that the lifetime risk of developing nonskin cancer in this country is 25%. Most familial aggregations of cancer are not due to a known mutation or to a known environmental etiology.

Approaches to investigation of familial cancer in populations

Although most familial aggregations of cancer occur without an identifiable genetic or environmental etiology, evidence that hereditary factors play a role in cancer etiology in the general population (i.e., in families without a known, single gene mutation) has accumulated gradually through increasingly sophisticated study of familial cancer occurrence in populations: genetic epidemiology. Both standard and original epidemiologic and statistical methods are used to identify nonmendelian traits and genetic factors in large groups of people; the large group may be a single large kindred or, more often, many unrelated small families in a specified population group. Several approaches to the genetic epidemiologic study of familial aggregation of cancer have been used, each originating with probands selected in a particular way.

Probands with cancers of many kinds

Investigators from the mid-nineteenth century through the early twentieth century attempted to evaluate the family history of malignancy of cancer patients with all types of malignancies considered together [reviewed in Reference 39]. Often, these reports consisted of nothing more than enumeration of hospital cases with a positive family history. In general, methods of data collection were unsystematic and incomplete, and adequate methods of data analysis were unavailable at that time. Not surprisingly, these investigators drew contradictory conclusions as to the heritability of cancer. However, during this time an appreciation began to develop that cancer is a common disease and can occur in several individuals in a single family by coincidence alone. The importance of coincidence and the necessity of comparing a familial incidence with that in the general population had been stressed by Broca in 1866, but this remained almost universally unrecognized by investigators of familial cancer for the next 75 years.

Probands with cancers of single sites

By the mid-twentieth century, investigators in Europe and the United States had begun to use more rigorous and sophisticated methods of study. Experiments with mice had demonstrated genetic susceptibility to specific types of tumors [40] and suggested that predisposition to cancer was not generalized but, in a given mouse strain, was almost always limited to a single anatomic site. Consequently, most investigators of this era limited their studies in human populations to cancer of a single site. In studies of this type, family histories of cancer were carefully collected from a large number of patients with cancer of a particular anatomic site, and also from noncancerous patients who served as a control group. Family-cancer experience of the two groups was then analyzed for statistically significant differences. Some investigators also made use of the cancer experience in the general population for a second comparison, as population data were becoming available for the first time in Europe and the United States during this period. In most studies, cancers reported in relatives were verified whenever possible by inspection of medical records or death certificates. Breast cancer was most widely studied, probably because of its frequency and ease of diagnosis compared to visceral malignancies [5, 31, 33, 41, 47–49]. Studies of the familial incidence of cancer of single sites other than breast cancer were less frequently done but included probands with malignancies of the stomach, uterus, bowel, and leukemia, among others [11, 14, 27, 32, 46, 48, 49]. Many of these large-scale studies [reviewed in References 13 and 39] reached remarkably similar conclusions, finding that occurrence of the particular cancer under investigation was increased two to three times in close relatives of probands compared to controls or to the general population. The risk for cancer in general, however, was not increased.

More recent studies have found that this overall risk figure can be further partitioned and refined by subclassifying aspects of either the cancer or the individuals under study. For instance, different histologic types of malignancies at the same anatomic site may have different implications for close relatives of the patient. Lehtola [25] in Finland found that relatives of patients with a differentiated, glandular type of stomach cancer had no significantly increased risk of developing gastric cancer but that relatives of patients with a diffuse, mucin-producing type of gastric cancer had seven times the risk of controls. Similarly, Gunz and colleagues [22], in a large study in Australia, found the usual two- to threefold increased risk to relatives of patients with leukemia of all types; however, when the types of leukemia were examined separately it was found that relatives of patients with chronic lymphocytic leukemia had the highest risk of leukemia development, relatives of acute leukemics had an intermediate level of risk, and there was no increased risk to relatives of patients with chronic myelogenous leukemia. D. E. Anderson [2, 3] subcategorized patients with breast cancer according to the patient's age at onset and whether the malignancy had occurred bilaterally; he then calculated the risks of breast cancer for sisters and daughters of probands in these subcategories. In general he found a greatly increased risk of breast cancer for female relatives of patients who had had premenopausal, bilateral disease, but virtually no increased risk to female relatives of patients with postmenopausal, unilateral disease. The usual two- to threefold increased risk to relatives could be derived from the pooled data but, in fact, did not describe the risk for breast cancer accurately for any of the relatives when age and laterality had been taken into account. These data have been confirmed recently in an epidemiologic survey by Bain et al. [6].

Healthy probands

The incidence of cancer in healthy probands' families rarely has been studied. Control probands in the large population surveys just discussed have tended to be noncancerous hospital patients; whether controls were hospitalized patients or healthy individuals, information obtained about the controls' families has tended to be seriously incomplete [10]. However, in 1977 Albert and Child [1] studied the family history of cancer of the first-degree relatives of parents of healthy women participating in an unrelated study—an original approach to subject selection. Their remarkably complete and well-analyzed data showed a nonrandom occurrence of epithelial cancers in families; in contrast, leukemias, lymphomas, central nervous system tumors, and sarcomas occurred randomly. Lynch and his colleagues [28] also have accumulated data concerning the incidence of cancer in families of large numbers of the general population seen in community cancer-screening programs. In general, both studies have showed that the familial occurrence of cancer tends to deviate from the Poisson distribution, in that more families than expected by chance have two or more affected, closely related members.

Probands selected with a genetic hypothesis

In the last decade, studies of the genetic epidemiology of cancer usually have not been general population surveys of familial cancer but have been organized to investigate a particular hypothesis about familial cancer incidence. For example, several studies have investigated the incidence of cancer among the relatives of individuals homozygous for one of the rare cancer-predisposing genes (reviewed elsewhere in this volume [20]), testing the hypothesis that carriers of these genes also would have an increased predisposition to develop cancer greater than that of the general population but less pronounced than that of the homozygote. A study of this type of families of ataxia-telangiectasia patients showed that the risk of dying of cancer is five times higher than normal for close relatives younger than 45 years old [43], and their elevated risk cannot be accounted for by a number of epidemiologic variables other than heterozygosity for the ataxia-telangiectasia gene [16].

Others have hypothesized that the characteristics of known genetic cancers can be used to identify individuals in the general population who have a prominent genetic component in the etiology of their cancers. Known genetic cancers, such as colon cancer in the multiple polyposis coli syndromes, are characterized by unusually early age at onset, frequent multiple primary sites, and several affected close relatives. The second characteristic—more than one primary cancer—was investigated in 1966 by Moertel [35], who examined the family history of cancer of nearly 800 patients with multiple primary neoplasms; he found that families of these patients had 26% more malignant disease than families of noncancerous control patients, indicating that occurrence of multiple primaries is a sign of increased genetic cancer predisposition not only in patients with known genetic cancers but also in a presumed genetically normal cancer population. We [38] have investigated the relationship between the first characteristic of genetic cancer—early age at onset of disease—and familial incidence of cancer in a general population of cancer patients. We hypothesized that patients with an unusually early age at the onset of malignancy would show a stronger genetic predisposition to cancer measured by higher cancer incidence among their close relatives than cancer patients diagnosed at more typical ages. Patients with a wide variety of malignancies were included and were assigned to young, intermediate, median, or old study groups after comparison of their ages at cancer diagnosis with the distribution of ages at diagnosis for their cancer sites compiled by the Third National Cancer Survey [15]. Family histories of cancer were collected by questionnaire. All the reported cancers and deaths from any cause in relatives were verified when possible with medical records and death certificates. The numbers of cancers expected among the first-degree relatives of probands were calculated using standard epidemiologic methods. Statistical analysis showed that a familial tendency to develop cancer did exist in this randomly selected population of cancer patients—that is, families with one case of any kind of cancer were more likely to have other cases of cancer in the family—but that this familial tendency existed regardless of the age of cancer onset in the proband. It appears that, in the general population, early age at diagnosis of cancer may indicate genetic predisposition to malignancy only in exceptional cases.

Evaluation of cancer aggregation in a family

Familiarity with methods and results of studies of familial cancer in populations can be useful to the clinician in evaluating the presence of hereditary cancer predisposition in a given family and estimating the risk of cancer development to individual family members.

First, it has consistently been the experience of investigators of familial cancer from the mid-nineteenth century to the present that reported information about family members is notoriously incomplete or inaccurate [10, 18, 23, 38, 39]. Informants commonly possess little or no useful information about relatives other than their first-degree rela-

Table 9–2 Kinds of familial aggregation.

Single-site cancer
Cancer of two or three sites or types
Cancer of several or many sites or types
Unusual sites or types and/or at unusual ages
Common types at common ages

tives; nearly half do not know the cause of death or age of death of their grandparents. Also, because of the fear and anxiety attached to a diagnosis of malignancy, informants may overreport or underreport incidence of cancer among their relatives. Finally, even the rare informant confidently possessed of the most detailed family history is not sophisticated enough medically to be a sufficiently reliable informant for the purpose of genetic evaluation. Therefore, it cannot be emphasized too strongly that verification of reported malignancies with hospital and vital records whenever possible is of critical importance. Fortunately, numerous studies over the past 50 years have shown repeatedly that cancer is the most accurately reported diagnosis on death certificates [7, 17, 19]. It is highly probable that more than one family member will have to be consulted to acquire sufficient information about a particular relative to be able to obtain his or her hospital records or death certificate.

Second, the nature of the cancer aggregation in the pedigree should be evaluated after as much information as possible has been verified. A useful classification based on the number of anatomic sites involved in a given family is outlined in Table 9–2. Epidemiologic data are most apt to be useful in evaluating the first and last types of aggregations listed: single-site cancer and common types of cancer occurring at common ages.

Kinds of familial aggregations of cancer

Single-site cancer

The most common kinds of site-specific aggregations of familial cancer coming to the attention of the clinician are breast cancer and colon cancer, discussed elsewhere in this volume [4]. Pedigrees of breast cancer like those in Figure 9–1 can be evaluated for ages of onset and bilaterality, and risk of breast cancer for individual members can be estimated using Anderson's data [2, 3]. However, these pedigrees may be somewhat problematical; note the typical age of onset in individual III-

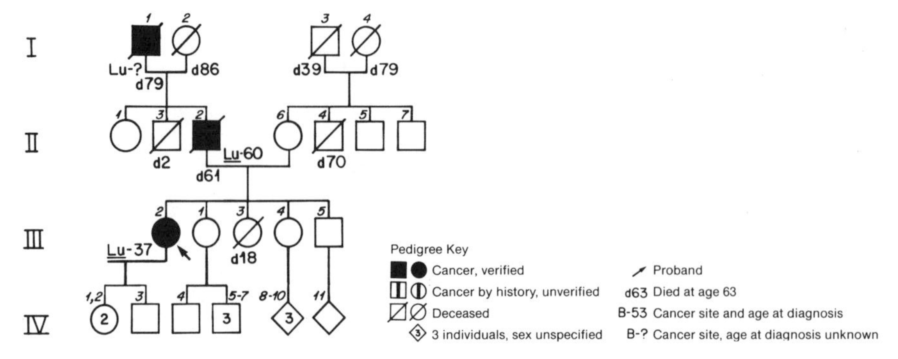

Figure 9–1 Familial breast cancer.

1 in Figure 9–1. Hill et al. [24] have pointed out that in familial aggregations of common cancers such as breast cancer, sporadic cases as well as those with a strong genetic determination may occur together in the same kindred, making genetic analysis more difficult.

A small number of single genes are known to predispose to site-specific cancer; the most common and well known of these are the genes for the polyposis coli syndromes with their great predisposition to colon cancer at an early age. In the evaluation of any familial cancer aggregation, known single-gene mutations should be ruled in or out as soon as possible. Establishing the presence of one of these "mendelizing" genes permits accurate risk assessment and an appropriate surveillance program. Mulvihill compiled a list of the known single-gene mutations with associated neoplasms [36] as German has done for the present volume [21].

In general, if the presence of a single-gene mutation has been ruled out, a site-specific aggregation of cancer can be considered an example of those included in the large population surveys described earlier, and a two- to threefold increased risk for that cancer can be assigned to close relatives of the proband. It is necessary, however, to be aware that such a risk figure is usually an average derived from pooled data and may not be individually accurate. When more specific classes of risk are available, such as for breast and gastric cancers, these obviously should be used. It is reasonable to assume that when other cancers also are examined in clinically and genetically meaningful subcategories, their two- to threefold increased risk figures will be modified and partitioned also. It should also be noted that single-site cancer in a family may be related to unusually intense exposure to environmental carcinogens (Figure 9–2). It perhaps is not surprising that a cigarette smoker who was also a mine worker died of lung cancer (II-2); however, the familial susceptibility to lung cancer is illustrated by I-1, a nonsmoking

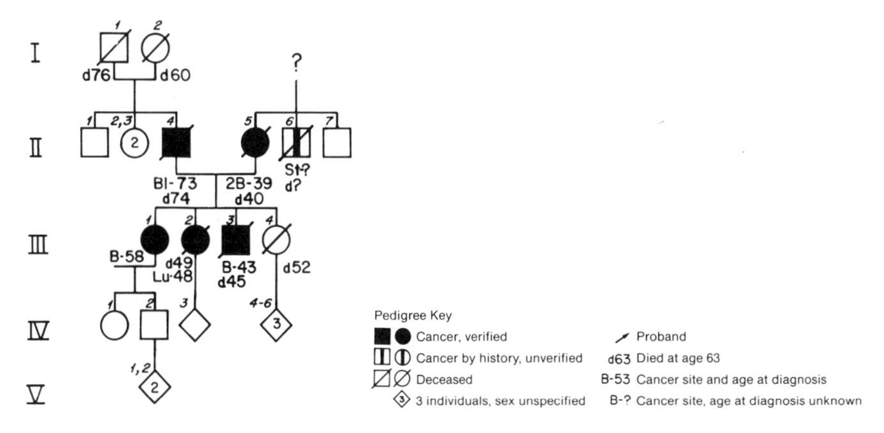

Figure 9–2 Familial lung cancer with prominent environmental components. I-1 and II-2 were mining engineers; II-2 and III-2 were cigarette smokers.

mining engineer, and by the proband, a smoker who died of lung cancer at an unusually early age.

Cancer of two or three specific sites

Cancers occurring alone or in combination at a restricted number of anatomic sites are likely to be part of a single-gene syndrome such as the multiple endocrine adenomatoses or may represent the so-called cancer-family syndrome or one of its variants [30]. In the cancer-family syndrome, colon and endometrial cancer occur together or singly in successive generations in more than 25% of the members of an affected kindred; the pattern of inheritance appears to be dominant, but is not strictly mendelian. Similar frequent associations of specific tumor types in families have been described [e.g., 29, 44].

Cancers of multiple anatomic sites

Individual families have been reported in which aggregations of cancers have occurred at unusual ages or in which unusual kinds of cancers have clustered. For example, four of nine siblings reported by Meisner et al. [34] died of lymphoma at age 3, of meningeal sarcoma at age 4, of osteogenic sarcoma at age 6, and of adenocarcinoma of the cecum at age 17. Lymphoma and osteogenic sarcoma are common childhood malignancies; however, meningeal sarcoma is rare, and cecal adenocarcinoma, a cancer that usually occurs in the fifth to seventh decades, is extremely rare at age 17. Other examples of clusters of cancers at various sites in young people are the pedigrees reported by Li and Fraumeni [26] and by Blattner et al. [8]. These kindreds usually have been subjected to extensive laboratory investigations in an effort to find either a clue to their etiology or a laboratory marker that could be used to identify other members at risk. In general, such investigations have been unrevealing, and these unusual aggregations of familial cancer remain unexplained.

The type of familial aggregation the clinician is most likely to be asked to evaluate is a familial aggregation of common kinds of cancer occurring at common ages (Figures 9–3 and 9–4). Often the aggregation will be seen in a sibship (Figure 9–3). Individual II-2 in Figure 9–3 had had more than 150 exposures to diagnostic X-rays over the months following injury in an accident, and these were almost certainly important in the etiology of his leukemia. Individual II-5 in Figure 9–4, who died of melanoma at an unusually early age, complicates evaluation of this family, particularly because her brother also developed melanoma but at a typical age. A familial predisposition to melanoma could be considered unlikely if the siblings of I-3 or I-4 and the children of II-5, 6, and 7 were unaffected. Because genetic epidemiology has shown increased familial cancer risk for common cancers to be site-specific (except in the case of the rare single-gene mutations referred to earlier), it is highly probable that the other cancers in Figure 9–4 are a random cluster. Familial aggregations such as Figure 9–3 and 9–4 usually have no known genetic or environmental associations and can, *with our present state of knowledge,* best be explained by coincidence. Although intellectually somewhat unsatisfying to the cancer geneticist seeking biologic mechanisms, the explanation of coincidence or random clustering of familial cancer aggregations is nevertheless statistically

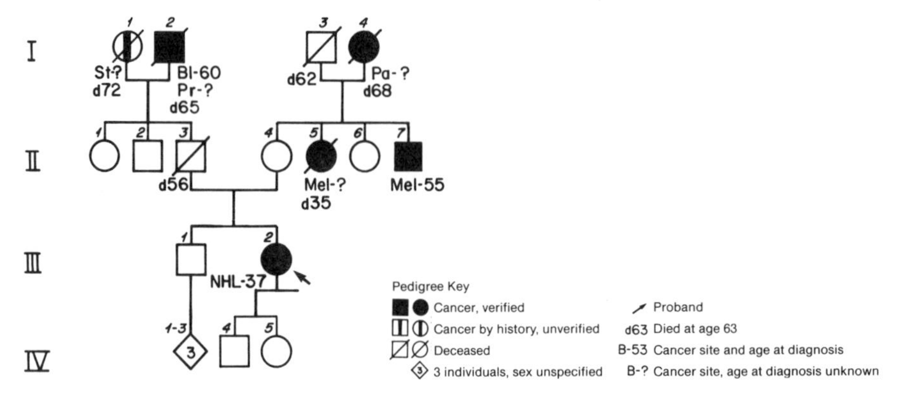

Figure 9–3 Familial aggregation of common cancers at typical ages in a sibship. III-2 had had more than 150 exposures to diagnostic X-rays following injury in an accident.

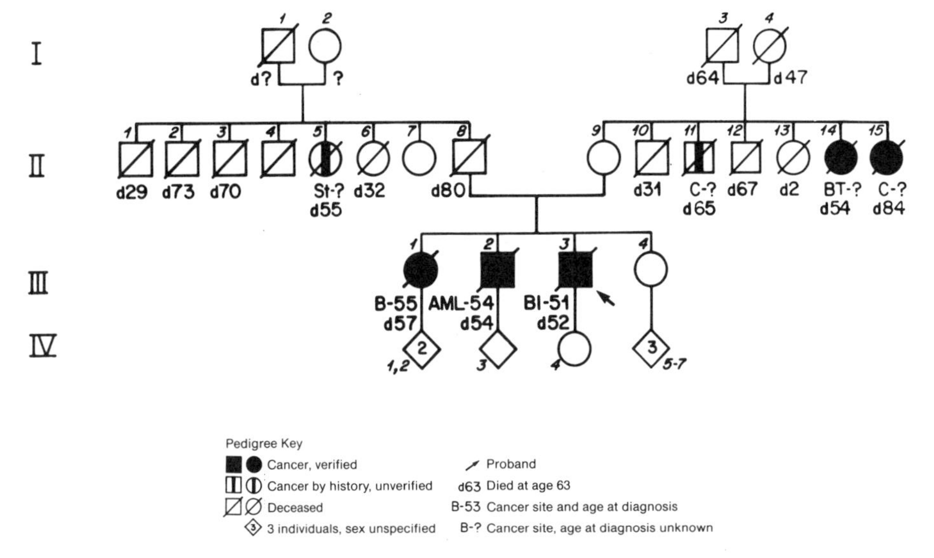

Figure 9–4 Familial aggregation of common cancers at typical ages with one exception (II-5).

and epidemiologically valid and is likely to be comforting to the individual concerned about his or her apparent genetic predisposition.

Conclusions

The conclusions of epidemiologic studies of familial cancer have shown that although cancer does occur nonrandomly in families, significant hereditary predisposition is present only in exceptional cases. Recent studies suggest that certain characteristics and not others of particular cancers may be associated with increased familial predisposition to malignancy.

In the evaluation of a particular familial aggregate of cancer it is important to document all reported cases with medical or vital records. In every case, unusual environmental exposures and known single-gene mutations should be considered. Aggregations of unusual cancers should be referred when possible for laboratory evaluations, with the realization that the results are likely to have more research value than clinical utility. The risk of single-site cancer to relatives can be considered to be increased two- to threefold (derived from population studies), unless more specific data are available as in the instances of breast and gastric cancers and leukemias. In most other cases, and indeed in the majority of cases of familial aggregation of cancer coming to the attention of clinicians, aggregation can best be ascribed to coincidence, and reassurance can be provided.

Literature cited

1. Albert S, Child M: Familial cancer in the general population. *Cancer* 40:1674–1679, 1977.
2. Anderson DE: Genetic study of breast cancer: Identification of a high risk group. *Cancer* 34:1090–1097, 1974.
3. Anderson DE: Breast cancer in families. *Cancer* 40:1855–1860, 1977.
4. Anderson DE, Williams WR: Familial cancer: Implications for healthy relatives. Chapter 15 in this volume.
5. Anderson VE, Goodman HO, Reed SC: Variables Related to Human Breast Cancer. Minneapolis, University of Minnesota Press, 1958.
6. Bain C, Speizer FE, Rosner B, Belanger C, Hennekens CH: Family history of breast cancer as a risk indicator for the disease. *Am J Epidemiol* 111:301–308, 1980.
7. Barclay THC, Phillips AJ: The accuracy of cancer diagnosis on death certificates. *Cancer* 15:5–9, 1962.
8. Blattner WA, McGuire DB, Mulvihill JJ, Lampkin BC, Hananian J, Fraumeni JF Jr: Genealogy of cancer in a family. *JAMA* 241:259–261, 1979.
9. Broca P: Traite' des Tumeurs. Asselin, Paris, 1866.
10. Busk T: Some observations on heredity in breast cancer and leukemia. *Ann Eugen* 14:213–229, 1948.
11. Brobech O: Heredity in Uterine Cancer. Aarhus, Universitetsforlaget, 1949.
12. Cancer Facts and Figures 1981. New York, American Cancer Society, 1981.
13. Clemmesen J: Statistical studies in the aetiology of malignant neoplasms. Review and results. *Acta Pathol Microbiol Scand* (Suppl) 174:1–543, 1965.
14. Crabtree JA: Observations on the familial incidence of cancer. *Am J Publ Health* 31:49–56, 1941.
15. Cutler SJ, Young JL: Third National Cancer Survey: Incidence data. *National Cancer Institute Monograph 41.* Washington, DC, U.S. Government Printing Office, 1975.
16. Daly MB, Swift M: Epidemiological factors related to the malignant neoplasms in ataxia-telangiectasia families. *J Chron Dis* 31:625–634, 1978.
17. Dorn HF, Horn JI: The reliability of certificates of deaths from cancer. *Am J Hyg* 34:12–23, 1941.
18. Elwood JM, Crawford GM, Werner M: Investigation of a family suspected of being at high risk for cancer. *Can Med Assoc J* 121:559–563, 1979.
19. Engel LW, Strauchen JA, Chiazze L, Heid M: Accuracy of death certification in an autopsied population with specific attention to malignant neoplasms and vascular diseases. *Am J Epidemiol* 111:99–112, 1980.
20. German J: Recessive inheritance in human cancer. Chapter 6 in this volume.
21. German J: Heritable conditions that predispose to cancer. Chapter 5 in this volume.
22. Gunz FW, Gunz JP, Veale AMO, Chapman CJ, Houston IB: Familial leukemia: A study of 909 families. *Scand J Haematol* 15:117–131, 1975.
23. Harris R, Linn MW, Hunter K: How valid are patient reports of family history of illness? *Meth Inform Med* 19:162–164, 1980.
24. Hill J, Carmelli D, Gardner E, Skolnick M: Likelihood analysis of breast cancer predisposition in a Mormon pedigree. *In* Genetic Epidemiology (Morton NE, Chung CS, eds) New York, Academic Press, 1978, 304–310.
25. Lehtola J: Family study of gastric carcinoma with special reference to histological types. *Scand J Gastroenterol* (Suppl 50) 13:12–54, 1978.

26. Li FP, Fraumeni JF Jr: Familial breast cancer, soft-tissue sarcomas, and other neoplasms. *Ann Intern Med* 83:833–834, 1975.
27. Lovett E: Family studies in cancer of the colon and rectum. *Br J Surg* 63:13–18, 1976.
28. Lynch HT, Brodkey FD, Lynch P, Lynch J, Maloney K, Rankin L, Kraft C, Swartz M, Westercamp T, Guirgis HA: Familial risk and cancer control. *JAMA* 236:582–584, 1976.
29. Lynch HT, Harris RE, Guirgis HA, Maloney K, Carmody LL, Lynch JF: Familial association of breast/ovarian carcinoma. *Cancer* 41:1543–1549, 1978.
30. Lynch HT, Krush AJ, Thomas RJ, Lynch J: Cancer family syndrome. *In* Cancer Genetics (Lynch HT, ed) Springfield IL, Charles C Thomas, 1976, 355–388.
31. Macklin MT: Comparison of the number of breast cancer deaths observed in relatives of breast cancer patients, and the number expected on the basis of mortality rates. *J Natl Cancer Inst* 22:927–951, 1959.
32. Macklin MT: Inheritance of cancer of the stomach and large intestine in man. *J Natl Cancer Inst* 24:551–571, 1960.
33. Martynova RP: Studies in the genetics of human neoplasms. Cancer of the breast, based on 201 family histories. *Am J Cancer* 29:530–540, 1937.
34. Meisner LF, Gilbert E, Ris HW, Haverty G: Genetic mechanisms in cancer predisposition. *Cancer* 43:579–689, 1979.
35. Moertel CG: Multiple Primary Malignant Neoplasms. Recent Results in Cancer Research, vol 7, New York, Springer-Verlag, 1966.
36. Mulvihill JJ: Genetic repertory of human neoplasia. *In* Genetics of Human Cancer (Mulvihill JJ, Miller RW, Fraumeni JF, eds) New York, Raven Press, 1977, 137–143.
37. Schimke RN: Dominant inheritance in human cancer. Chapter 7 in this volume.
38. Schneider NR, Chaganti SR, German J, Chaganti RSK: Familial predisposition to cancer and age at onset of disease in randomly selected cancer patients. *Am J Hum Genet* 35:454–467, 1983.
39. Schneider NR, Williams WR, Chaganti RSK: Genetic epidemiology of familial aggregation of cancer (in preparation).
40. Slye M: Biological evidence for the inheritability of cancer in man: Studies in the incidence and inheritability of spontaneous tumors in mice. Eighteenth report. *J Cancer Res* 7:107–147, 1922.
41. Smithers, DW: Family histories of 459 patients with cancer of the breast. *Br J Cancer* 2:163–167, 1948.
42. Sokoloff B: Predisposition to cancer in the Bonaparte family. *Am J Surg* 40:673–678, 1938.
43. Swift M, Sholman L, Perry M, Chase C: Malignant neoplasms in the families of patients with ataxia-telangiectasia. *Cancer Res* 36:209–215, 1976.
44. Thiessen EU: Concerning a familial association between breast cancer and both prostatic and uterine malignancies. *Cancer* 34:1102–1107, 1974.
45. Tokuhata GK: Familial factors in human lung cancer and smoking. *Am J Publ Health* 54:25–32, 1964.
46. Videback A: Heredity in Human Leukemia. Copenhagen, Busek, 1947.
47. Wainwright JM: A comparison of conditions associated with breast cancer in Great Britain and America. *Am J Cancer* 15:2610–2645, 1931.
48. Wassink WF: Cancer et heredite. *Genetica* 17:103–144, 1935.
49. Woolf CM: Investigations on genetic aspects of carcinoma of the stomach and breast. *Univ Calif Publ In Public Health* 2:265–349, 1955.

PART TWO

GENETICS AND THE CLINICAL ONCOLOGIST

Part One of this book summarized for the clinician much of the current understanding of the neoplastic process and what is known about heritable factors that predispose to neoplastic transformation. Part Two is devoted to certain applications of this basic information in clinical medicine.

The chapters in this second part address the following matters. First, over the past half-century considerable effort has been expended in typing persons for various genetically determined traits in an effort to identify cancer-prone persons. It is important for the physician to know the predictive value of such typing—or, as is more often the case, the disappointing lack of value. Chapter 10 mentions the more often discussed "genetic markers." Second, analysis of the chromosome complement of neoplastic tissues, long a science-art practiced by the research-laboratory investigator, has taken on major diagnostic importance in clinical medicine. Chapter 11 summarizes the use of cytogenetics in relation to the leukemias, and Chapter 12 discusses its use in relation to lymphomas and other solid tumors. Next, two chapters, 13 and 14, address a variety of difficult problems the physician may encounter in dealing with persons once they are recognized as being predisposed to cancer on a genetic basis. The handling of the relatives of persons with cancer also is considered. In these chapters the authors write from personal clinical experience. Finally, Chapter 15 deals with the methods for determining the risk that relatives of persons affected with two common types of cancer, breast and colon, have of developing the same disease.

In the clinical application of genetic knowledge, significant feedback from the clinic to the basic geneticist or cell biologist often has occurred and will continue to do so. The chance that this will take place is increased when the clinician is an astute observer and sophisticated scientifically. Some examples of such feedback—clinicial to basic investigator—are the following. Epidemiological and population studies, while providing valuable risk figures for the clinician and genetic counselor, simultaneously provide the tumor biologist with important clues regarding cancer etiology. The delineation and detailed observation of increasing numbers of the so-called cancer families should do the same. Also, the evaluation of persons with cancer proneness on a genetic basis because of defective DNA metabolism, defective DNA-repair ability in response to DNA-damaging agents, or a defective immune response have shed new light repeatedly on fundamental biological processes. Finally, the cytogenetic studies of human cancers have repeatedly provided new observations of basic significance. Edwards commented on this important interaction and interdependence two decades ago in the following way: "Knowledge may be utilized directly, or it may be invested by research, or disseminated by teaching. Ideally these three activities should be combined, for each will contribute to the other two." [J. H. Edwards, "The application of knowledge." In *New Directions in Human Genetics* (ed. D. Bergsma and J. German), Birth Defects: Original Articles Series 1(2):75–84, 1965.] —*J. G.*

10.

Genetic indicators of cancer predisposition

R. S. K. CHAGANTI

Genetic determination of cancer predisposition implies that, in principle, gene carriers in affected families can be identified without reference to the morbidity phenotype (clinical cancer) if either the mutation itself manifests an unambiguous and easily recognizable phenotype or the locus in question exhibits close linkage to a gene with an easily recognizable phenotype. There has been considerable interest in searching for such phenotypes and linkages associated with genetic cancer because of their obvious value in the counseling and management of families. A second reason for attempting to identify indicators of cancer predisposition has been the hope that they may enable detection of individuals in the population who are at high risk for cancer development because of inherited factors. Unfortunately, with the exception of certain specific patterns of chromosome breakage and a DNA-repair defect that comprise the cellular phenotypes of a handful of rare and recessively inherited disorders, no unambiguous indicator of genetic cancer predisposition currently is known. Even the specific phenotypes mentioned above are useful only in establishing the diagnosis of a clinical syndrome determined by the homozygous state of a mutant gene rather than for the identification of gene carriers in the populations.

The traits that have been suggested as having actual or potential predictive value are briefly discussed below. The list of traits presented here is not meant to be comprehensive; rather, it is a summary and indicates some of the difficulties associated with attempts to develop methods to detect individuals at high risk for cancer development due to genetic causes.

Cellular phenotypes of cancer-predisposing genes as indicators of cancer predisposition

The first group of traits to be discussed will comprise certain cellular and other phenotypes of cancer-predisposing mutations that can be useful in identifying gene carriers. Some of them (e.g., chromosome breakage) are now used in establishing the diagnosis of patients with certain syndromes; on occasion they have been used to establish phenotypes prenatally.

Chromosome instability

Spontaneous chromosome breakage in cultured cells occurring at a level significantly higher than in normal cells is characteristic of Fanconi anemia (FA), Bloom syndrome (BS), ataxia-telangiectasia (A-T), and Werner syndrome (WS), which comprise the so-called chromosome breakage syndromes [19, 20]. The pattern of chromosome breakage in each disorder is specific and diagnostic [19, 38]. FA is characterized mainly by increased incidence of chromatid breaks, gaps, and exchanges; A-T by increased incidence of chromosome-type rearrangements (including dicentric chromosomes) and clonal expansion of T lymphocytes with specific chromosome rearrangements; WS by clonal expansion of fibroblasts in long-term culture that are characterized by chromosome rearrangements; and BS by increased incidence of homologous sister and nonsister chromatid exchanges. Although a number of reports have appeared in the literature suggesting that chromosome breakage is a trait of several other inherited disorders with increased cancer predisposition [24], in general these data either are limited or lack confirmation.

Tetraploidy and hyperdiploidy in cultured fibroblast cells

Cultured fibroblast cells from patients with certain dominantly inherited cancer-predisposing disorders have been reported to exhibit significantly increased numbers of cells with tetraploid or hyperdiploid chromosome numbers compared to cells from normal individuals [13, 23]. Thus, increased tetraploidy has been suggested to characterize some, but not all, forms of familial colon cancer, and hyperdiploidy has been suggested to be associated with tumors of the breast, melanoma, ovary, and pancreas [13, 23]. While these findings are of interest and potentially significant in understanding the cellular defects caused by the presumptive cancer-predisposing mutations, their diagnostic utility in identifying gene carriers has not been established. Furthermore, familial large bowel cancer exhibits considerable heterogeneity of phenotype that is probably genetically based [1, 2]. Therefore, any attempts to correlate cellular markers with the inheritance of these disorders have to take cognizance of the heterogeneity.

Immunodeficiency

Specific defects in humoral as well as cellular immunity define the primary or inherited immunodeficiencies, a group of disorders that exhibit increased predisposition to lymphoid and other malignancies [19, 21, 22]. While an inherited immunodeficiency almost certainly places an individual at increased risk for neoplastic disease, the nosology and genetics of these rare disorders are complex [20, 22]. The risk of malignant disease is variable among the different disorders [20, 46], and in some cases (e.g., severe combined immunodeficiency) risk of lethal infections may be more significant than risk of malignant disease.

Cellular and chromosomal hypersensitivity to ionizing radiation and radiomimetic drugs

The classical example of cellular hypersensitivity to ionizing radiation and radiomimetic drugs is A-T [4, 25, 37]. In a number of studies, heterozygous cells also have been shown to exhibit sensitivity expressing to a level that is intermediate between homozygous and normal cells [37, 43]. This hypersensitivity phenotype is of value in diagnosing the affected and gene carriers in families. Its usefulness in detecting heterozygotes in the population, however, has not been established. This is particularly important to note because A-T heterozygotes have been suggested as being at increased risk for neoplastic disease [45]. Chromosome breakage in cultured A-T cells is significantly increased over normals when treated with ionizing radiations or radiomimetic drugs providing another aid in diagnosing the affected in families [4, 37]. A point of caution here is that possible genetic heterogeneity for this phenotype has been reported recently [16]. It has been suggested that several other cancer-predisposing disorders exhibit increased cellular or chromosomal radiosensitivity, such as Gardner syndrome and retinoblastoma [11, 25]. However, neither is the level of radiosensitivity in these cases as high as that exhibited by A-T cells nor are the data themselves as unequivocal as those of A-T, thus limiting their usefulness in the detection of gene carriers [9, 23].

Cellular hypersensitivity to ultraviolet (UV) radiation

Xeroderma pigmentosum (XP) comprises the best-known example of genetically determined hypersensitivity to UV radiation [3, 11, 25]. The exquisite sensitivity of cells to the lethal effects of UV and the hypersensitivity of patients to sunlight and their susceptibility to developing skin cancers has been well known [3, 11]. The biochemical basis for the disorder has been reviewed elsewhere [3, 11]. The cellular phenotype is easily recognized by the inability to perform unscheduled DNA synthesis normally associated with removal of the thymidine dimers generated by UV exposure [3]. Based on somatic cell hybridization studies, eight complementation groups of XP have been recognized [3, 11].

Cellular hypersensitivity to DNA-crosslinking and alkylating agents

If A-T and XP are the classical examples of cellular hypersensitivity to ionizing radiations and UV, then FA is the classical example of cellular hypersensitivity to DNA-crosslinking agents [11, 25]. This hypersensitivity is easily recognized in cytotoxicity as well as significantly enhanced chromosome breakage when cells are treated with such chemicals as mitomycin C and diepoxybutane [5, 8]. Significantly increased incidence of a characteristic chromosome breakage pattern (multiple chromatid exchanges) following treatment with these drugs is now used in prenatal as well as postnatal diagnosis of the affected in families in which the mutation is segregating [5]. There is some indication that the chromosomes of obligate FA heterozygotes in families may also exhibit a slight but significant hypersensitivity to these agents [5, 34]. However, the heterogeneity in the heterozygote response observed [5, 12] precludes its usefulness in detecting gene carriers.

Chromosomal hypersensitivity to clastogenic (chromosome-breaking) factors produced by mutant cells

A number of studies have suggested that BS, FA, and A-T cells may produce certain diffusable chemical factors that act as clastogenic agents on normal cells [38]. While data on such factors produced by FA and BS cells are equivocal [38], a low-molecular-weight (500 to 1000 dalton) protein factor has been identified from the serum of A-T patients, the amniotic fluid of affected A-T fetuses, and the supernatent medium from cultured A-T fibroblasts [40, 41, 42]. The ability to produce the clastogenic factor has been used in prenatal and postnatal diagnosis of A-T in families in which the mutation has been known to be segregating [42].

In vitro cell transformation

Because increased transformation in vivo is a key phenotype of mutations that predispose to neoplastic development, it is reasonable to expect that cells derived from gene carriers would be more readily amenable to induction of transformation by carcinogens in vitro. Many investigators have tested the in vitro transformability of cells from individuals with a number of cancer-predisposing syndromes; so far, FA and the inherited adenomatosis of colon and rectum (ACR) syndrome cells yielded positive results. Thus, both FA and ACR fibroblasts have been shown to be more susceptible to T-antigen display and growth transformation induced by simian virus 40 (SV40) than control cells [32, 47]. In addition, ACR cells also have been shown to be more susceptible to induction of growth transformation by the Kirsten murine sarcoma virus (KiMSV) [23, 31]. Although these data are interesting and potentially important in terms of identification of gene carriers, neither sufficient data from gene carriers nor baseline data from sufficient numbers of non-gene-carrier "control" individuals are available in order to evaluate rigorously the value of these traits as diagnostic indicators of the carrier state. Furthermore, available data indicate that familial large bowel cancer is a highly heterogeneous entity and most probably involves several mutations in its etiology [2, 39].

Cell growth in medium with low serum

ACR fibroblasts have been shown to be capable of growth in vitro under conditions of low serum concentration in the medium, a trait generally associated with transformed cells [31]. The usefulness of this trait as a general diagnostic indicator of carrier state has not been established; it falls in the same category as susceptibility to transformation by viruses.

Inherited traits that exhibit linkage relationships with cancer-predisposing genes

In contrast to the above, a number of inherited traits have been shown to exhibit possible or probable association or genetic linkage with inherited cancer. When the allelic frequencies at a given locus differ significantly between a population of patients with a specific cancer type and normal controls, association between the locus and the disease state is indicated. The association may reflect direct or indirect etiological relationship between the two [30]. Relative risk for a number of cancer types associated with alleles of a number of polymorphic red blood cell group and serum protein markers has been summarized by King and Petrakis [30]. However, association data are of limited predictive value for the individual or family; therefore, they are not discussed further here. Close linkage of an inherited cancer phenotype to an unrelated interited trait with a clearcut and easily recognizable phenotype, on the other hand, is of predictive value. Linkage relationships of a number of inherited traits with inherited cancer phenotypes have been investigated; the results of these studies are summarized below and their value commented upon.

Major histocompatibility complex (HLA)

Comparison of HLA haplotypes of populations of Hodgkin's disease (HD) patients and controls suggested an association between HD and HLA [14, 26] which led to investigation of possible linkage between the two. However, the clustering of Hodgkin's disease in families does not follow a recognizable Mendelian pattern, thereby ruling out a classical linkage analysis. The analyses mainly have tested the significance of deviation from expected of haplotype sharing by affected relatives. The data show strong but not complete haplotype sharing, indicating a relationship, as well as heterogeneity in the relationship [14, 26]. Possible linkage between HLA and malignant melanoma also has been suggested by a number of investigators [14]. More data are needed to confirm these presumed linkages.

Glutamate-pyruvate transaminase (GPT)

The report of a lod score reaching +1.95 [28, 29] between breast cancer and the locus for GPT, a gene mapped to chromosome 16 [36], suggesting possibility of linkage, elicited considerable interest because it was the first report of a potential linkage between an inherited cancer phenotype and an unrelated enzyme marker. However, a number of other investigators have been unable to confirm this association [7, 10, 35]. It should be noted that many of the pedigrees in which the initial positive lod score was observed included individuals with ovarian cancer, a possible genetic subset of familial breast cancer [33]. More data are needed both to properly define the heterogeneity in familial breast cancer and to establish the possible linkage of GPT with one or more forms of it. (See also Chapter 8 by King [29] in this volume for further discussion of these data.)

Inherited chromosome rearrangements

Three clinical entities are recognized in which a germ-line chromosome abnormality shows almost complete linkage with cancer development. The first is a single family with an inherited translocation between chromosomes 3 and 8 in which all carrier individuals developed renal cell carcinoma [12]. The second is deletion of band 13q14 associated with retinoblastoma [23, 49]. The third is deletion of band 11p13 associated with aniridia-Wilms' tumor association [17, 23]. Although these abnormalities are diagnostic of the respective syndromes, they were identified through ascertainment of the affected; therefore, it cannot be stated categorically that fortuitous observation of one of these rearrangements, if made in an unaffected individual in the general population, is a prognostic finding. For example, study of several additional families with renal cell carcinoma did not reveal the 3/8 translocation [23].

Fragile sites on chromosomes

Fragile sites are specific sites on chromosomes at which gaps and breaks are expressed when cells, especially blood mononuclear cells, are cultured in medium deprived of folic acid and thymidine [15]. Among a large number of such sites recognized, 17 have been called by Yunis and Soreng [50] as heritable (h-fra) and over 50 as constitutive (c-fra), the heritability status of the latter being unknown. Recently, it has been shown that the positions of 8 of the 18 cellular oncogenes that map to a specific chromosome band or region also are at or near c-fra sites [50]. In addition, 20 of 51 c-fra sites and 6 of 16 h-fra sites appear to map at or close to breakpoints seen in 26 of 31 specific structural chromosome alterations recognized in cells from histologically diverse tumors [50]. These data suggest a relationship between tumor-specific chromosome breaksites and c-fra sites. If c-fra expression indeed is genetically determined, then inheritance of some of these sites may possibly indicate predisposition. Clearly, more data are needed to establish the relationship between c-fra sites and cancer predisposition. Currently available data do not permit use of c-fra sites as markers for genetic cancer predisposition.

Restriction fragment length polymorphisms

Restriction endonucleases recognize and catalyze cleavages at specific sequences in DNA, resulting in fragments of defined lengths. Cleaving DNA from random individuals with a given enzyme yields fragments of different lengths whose molecular weight differences can be detected by electrophoresis. Such polymorphisms in DNA are referred to as RFLPs [6, 44, 48]. The human genome is endowed with a wealth of RFLPs, many of which have been shown to be inherited [44, 48]. A great deal of effort currently is being expended in mapping RFLPs to specific chromosomal regions using somatic cell hybridization or in situ molecular hybridization methods. Since RFLPs occur at specific as well as arbitrary loci that are distributed throughout the genome, in principle they are the most appropriate markers to establish linkage relationships of any inherited phenotype. Although no cancer-predisposing genes have so far been mapped using RFLPs, they comprise the most promising genetic markers for linkage studies, and significant results can be expected in the near future from the work of a number of investigators who are studying the linkage relationships between RFLPs and cancer-predisposing genes.

Conclusions

The data discussed above show that, in spite of an extensive literature that describes phenotypes associated with cancer-predisposing mutations, very few examples of reliable diagnostic indicators are available. Furthermore, not a single cellular or other phenotype is known at present with which unaffected but predisposed individuals can be identified by screening the general population.

Literature cited

In order to provide the reader with as wide a coverage as possible of the rather large literature in this area, publications cited here mainly comprise recent comprehensive reviews rather than primary reports.

1. Anderson DE: Familial cancer and cancer families. *Semin Oncol* 5:11–16, 1978.
2. Anderson DE, Williams WR: Familial cancer: Implications for healthy relatives. Chapter 15 in this volume.
3. Andrews AD: Xeroderma pigmentosum. *In* Chromosome Mutation and Neoplasia (German JL, ed) New York, Alan R Liss, 1985, 63–83.
4. Arlett CF, Lehman AR: Human disorders showing increased sensitivity to the induction of genetic damage. *Annu Rev Genet* 12:95–115, 1978.
5. Auerbach AD, Adler B, Chaganti RSK: Prenatal and postnatal diagnosis and carrier detection of Fanconi anemia by a cytogenetic method. *Pediatrics* 67:128–135, 1981.
6. Botstein D, White R, Skolnick M, Davis R: Construction of a genomic linkage map in man using restriction length fragment polymorphisms. *Am J Hum Genet* 32:314–331, 1980.
7. Cannon LA, Bishop DT, McLellan T, Skolnick MH: Pedigree and linkage analysis of breast cancer in eleven Utah kindreds: nonlinkage of breast cancer to GPT. *Am J Hum Genet* 35:60A, 1983.
8. Cervanka J, Arthur D, Yasis C: Mitomycin C test for diagnostic differentiation of idiopathic aplastic anemia and Fanconi anemia. *Pediatrics* 67: 119–127, 1981.
9. Chaum E, Doucette LA, Ellsworth RM, Abramson DH, Haik BG, Kitchin FD, Chaganti RSK: Bleomycin-induced chromosome breakage in G2 lymphocytes of retinoblastoma patients. *Cytogenet Cell Genet* 38:152–154, 1984.

10. Cleton FJ: Genetic markers of cancer susceptibility. *In* Proceedings, 13th International Cancer Congress, Part C (Mirand EA et al., eds) New York, Alan R Liss, 1983, 383–389.
11. Cleaver JE: Regulation of the responses to DNA damage in the hypersensitivity diseases and chromosome breakage syndromes. *In* Chromosome Mutation and Neoplasia (German J, ed) New York, Alan R Liss, 1983, 235–249.
12. Cohen MM, Simpson SJ, Honig GR, Maurer HS, Nicklas JW, Martin AO: The identification of Fanconi anemia genotypes by clastogenic stress. *Am J Hum Genet* 34: 794–810, 1982.
13. Danes BS: In vitro expression of cancer genes for heritable human tumors: numerical alterations in chromosome complement. *Disease Markers* 2:371–380, 1984.
14. Dausset J, Colombani J, Hors J: Major histocompatibility complex and cancer, with special reference to human familial tumours (Hodgkin's disease and other malignancies). *In* Inheritance of Susceptibility to Cancer in Man (Bodmer WF, ed) New York, Oxford University Press, 1983, 119–147.
15. de la Chapelle A, Berger R: Report of the committee on chromosome rearrangements in neoplasia and on fragile sites. *Cytogenet Cell Genet* 37:274–311, 1983.
16. Doucette LA, Pahwa S, Chaganti RSK: An ataxia-telangiectasia variant is not hypersensitive to the clastogenic effect of bleomycin. *Am J Hum Genet* 35:130A, 1983.
17. Francke U, Holme LB, Atkins LB, Riccardi VM: Aniridia-Wilms' tumor association: evidence for specific deletion of 11p13. *Cytogenet Cell Genet* 24:185–192, 1979.
18. Gatti RA, Hall K: Ataxia-telangiectasia: search for a central hypothesis. *In* Chromosome Mutation and Neoplasia (German J, ed) New York, Alan R Liss, 1983, 23–41.
19. German J: Genes which increase chromosomal instability in somatic cells and predispose to cancer. *Prog Med Genet* 8:61–102, 1972.
20. German J: Recessive inheritance in human cancer. Chapter 6 in this volume.
21. Good RA: Relations between immunity and malignancy. *Proc Natl Acad Sci USA* 69:1026–1032, 1972.
22. Good RA, Day NB, Cunningham-Rundles C, O'Reilly RJ, Auerbach AD, Fernandez G: Genetics of the primary immunodeficiency diseases. *In* Menarini Series on Immunopathology (Miescher PA et al., eds) Basel, Schwabe and Co, 1981, 184–240.
23. Harnden D, Morten J, Featherstone T: Dominant susceptibility to cancer in man. *Adv Cancer Res* 41:185–255, 1984.
24. Hecht F, McCaw BK: Chromosome instability syndromes. *In* Genetics of Human Cancer (Mulvihill JJ et al., eds) New York, Raven Press, 1977, 105–113.
25. Heddle JA, Krepinsky AB, Marshall RR: Cellular sensitivity to mutagens and carcinogens in the chromosome breakage and other cancer-prone syndromes. *In* Chromosome Mutation and Neoplasia (German J, ed) New York, Alan R Liss, 1983, 203–234.
26. Hors J, Dausset J: HLA and susceptibility to Hodgkin's disease. *Immunological Rev* 70:167–192, 1983.
27. King MC: Genetic epidemiology of human cancer: Application to familial breast cancer. Chapter 8 in this volume.
28. King MC, Petrakis NL: Genetic markers and cancer. *In* Genetics of Human Cancer (Mulvihill JJ et al., eds) New York, Raven Press, 1977, 281–290.
29. King MC, Go RCP, Elston RC, Lynch HT, Petrakis NL: Allele increasing susceptibility to human breast cancer may be linked to the glutamate-pyruvate transaminase locus. *Science* 208:406–408, 1980.
30. King MC, Go RCP, Lynch HT, Elston RC, Terasaki PI, Petrakis NL, Rodgers GC, Lattanzio D, Bailey-Wilson J: Genetic epidemiology of breast cancer and associated cancers in high-risk families. II Linkage analysis. *J Natl Canc Inst* 71:463–467, 1983.
31. Kopelovich L: Adenomatosis of the colon and rectum: relevance to inheritance and susceptibility mechanisms in human cancer. *In* Inheritance of Susceptibility to Cancer in Man (Bodmer WF, ed) New York, Oxford University Press, 1983, 71–91.

32. Lubiniecki AS, Blattner WA, Dosik H, McIntosh S, Wertelecki W: Relationship of SV40 T-antigen expression in vitro to disorders of bone marrow function. *Am J Hematol* 8:389–396, 1980.
33. Lynch HT: Genetic heterogeneity and breast cancer: variable tumor spectra. *In* Genetics and Breast Cancer (Lynch HT, ed) New York, Van Nostrand Reinhold, 1981, 134–173.
34. Marx MP, Smith S, Heyns AP, van Tonder IZ: Fanconi's anemia: a cytogenetic study on lymphocyte and bone marrow cultures utilizing 1,2:3,4-diepoxybutane. *Cancer Genet Cytogenet* 9:51–60, 1983.
35. McLellan T, Cannon LA, Bishop DT, Skolnick MH: The cumulative LOD score between a breast cancer susceptibility locus and GPT is −3.86. *Cytogenet Cell Genet* 37:536–537, 1984.
36. McKusick VA: *Mendelian Inheritance in Man*, Baltimore, Johns Hopkins University Press, 1982.
37. Paterson MC, Smith PJ. Ataxia telangiectasia: an inherited disorder involving hypersensitivity to ionizing radiation and related DNA-damaging chemicals. *Annu Rev Genet* 13:291–318, 1979.
38. Ray JH, German J: The cytogenetics of the "chromosome-breakage syndromes." *In* Chromosome Mutation and Neoplasia (German J, ed) New York, Alan R Liss, 1983, 135–167.
39. Schimke RN: Dominant inheritance of human cancer. Chapter 7 in this volume.
40. Shaham M, Becker Y, Cohen MM: A diffusable clastogenic factor in ataxia-telangiectasia. *Cytogenet Cell Genet* 27:155–161, 1980.
41. Shaham M, Becker Y: The ataxia telangiectasia clastogenic factor is a low molecular weight peptide. *Hum Genet* 58:422–424, 1981.
42. Shaham M, Kohn G, Yarkoni S, Becker Y, Voss R: Prenatal diagnosis of ataxia telangiectasia in cell free amniotic fluid. *J Pediatrics* 100:134–137, 1982.
43. Shiloh Y, Tabor E, Becker Y: Cellular hypersensitivity to neocarzinostatin in ataxia-telangiectasia skin fibroblasts. *Cancer Res* 42:2247–2249, 1982.
44. Skolnick MH, Willard HF, Menlove LA: Report of the committee on human gene mapping by recombinant DNA techniques. *Cytogenet Cell Genet* 37:210–273, 1983.
45. Swift M: Disease predisposition of ataxia-telangiectasia heterozygotes. *In* Ataxia-Telangiectasia: A Cellular and Molecular Link between Cancer, Neuropathology, and Immune Deficiency (Harnden DG, Bridges BA, eds) New York, John Wiley, 1982, 355–361.
46. Spector BD, Perry GS III, Kersey JH: Genetically determined immunodeficiency diseases (GDID) and malignancy: report from the immunodeficiency cancer registry. *Clin Immunol Immunopathol* 11:12–29, 1978.
47. Todaro GJ, Green H, Swift MR: Susceptibility of human diploid fibroblasts to transformation by SV40 virus. *Science* 153:1252, 1966.
48. White R: DNA polymorphism: new approaches to the genetics of cancer. *In* Inheritance of Susceptibility to Cancer in Man (Bodmer WF, ed) New York, Oxford University Press, 1983, 175–186.
49. Wilson MG, Abbin AJ, Towner JW, Spencer WH: Chromosome anomalies in patients with retinoblastoma. *Clin Genet* 12:1–8, 1977.
50. Yunis JJ, Soreng AL: Constitutive fragile sites and cancer. *Science* 226:1199–1204, 1984.

11.

Chromosome changes in leukemia

JOSEPH R. TESTA SHINICHI MISAWA

During the last decade there has been extraordinary progress in the understanding of patterns of chromosome changes in leukemia. The advances have come about with the application of cytogenetic banding techniques, which enable us to identify each human chromosome. Nonrandom patterns of chromosome alterations have been well-documented in leukemia. In acute leukemia, banding analysis of routine metaphase chromosome preparations have revealed clonal karyotypic abnormalities in bone-marrow cells from approximately 50% of patients with acute nonlymphocytic leukemia (ANLL) [24, 50, 86, 102] and in about 50 to 65% of those with acute lymphoblastic leukemia (ALL) [86, 103, 104]. Most of the data presented in this chapter were obtained between 1974 and 1980. Many of the studies made during this period used chromosomes that were relatively contracted, and the banding pattern often was poorly defined. Newer high-resolution chromosome banding methods can yield elongated, finely banded mitotic chromosomes that allow us to detect chromosome rearrangements that might have been overlooked previously [113]. Recent investigations utilizing this improved technology now indicate that most patients with acute leukemia have cytogenetic changes in their bone-marrow cells [100, 115]. Furthermore, during the last 2 years there has been considerable excitement as a result of recent findings linking human cellular oncogenes to chromosome changes in malignancy. There appears to be some concordance between the chromosome location of cellular oncogenes and the breakpoints involved in specific chromosome rearrangements in various malignant diseases, including leukemia. This evidence appears to support the proposal that chromosome alterations have a fundamental role in neoplasia.

Along with these advances, correlations between chromosome changes and clinical findings have been documented in several hematologic malignancies. In ANLL and ALL it is now possible to recognize specific subgroups, each characterized by a particular chromosome rearrangement and distinctive clinical course.

Methodology

In order to examine the karyotype of leukemic cells, cytogenetic analysis is ordinarily done on a bone-marrow aspirate or, in some cases, on biopsied bone core. If there are a considerable number of circulating leukemic blasts, a peripheral blood sample can be cultured without adding any mitogen; the karyotype of the mitotic cells is generally similar to that seen in the marrow. In many cytogenetics laboratories these specimens are processed by the traditional "direct" technique after a 1- to 2-hour exposure to a mitotic arresting agent such as Colcemid [105]. In some laboratories, specimens are cultured without mitogen stimulation for 1 to 3 days prior to harvesting the dividing cells.

The chromosome changes discussed here are clonal abnormalities. An abnormal clone is defined as two or more metaphase cells with an identical structural rearrangement or extra chromosome, or three or more cells lacking the same chromosome [24]. Patients are considered to have an abnormal karyotype if a clonal chromosome change is found. Patients are considered to be karyotypically normal if their marrow shows no chromosome alterations or if the abnormalities involve different chromosomes in different cells. Sporadic changes may be due to technical artifacts or random mitotic errors.

Chromosomes are identified according to the Paris Conference nomenclature [63] and the International System for Human Cytogenetic Nomenclature [37, 38], and karyotypes are expressed as recommended under these systems. The total chromosome number is indicated first, followed by the sex chromosome constitution, and then by any gains, losses, or rearrangements of the autosomes. A plus sign (+) or minus sign (−) before a number indicates gain or loss, respectively, of a whole chromosome. A plus or minus sign after a number indicates gain or loss of a part of a chromosome. The following abbreviations are used in the subsequent text: p and q are the short and long arms of the chromosome, respectively; i = isochromosome; del = deletion; inv = inversion; t = translocation. Abbreviations for structural rearrangements are followed by the chromosome(s) involved in the first set of parentheses; the chromosome bands in which the breaks occurred are given in the second set of parentheses.

When leukemic cells show a chromosome pattern compatible with a constitutional abnormality and normal phenotype (e.g., a balanced translocation), it is important to analyze cells from normal tissues. In most instances, cells from unaffected tissues such as phytohemagglutinin (PHA)-stimulated lymphocytes or skin fibroblasts have a normal chromosome pattern. Thus, the karyotypic abnormalities observed in leukemic cells represent acquired, somatic mutations.

Chronic myelogenous leukemia

Chronic phase

Nowell and Hungerford [59] reported the first consistent chromosome change associated with a human neoplasm in 1960. They described an abnormally small G-group chromosome, referred to as the Philadelphia chromosome or Ph^1, in the leukemic cells of patients with chronic myelogenous leukemia (CML). Subsequent studies have shown that the Ph^1 is present in bone-marrow cells from about 85 to 90% of patients with CML [87].

With the application of banding techniques, Rowley [71] showed that the "missing" distal segment of the long arm of the Ph^1 was translocated to chromosome 9. The breakpoints are at band q34 in chromosome 9 and at q11 in chromosome 22 (Figure 11–1*a*). Of considerable interest is the recent finding by de Klein et al. [22] that the human cellular oncogene c-*abl* is located at band 9q34, the affected site on chromosome 9 in the 9;22 translocation. Furthermore, the investigators were able to demonstrate that c-*abl* was translocated to the Ph^1(22q−), which provides evidence for the reciprocal nature of the rearrangement.

Karyotypes of more than 1100 patients with Ph^1-positive (Ph^1+) CML examined with banding techniques have been reported by various investigators [summarized in reference 80]; 92% of these cases had the typical 9;22 translocation. The remaining 8% had variant translocations or, in a few patients, a Ph^1 but with no apparent translocation. Of the patients with variant translocations, approximately half had a simple translocation involving chromosome 22 and some chromosome other than 9. The others had a complex rearrangement involving chromosome 22 and two or more other chromosomes (one of which is usually chromosome 9).

Ph^1+ CML patients have a much better prognosis than Ph^1-negative patients. The median survival of Ph^1+ patients is about 30 to 40 months, compared to only 12 to 15 months for Ph^1-negative patients [109]. At the present time the clinical significance of variant Ph^1 translocations is controversial. Sandberg [85] has collected multicenter data, which indicate that the survival curves for patients with variant translocations do not differ from those for patients with the typical t(9;22). However, in another investigation the "benign" phase in patients who had variant Ph^1 translocations was significantly shorter than in the standard type of translocation [68]. To resolve this issue, it will be necessary to study more patients for a longer period of time.

The Ph^1 is usually seen in all of the bone-marrow cells examined,

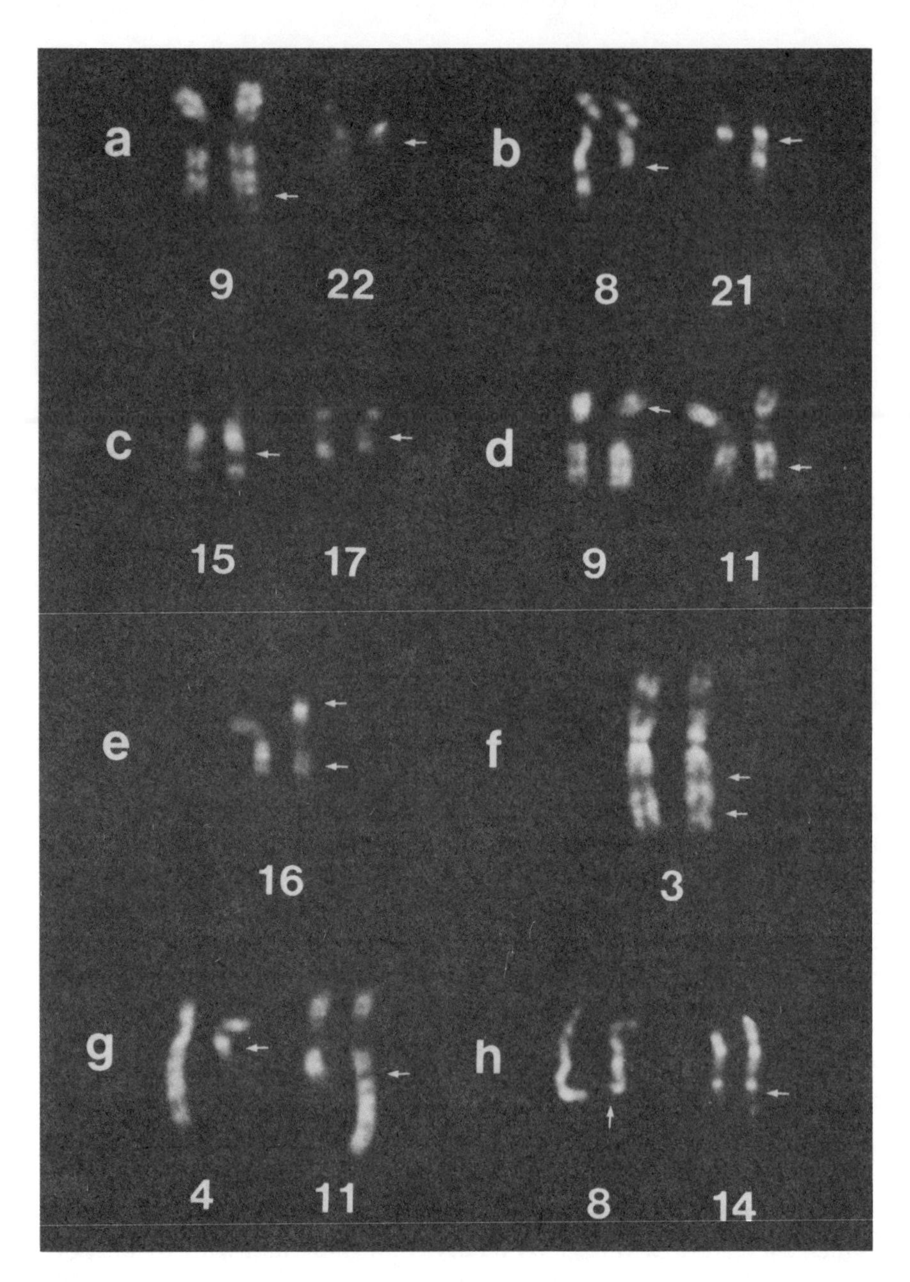

Figure 11–1 Selected Q-banded chromosomes showing rearrangements frequently observed in leukemia: t(9;22)(q34;q11) seen in a patient with Ph_1+ CML (*a*); t(8;21)(q22;q22) in AML-M2 (*b*); t(15;17)(q22;q12) in APL-M3 (*c*); t(9;11)(p21;q23) in AMoL-M5 (d); inv(16)(p13q22) in AMMoL-M4 (e); inv(3)(q21q26.2) in a case of CML in blastic crisis with abnormal megakaryocytes (*f*); t(4;11)(q21;q23) in ALL-L2, although there is controversy regarding this diagnosis (g); and t(8;14)(q24;q32) in non-African Burkitt's lymphoma cell line (*h*). Arrows indicate breakpoints of rearrangements.

even when the patient is in remission. There is evidence that patients who have some normal cells have a better prognosis than patients who have only Ph^1+ cells [81]. There have been some attempts to use intensive forms of therapy to eradicate the Ph^1+ cell line. In one study 12 of 37 patients (33%) who were initially 85 to 100% Ph^1+ converted to 0 to 30% Ph^1+ cells after receiving such therapy [20]. The decline in the percentage of Ph^1+ cells lasted less than 4 months in most patients. Two patients with slowly progressive disease had a sustained decline in Ph^1+ cells for more than 3 years. With this therapeutic approach, it was not possible to prevent blastic crisis or to prolong the overall survival of the group. However, patients who responded had a longer median survival.

Karyotypic changes, in addition to the Ph^1 translocation, may be found in about 10 to 30% of CML patients in the chronic phase [12, 24]. These changes include a +8, i(17q), a second Ph^1, loss of the Y chromosome, and various structural rearrangements. Many of these abnormalities are identical to those found in the terminal acute phase of the disease. Patients in whom additional chromosome alterations are detected at the time of diagnosis of the chronic phase of CML do not appear to have a substantially poorer prognosis than those who have the Ph^1 alone [109].

Blastic crisis

At the onset of the acute phase of CML (blastic crisis), approximately 20% of the patients show no further change in the karyotype, whereas 80% have additional chromosome abnormalities superimposed on the Ph^1+ cell line [74]. Usually, a change in the karyotype is considered to be a grave prognostic sign; the median survival from the time of change until death is about 2 to 5 months [109]. In some cases, evolution of the karyotype may precede the clinical signs of blastic crisis by several months [87].

More than 300 patients with CML who had a change in their karyotype at the time of blastic crisis have been reported by various investigators [summarized in Reference 80]. In 30% of these cases the change in karyotype was not reflected in a change in the modal chromosome number of 46. For instance, Rowley studied one patient with new abnormalities (−7 and +8) at the onset of blastic crisis that would not have been detected without banding analysis [80]. Nearly 65% of all patients who had a change in karyotype in the acute phase had hyperdiploid clones, most frequently with modal chromosome numbers of 47 to 50. Hypodiploid clones were observed in only 6% of all cases summarized. There is some evidence that patients whose marrow cells have a hypodiploid modal number in blastic crisis may respond better to therapy with vincristine and prednisone than patients with other karyotypic changes [17].

There is a distinctly nonrandom pattern of karyotypic evolution in the acute phase of CML. The most frequent changes include a +8 (about 40% of cases), a gain of second Ph^1 (about 40%), an i(17q) (about 25%), and a +19 (about 15%) [80]. In a given patient, these changes occur singly or in combinations.

There are conflicting data regarding the prognostic significance of karyotypic evolution in the acute phase of CML. Prigogina and Fleischman [69] observed a higher remission rate and a longer survival in patients who retained the $46,Ph^1+$ cell line unchanged compared with those whose marrow cells acquired additional changes. However, other investigators have reported that the survival time of patients who developed additional chromosome abnormalities was similar to that of patients whose karyotype did not change [2, 93]. Alimena et al. [2] did observe that patients with a lymphoid rather than a myeloid blastic crisis survived longer.

Acute nonlymphocytic leukemia de novo

Between 1976 and 1983 there were 14 reports on banding studies of 25 or more ANLL patients, mostly adults, whose leukemic cells were processed primarily by conventional cytogenetic methods [1, 10, 14, 25, 30, 46, 52, 62, 67, 70, 73, 79, 92, 94]. The incidence of cases with an abnormal karyotype ranged from 42% [62, 67, 94] to 73% [92], with an overall incidence of about 50%. Cells from most of the patients reported in these papers were processed by the direct technique, whereas cells from some patients were processed by the unsynchronized short-term culture method. Patients examined by the high-resolution banding method with methotrexate are not presented in this section, except for some of the cases included in the report by Hagemeijer et al. [30].

Modal chromosome numbers in ANLL patients are primarily distributed in the diploid range of 45 to 47 (approximately 85% of ANLL patients with abnormal karyotypes). In contrast to ALL, in ANLL modal numbers greater than 48 are relatively uncommon. Modal numbers up to the tetraploid range have been reported in ANLL on very rare occasions [101].

Banding patterns in adults

Although chromosome changes in patients with ANLL show considerable variation, certain nonrandom patterns are evident. The numerical and structural chromosome changes in 282 karyotypically abnormal adult patients reported by various investigators have been summarized previously by Misawa and Testa [50]. A gain of chromosome 8, the most frequent abnormality seen in ANLL, was found in 66 patients (23.4% of the 282 abnormal cases), in 33 of whom a +8 was the sole alteration present. Similarly, loss of chromosome 7, another common numerical change, was observed in 32 patients, 14 of whom showed only this change in karyotype. As with chromosome gains and losses, the nonrandom distribution of structural rearrangements is especially evident in the cases with a single abnormality. The most frequently rearranged chromosome was 17, which was observed in 51 patients (18.1% of the 282 cases summarized), 30 of whom showed only one abnormality. Likewise, the majority of cases with rearrangements of chromosomes 8, 15, or 21, the next three most commonly altered chromosomes, involved a single abnormality or, in some of the cases with a t(8;21), showed loss of a sex chromosome as the only accompanying change. Thirty-three of the cases with rearrangements of chromosomes 15 and 17 involved an apparently identical 15;17 translocation. An 8;21 translocation was observed in 37 patients, 17 of whom showed a loss of an X or Y as the only other change. Rearrangements of chromosomes other than chromosomes 15, 17, 8, and 21, rarely occurred alone in ANLL.

As in CML, karyotypic evolution may develop during the disease course of ANLL [25, 99]. Testa et al. [99] reported evolution of the karyotype in 17 of 60 ANLL patients (28%) for whom serial samples of leukemic cells were obtained. The pattern of cytogenetic evolution was nonrandom. The most common change was a gain of a chromosome 8, which was found in 10 of 17 patients (59%) whose karyotype evolved. In all but one of the initially abnormal patients, karyotypic evolution involved the original cytogenetically abnormal clone.

8;21 translocation

Kamada et al. [40] were the first to recognize that a group of patients with acute myeloblastic leukemia (AML) may be characterized by an abnormality most likely representing a translocation between a C- and a G-group chromosome. Rowley [72] used the quinacrine (Q)-banding technique to determine that this abnormality involved a translocation between chromosomes 8 and 21, t(8;21)(q22;q22) (Figure 11–1*b*). Interestingly, Neel et al. [57] recently have mapped the oncogene c-*mos* to band 8q22, the break site of chromosome 8 in the 8;21 translocation. The frequency with which this translocation occurs may vary from one laboratory to another, but it amounted to 13% (37/282) of the abnormal cases summarized by Misawa and Testa [50]. The abnormality is usually restricted to patients with the M2 category of ANLL. At the Second International Workshop on Chromosomes in Leukemia

(SIWCL) [91], all 43 cases with a t(8;21) and adequate bone-marrow material available for cytologic review had a diagnosis of AML with maturation (type M2 according to the French-American-British (FAB) classification) [6]. Overall, the t(8;21) was seen in 9.3% of all patients with M2-type leukemic cells. Patients with a t(8;21) have also been reported to have a low leukocyte alkaline phosphatase level [41] and a high incidence of Auer rod-positive cells [41, 106]. These patients also have a relatively long median survival [41, 83, 106]. Data on survival of 48 patients with an 8;21 translocation were reviewed at the SIWCL; the median survival of the whole group was 11.5 months, which was longer than for patients with other chromosome abnormalities.

The t(8;21) is also of interest for two other reasons. First, chromosomes 8 and 21 can participate in three-way rearrangements similar to those involving chromosomes 9 and 22 in CML. In five reported cases with complex 8;21 rearrangements the third chromosome involved was 1 (twice), 2, 11, or 17 [45, 49, 60, 65]. Second, as mentioned earlier, the t(8;21) often is accompanied by the loss of a sex chromosome; of the cases reviewed at the SIWCL 32% of the males with a t(8;21) were −Y, and 36% of the females were missing one X [91]. This association is particularly noteworthy because sex-chromosome abnormalities otherwise rarely are seen in ANLL patients.

15;17 translocation

A consistent structural rearrangement involving chromosomes 15 and 17 in acute promyelocytic leukemia (APL, or ANLL type M3) was first recognized by Rowley et al. [77]. There has been considerable confusion with regard to the breakpoints in this rearrangement. Recently, Hagemeijer et al. [32] performed detailed banding analysis on elongated chromosomes from four patients with a t(15;17), and the breakpoints (15q2200 and 17q12) appeared to be identical in each case (Figure 11–1*c*). Using somatic cell-hybridization techniques to perform genetic analysis of the 15;17 translocation, Sheer et al. [90] similarly proposed breakpoints at 15q22 and proximally in 17q21 (near to 17q12). Of the 80 patients with APL who were reviewed at the SIWCL, 33 had a t(15;17), 7 had other types of chromosome changes, and 40 had a normal karyotype [91]. The t(15;17) was not found in patients with any other type of acute leukemia.

In some cases of APL with a t(15;17), the granules typically seen in the leukemic promyelocytes cannot be seen by light microscopy although they are evident when the cells are examined ultrastructurally [27, 97]. The FAB Co-operative Group has since recognized that not all APL patients have coarse granules, and has thus added a category called the M3 variant [7]. The variant subgroup was identified largely on the basis of the clinical features and the presence of the 15;17 translocation.

Of further interest, variant simple or complex forms of the t(15;17) have been reported recently. The variant simple translocations involved a chromosome 17 and either a 1 or a 7 [111]. Two cases with complex translocations involved chromosomes 15, 17, and either chromosome 2 or 3 [11]. Thus, a similar pattern of variation of a specific translocation can involve the t(15;17) as well as the t(9;22) and t(8;21).

An unusual geographical distribution of the APL cases with a t(15;17) has been noted [96]. For example, the translocation has been observed in 23 of 23 patients in Chicago [44], 6 of 6 patients in Baltimore [50], 18 of 19 patients in Japan [84], 11 of 16 patients in Belgium [107], and 0 of 12 patients in Finland and Sweden [96]. The reason or reasons for this variation are not yet entirely certain, but the particular cytogenetic methodology used appears to be a likely factor. Berger et al. [9] have provided evidence that APL patients who appear to be chromosomally normal based on direct marrow preparations actually may show the t(15;17) in preparations from cells cultured for 24 to 48 hours. Subsequently, a number of other investigators have reported similar findings [26, 44, 50, 112].

Other specific structural rearrangements

Recently, several additional, specific chromosome rearrangements have been associated with distinctive clinical and hematologic features in patients with ANLL. Each of these rearrangements is rather subtle, and the identification of such changes has coincided with recent improvements in cytogenetic methodology.

One of these chromosome abnormalities is an apparently balanced rearrangement, between chromosomes 9 and 11, which involves reciprocal translocation of small chromosome segments of about the same lengths. Hagemeijer et al. [31] found a t(9;11)(p21;q23) in three cases of acute monoblastic leukemia (AMoL, or ANLL type M5), and proposed a new specific association between this translocation and acute leukemia of the M5 type (Figure 11–1*d*). Preferential involvement of band 11q23 has been observed in cases with acute myelomonocytic leukemia (AMMoL, or ANLL type M4) or AMoL, and a deletion at band 11q23 or translocations between chromosomes 11 and either 6, 9, 10, 17, or 19 have also been reported [23].

Arthur and Bloomfield [3] identified another specific abnormality, del(16)(q22), in 5 of 60 patients with ANLL. In each of the 5 cases this deletion was the only chromosome change found. The unique hematologic feature of these patients was bone-marrow eosinophilia (range 8 to 54%). None of the other patients in their series had more than 4%

marrow eosinophils. All 5 patients with the deleted 16 had blasts with monocytoid morphologic features; however, 3 patients had results with cytochemical stains that were most consistent with an M2 type. Thus, 3 patients were designated as having M2 and 2 as having M4. Le Beau et al. [48] subsequently have identified an apparently related cytogenetic-clinicopathologic correlation. They found an association of a pericentric inversion of chromosome 16, inv(16)(p13q22), with abnormal marrow eosinophils in 18 patients who had a diagnosis of M4 (Figure 11–1*e*). The atypical eosinophils demonstrated unusually coarse cytoplasmic granulation, with a mixture of eosinophilic and basophilic staining characteristics. Clinically, a complete remission was achieved in 13 of 17 patients who received chemotherapy, and they appear to have a better prognosis than patients with other types of ANLL [46, 48]. Recently, we have identified a translocation involving both chromosome 16 homologues, t(16;16)(p13;q22), in a patient with AMMoL and abnormal marrow eosinophils [98]. In all of these chromosome 16 rearrangements, the consistent finding is a break at band 16q22. The consistent involvement of this site suggests that genetic sequences may reside there that are important in eosinophilic differentiation and that alteration of this region in a myeloid precursor may be associated with proliferation of abnormal eosinophils.

Bernstein et al. [13] have recently reported rearrangements of chromosome 3 in four patients with ANLL (M1 or M2 types), each of whom showed abnormalities of thrombopoiesis, including the presence of abnormal, hypolobulated micromegakaryocytes in the marrow. Three of their patients had an inv(3) with consistent breakpoints at q21 and q26 (Figure 11–1*f*). Their fourth patient had a translocation, t(3;3)(q2?6;q2?9).

In 1976 Golomb et al. [28] described a translocation between chromosomes 6 and 9, t(6;9)(p23;q34), in two patients with AML. More recently, a t(6;9)(p23;q34) or a t(6;9)(p21;q23) have been reported in three patients with AML (type M2), and in three others with either AMMoL, AML, or CML [88, 89, 108]. Overall, the group of patients with ANLL was relatively young, and only two of the seven achieved a complete remission. Recently, Pearson et al. [66] described an association between the t(6;9) and bone marrow basophilia in ANLL. They reviewed bone marrow cytology on 8 patients (4 from their laboratory) and found an increased percentage of bone marrow basophils in 7 of the 8 patients; the percentage of marrow basophils ranged from 1.5% to 12% (normal value is less than 0.2%). The investigators concluded that the association of t(6;9) in ANLL with basophilia may point to the chromosomal location of genes regulating the production of basophils [66].

Banding patterns in children

Eighty-seven children and adolescents with ANLL who showed clonal karyotypic abnormalities are summarized here [16, 80]. Although the number of children examined cytogenetically is relatively small compared with that for adults, it is possible to make some preliminary comparisons between the patterns seen in these two groups. Whereas specific abnormalities such as t(8;21), t(15;17), and +8 are common to each age group, there appear to be differences with regard to the incidence of these changes in adults and children. Thus, a +8 has been reported in only 7 of 87 (8%) children with abnormal karyotypes. In adults, 23.4% of the 282 patients with abnormal karyotypes from our summary had a +8 [50]. In childhood ANLL, a gain of chromosome 19 was a relatively common finding, being seen in 8 cases (9.2%), but in adults a +19 was rare (3.9% of abnormal cases). Loss of chromosome 7 was observed in 5 children; however, 3 of these were reported as selected cases. In four series describing ANLL in children [5, 16, 34, 54], only 2 of 66 cytogenetically abnormal patients were missing a chromosome 7. Thus, unlike in adults, a −7 may not be a common occurrence in children. Abnormalities of chromosome 5 were present in only 3 of 87 children and adolescents (3.4%) but in 33 of 282 adults (11.7%). Rearrangements of chromosome 11, like gain of 19, appear to be more common in children. Thus, 12 children had a rearranged 11 in contrast to 4 (1.4%) of the adults.

A t(8;21) was the most common abnormality in children with ANLL, being reported in 17 of 87 (19.5%) karyotypically abnormal cases. Six of 8 males with the t(8;21) were missing a Y, whereas 4 of 9 females with this rearrangement were −X. Six children with APL had a 15;17 translocation.

Prognostic significance of karyotype in ANLL

Sakurai and Sandberg [82] using conventional staining methods, were the first to show that the karyotypic pattern of the bone-marrow cells is correlated with survival in patients with ANLL. Patients with only normal mitoses (NN) had a median survival of 11.5 months from the onset of symptoms, compared to 10.3 months for patients with a mixture of normal and abnormal mitoses (AN) and 3.2 months in those with only abnormal mitoses (AA). More recent studies, based on a series of ANLL patients studied with routine banding techniques, have shown similar results [10, 29, 36, 46, 47, 58]. Of the AML (M1 and M2 according to the FAB classification) patients reviewed at the FIWCL, a substantially longer median survival (8 months) was found for NN patients compared to those who were AA (3.5 months) [24]. Patients who were AN had an intermediate survival (5 months). No such differences were found in patients with the M4 or M5 leukemic types. Benedict et al. [5] demonstrated that the karyotypic pattern can also correlate with prognosis in childhood ANLL. They reported a median survival of 20.5 months in patients who were NN, whereas those with chromosome changes had a median survival of only 7.1 months. Three of their patients were AA; none of the three survived longer than 4.5 months from diagnosis. It remains to be seen whether or not such correlations will hold up as newer cytogenetic methods are used to study karyotypes in ANLL (see the section on High-Resolution Banding Analysis of Acute Leukemia).

ANLL as a second malignancy

ANLL can be a late complication in malignant and nonmalignant diseases treated with radiation or chemotherapy, or both, which are mutagenic and potentially carcinogenic agents. Rowley et al. [78] reported on the karyotypes of bone-marrow cells from 27 ANLL patients who had received prior treatment for a primary neoplasm or, in one case, for a renal transplant. Fifteen of the patients had previously had both radiotherapy and chemotherapy, 8 had had only chemotherapy, and 4 had had only radiotherapy. The median times from diagnosis of the initial disease to the onset of ANLL for these three treatment groups were 61, 59, and 59 months, respectively. Abnormal karyotypes were observed in 26 of 27 cases (96%). Most (19 of 27) of the patients had a hypodiploid clone. One or both of two consistent chromosome changes were found in marrow cells from 23 of the 26 karyotypically abnormal cases: loss of chromosome 7 (18 patients) or part of the long arm of 7 (1 patient); and loss of 5 (11 patients) or part of the long arm of 5 (3 patients). This karyotypic pattern is similar to that found in about 25% of aneuploid patients with ANLL de novo.

The data regarding nonrandom changes of chromosomes 5 and 7 in secondary ANLL seem to be especially pertinent to the question of whether or not patients with ANLL de novo can be identified who may have been exposed to environmental mutagens. Although this question cannot be answered at this time, two lines of evidence suggest that the answer may be positive. The first of these comes from a retrospective study of the correlation between karyotype and occupational exposure in ANLL, and the second from studies of childhood ANLL.

Mitelman et al. [51] reported on a retrospective study of 56 patients with ANLL de novo; 23 had a history suggesting occupational exposure to chemical solvents, insecticides, or petroleum products, whereas 33 had no such known exposure. Only 24% of the nonexposed patients had clonal chromosome abnormalities, compared to nearly 83% in the exposed group. A clearly nonrandom pattern of chromosome changes was observed in the exposed group, with 84% of these cases having at least one of four specific changes: −5 (or 5q−), −7 (or 7q−), +8, or +21. In the nonexposed group, only two patients had any of these abnormalities: one was −7 and the second was +21.

Four series have described the karyotypic pattern in ANLL de novo in children [5, 16, 34, 54]. Only 2 of 66 patients with abnormal karyotypes from these investigations were missing a chromosome 7. Five patients had partial deletions of 7q, but only two had a deletion at 7q22, which is the usual abnormality seen in ANLL de novo in adults and in secondary ANLL. Only one of these children was −5, and none had a 5q−.

The consistent finding of a loss of all or part of chromosomes 5 or 7 in patients with secondary ANLL, in conjunction with the frequent occurrence of these same changes in patients with ANLL who may have an occupational exposure to mutagens, and the seeming paucity of these aberrations in childhood ANLL all provide support for the proposal that these specific chromosome changes may identify ANLL associated with exposure to mutagenic agents [78].

Acute lymphoblastic leukemia

Banding patterns

Chromosomes can be quite fuzzy in patients with ALL, and often it is difficult to obtain finely banded chromosomes that are suitable for detailed analysis. Consequently, there have been fewer banding studies in ALL than in ANLL. Nevertheless, important advances have been made during the past few years.

In early banding studies, karyotypic abnormalities were reported in approximately one half of patients with ALL [19, 61, 110]. More recently, the Third International Workshop on Chromosomes in Leukemia (TIWCL) [103, 104] evaluated data on 330 patients (173 adults, 157 children) with ALL whose marrow chromosomes were analyzed with banding techniques at diagnosis; clonal chromosome abnormalities were found in 66% of the cases. Among these abnormal cases, pseudodiploidy was found most frequently (51%); hypodiploidy was rare

(12%). Hyperdiploid cases (37%) were thought to be underrepresented owing to technical difficulties in obtaining analyzable banded preparations in such cases.

Among the chromosome abnormalities noted at the TIWCL, structural rearrangements predominated (87% of the cases), either alone (44%) or in combination with numerical changes (43%). Among the nonrandom numerical changes, a gain of chromosome 21 was the most common (23% of all abnormal cases). Others preferentially gained were chromosomes 6, 8, and 18. Those preferentially lost included chromosomes 7 and 20. Of the cytogenetically abnormal cases examined, 46% had one of the following specific structural aberrations: Ph^1 translocation, 4;11 translocation, 8;14 translocation, 14q+, and 6q− (listed in descending order of frequency).

Ph^1-positive ALL

At the TIWCL, the incidence of ALL patients with a Ph^1 was 5.7% for children, compared to 17.3% for adults. The incidence previously reported was 2.0% for children [18] and 25% for adults [15]. Of 39 untreated patients with Ph^1+ ALL evaluated at the TIWCL [104], 36 had the standard 9;22 translocation, t(9;22)(q34;q11). The remaining 3 cases had variant translocations; thus, the incidence of variant types was 7.7%, which is similar to that observed in patients with Ph^1+ CML [80]. Approximately half of the patients showed chromosome changes in addition to the Ph^1. The most common numerical changes were an additional Ph^1 or chromosome 8, which are the same extra chromosomes often found in Ph^1+ CML. Most notable in the ALL patients, however, was the absence at diagnosis of an i(17q), which, when it occurs in CML, is a reliable indication of imminent blastic crisis.

Each of the Ph^1+ cases examined at the TIWCL had non-T, non-B ALL. This group had a poor prognosis, with a median survival of 9 months.

4;11 translocation

A t(4q−;11q+), usually involving chromosome bands 4q21 and 11q23, was reported in 18 of 330 ALL patients (5.5%) at the TIWCL [104] and in 7 of 77 patients (9.1%) studied by Arthur et al. [4] (Figure 11–1g). Most of these cases were classified as FAB type L2 or L1 morphologically and non-T, non-B ALL immunologically [4, 56, 64, 103]. However, detailed studies on the nature of the leukemic cell by morphologic, cytochemical, ultrastructural, and immunologic methods now indicate that t(4;11)-associated acute leukemia involves an early myeloid precursor cell that has a lymphoid appearance by light microscopy [56, 64].

Clinically, one half of the reported patients were adults and the other half were children, most of whom were less than 1 year old [4, 55, 56]. The leukocyte count was very high (median WBC count 214 $\times$ 10^9/liter in children and 127 $\times$ 10^9/liter in adults) [103]. Most of the patients had hepatosplenomegaly or lymphadenopathy, or both [55, 56]. Overall, these patients had a very poor prognosis, with a median survival of 7 months [103] or 10.5 months [4], even though a complete remission was achieved in 16 of 24 reported cases [4, 55, 56].

8q−;14q+ translocation and other t(8)

A t(8;14)(q24;q32) has been observed in ALL patients with B-cell markers and in patients whose marrow cell morphology is of the FAB type L3 (Figure 11–1*h*) [8, 53]. An apparently identical chromosome rearrangement is frequently found in Burkitt's lymphomas of both African and non-African origin [39, 117]. Thus, Burkitt's lymphoma and most B-cell ALL of the L3 type probably are different manifestations of the same disease. Fourteen ALL patients reviewed at the TIWCL had a t(8q−;14q+). Two others had a translocation involving chromosome 8 (break in 8q23 or 24) and some chromosome other than 14. With one exception, all the tested cases with a t(8q−;14q+) or an 8q+ had B-cell markers and were FAB type L3. In the exceptional case, the leukemic cells showed a pre-B-cell phenotype and were of the L1 type [42]; the morphology of the leukemic cells, however, changed to L3 type at relapse [80]. Among the 16 cases with a t(8q−;14q+) or an 8q+ that were reviewed at the TIWCL, there was an excess of males over females and of adults over children. The group had a higher frequency (33%) of central nervous system involvement at diagnosis and a poorer prognosis (median survival, 5 months) than any other group of patients classified according to various karyotypic patterns.

The c-*myc* oncogene is located on chromosome 8 at band 8q24 [21, 57, 95], which is the specific band involved in the 8;14 translocation. The break in chromosome 14 at band 14q32 is at the locus for the immunoglobulin heavy chain gene [43]. Although the break within the heavy chain gene appears to be variable among different Burkitt's lymphoma cell lines, the c-*myc* oncogene is translocated from its native site on chromosome 8 to a new position adjacent to the heavy chain locus on 14 [21, 95].

14q+, not involving chromosome 8

Fifteen patients with ALL reviewed at the TIWCL had a 14q+ that did not involve a translocation with the terminal segment of 8q. An excess of males over females and of adults over children is similar to that in the patients with a t(8;14)(q24;q32). Approximately half of the patients, however, had non-T, non-B ALL, and the other half had B-ALL. Of the 13 cases classified by FAB type, 7 were L2, 3 were L1, and 3 were L3. The overall median survival of the group was 9 months. The donor chromosome involved in the translocation to chromosome 14 was identified in six patients, four of whom had a t(11;14)(q23;q32). A t(11;14) is one of the common abnormalities seen in malignant lymphoma of the poorly differentiated lymphocyte type [76], suggesting a possible relationship between it and ALL with a t(11;14).

6q−

Thirteen patients at the TIWCL had a 6q−; in five cases this was the only abnormality detected. There was an excess of children over adults in patients with a 6q−. None of these cases were L3 or had B-cell markers. The breakpoint in chromosome 6 varied from 6q15 to 6q25. Of possible relevance to this, Harper et al. [35] have recently mapped the oncogene c-*myb* to chromosome 6 at bands q22 to q24, which is within the region of breakpoints of the 6q− abnormality.

Near-haploid ALL

Near-haploidy in ALL is a rare but interesting occurrence. Seven ALL cases with near-haploid leukemic clones have been reviewed by Rowley and Testa [80], including two cases presented at the TIWCL. These seven individuals showed a remarkably consistent chromosome pattern. The chromosome number of the near-haploid clone ranged from 26 to 36. In addition to a haploid set, +21 was seen in all seven cases, +10 and +18 in six, +X or +Y in five, +6 in four, and +1, +19, and +22 in three each. Four of the patients had a second clone with twice the number of chromosomes as the near-haploid clone.

Hyperdiploidy with >50 chromosomes

Modal numbers in this group of patients usually range from 51 to 60 chromosomes. Among 31 cases (22 children, 9 adults) evaluated at the TIWCL, 19 had clones with 51 to 55 chromosomes, and the remainder had from 56 to 92 chromosomes. Twenty-two patients had unidentified markers, and 8 had other rearrangements. Certain chromosome gains were commonly observed. The most frequent change was +21, which was seen in 21 patients. This was particularly noteworthy in children because +21 occurred in 19 of 22 children in this group. Other common extra chromosomes included +6 in 14 patients, +18 in 13, +14 in 11, +4 in 10, and +10 in 9. If one compares the chromosome gains common in this group with the additional chromosomes in the seven cases reviewed in the previous section that showed near-haploidy, the similarities are notable since the most common in the latter group are +21, +10, and +18.

The median age of the 31 patients with hyperdiploidy with more than 50 chromosomes was 5 years, which was younger than that of patients with other abnormalities. All of the patients in the hyperdiploid group had non-T, non-B ALL. The L1 and L2 types were observed with equal frequency. Presenting leukocyte counts in the group was low, with a median of 6×10^9/liter. Patients in this group had a good prognosis (median survival, 34 months). In fact, the median survival of patients who had hyperdiploidy with more than 50 chromosomes was longer than that of patients with a normal karyotype.

Prognostic significance of karyotype in ALL

Patients reviewed at the TIWCL were classified into ten groups according to karyotype: no abnormalities; one of the common rearrangements discussed earlier (e.g., 14q+ or 6q−); or, in the remaining cases with abnormalities, the modal number (i.e., <46, 46, 47–50, >50). Therapeutic response (achievement of complete remission and remission duration) and survival differed significantly among the various karyotype groups. The best responses to treatment were seen in the patients with a modal chromosome number greater than 50. Patients with a normal karyotype or with a 6q− also did well. In contrast, patients with a t(4;11) or a t(8;14) responded very poorly to treatment and had the worst prognosis. The TIWCL study was the first rigorously to demonstrate that the chromosome pattern is an important independent risk factor in ALL. Multivariate analysis showed that the karyotype was an independent prognostic factor because it correlated with survival even when other major risk factors in ALL, such as age, initial leukocyte count, and the FAB type were taken into account.

Survival curves could be superimposed for children and adults who had karyotypes with the same specific structural rearrangements, for example, the t(9;22) or t(4;11). This was thought to be of considerable interest because karyotype may represent the first biologic marker to identify a type of ALL that has identical survival curves in children and adults.

High-resolution banding analysis of acute leukemia

An analysis of the karyotype of leukemic cells may be hampered by poor chromosome morphology and fuzzy, indistinct banding patterns.

The problems can be overcome with newer techniques, particularly the use of methotrexate to synchronize dividing cells. These procedures also provide cells with elongated chromosomes that have a larger number of transverse bands when stained according to various chromosome banding methods [33, 113].

Yunis et al. [115] examined high-resolution banded chromosomes from methotrexate-synchronized cultures from 24 patients with ANLL and detected karyotypic changes in every case. Karyotypic abnormalities were observed in 18 of 24 patients with the direct technique. Thus, with elongated chromosomes, an abnormal clone was detected in 6 patients who had seemed to have a normal karyotype on the basis of the direct preparation.

Although Yunis et al. [115] have proposed that all patients with ANLL may have a chromosome abnormality, Rowley [75] has stated that at least some of these patients will likely be found to have a normal karyotype, even when a large number of cells with elongated chromosomes are examined. She suggested that leukemia in such patients may be related to an alteration in the control of a normal cellular gene as a result of an insertion of DNA sequences adjacent to it; the amount of added DNA would be too small to be detected with light microscopy. We recently have reported data on 29 patients with acute leukemia that indicate that about 10 to 20% of these patients may have a normal karyotype even when high-resolution methods are used (see below). Moreover, Yunis [114] recently reported that further studies in his laboratory also have revealed some patients with a normal karyotype.

We examined 24-hour marrow cell cultures both with and without methotrexate synchronization from 29 adults (26 at diagnosis and 3 in relapse) with acute leukemia [50, 100]. In 10 of these patients we also examined direct preparations. Twenty-three of the patients had ANLL, 2 had ANLL secondary to treatment for a previous malignancy, and 4 had ALL. Overall, clonal karyotypic abnormalities were detected in 25 of 29 patients (86%). Of the 23 patients with ANLL de novo, 19 (83%) were cytogenetically abnormal. No clonal abnormalities were detected in 4 patients with ANLL de novo, even though 42 to 89 cells were fully karyotyped in each case. Abnormal karyotypes were observed in both patients with secondary ANLL and in all 4 with ALL. The results obtained with unsynchronized and methotrexate-synchronized culture methods were generally similar. However, in one patient a small clone was detected in synchronized cells but not in the unsynchronized cells. Moreover, synchronized cultures often yielded more mitoses with high-quality, elongated chromosomes than unsynchronized cultures. In two patients the only abnormality found was a tiny deletion or structural rearrangement, which would have been difficult to detect if not for the enhanced resolution of elongated chromosomes. Two other patients had small clones that were recognized only after a considerable number of mitoses (>40) were fully karyotyped in each case. In the ten patients who had a direct marrow examination, clonal abnormalities were detected in five patients with the direct method, compared with eight with culture methods.

Our results and those of Yunis et al. [114, 115] clearly indicate that chromosome abnormalities now can be detected in a much higher proportion of patients with acute leukemia than was previously thought. Since very few patients with ANLL may actually have a normal karyotypic pattern, the classification of patients based simply on normal or abnormal karyotype will be of little clinical value.

Results from studies with improved methods may eventually provide us with more reliable prognostic indicators. Already there is evidence that the t(8;21) is not associated with a poor prognosis [46, 91]. Similarly, patients with any of the related abnormalities involving 16q22 appear to have an especially favorable survival outlook [3, 48]. Recently, Yunis et al. [116] summarized their data on 105 patients with ANLL whose marrow chromosomes were studied with high-resolution banding techniques. They identified 17 karyotypic categories; 3 of these were found to have independent prognostic significance. Patients with an inv(16) had the best prognosis with a median survival of 25 months. In contrast, patients with complex karyotypes with a 5q−, monosomy 7, and/or other abnormalities had a very poor prognosis (median survival, 2.5 months). A third group with trisomy 8 as the single abnormality had an intermediate prognosis and a median survival of 10 months.

Conclusions

The application of chromosome banding techniques to the study of human leukemia has yielded many significant findings concerning cytogenetic patterns and clinical correlations. With newer cytogenetic procedures, including high-resolution banding methods, chromosome changes are now detectable in a much higher proportion of patients with acute leukemia than was previously assumed. As a result of these technical improvements, we have begun to identify subtle chromosome defects and recognize cytogenetic-clinicopathologic associations that heretofore were overlooked. As more patients with acute leukemia are examined with improved cytogenetic methods, it may be possible to find more reliable diagnostic and prognostic indicators. Moreover, specific chromosome rearrangements in leukemia and other malignancies have recently been linked to human cellular oncogenes. Further studies along this line will provide us with a greater understanding of the role of chromosome changes in leukemia.

Acknowledgment

J. R. Testa is a Scholar of the Leukemia Society of America, Inc.

Literature cited

1. Alimena G, Annino L, Balestrazzi P, Montuoro A, Dallapiccola B: Cytogenetic studies in acute leukaemias. Prognostic implications of chromosome imbalances. *Acta Haematol* 58:234–242, 1977.
2. Alimena G, Dallapiccola B, Gastaldi R, Mandelli F, Brandt L, Mitelman F, Nilsson PG: Chromosomal, morphological and clinical correlations in blastic crisis of chronic myeloid leukaemia—A study of 69 cases. *Scand J Haematol* 28:103–117, 1982.
3. Arthur DC, Bloomfield CD: Partial deletion of the long arm of chromosome 16 and bone marrow eosinophilia in acute nonlymphocytic leukemia: A new association. *Blood* 61:994–998, 1983.
4. Arthur, DC, Bloomfield CD, Lindquist LL, Nesbit ME Jr: Translocation 4;11 in acute lymphoblastic leukemia: Clinical characteristics and prognostic significance. *Blood* 59:96–99, 1982.
5. Benedict WF, Lange M, Greene J, Derencsenyi A, Alfi OS: Correlation between prognosis and bone marrow chromosomal patterns in children with acute nonlymphocytic leukemia: Similarities and differences compared to adults. *Blood* 54:818–823, 1979.
6. Bennett JM, Catovsky D, Daniel M-T, Flandrin G, Galton DAG, Gralnick HR, Sultan C: French-American-British (FAB) Cooperative Group: Proposals for the classification of the acute leukaemias. *Br J Haematol* 33:451–458, 1976.
7. Bennett JM, Catovsky D, Daniel M-T, Flandrin G, Galton DAG, Gralnick HR, Sultan C: French-American-British (FAB) Cooperative Group: A variant form of hypergranular promyelocytic leukaemia (M3). *Br J Haematol* 44:169–170, 1980.
8. Berger R, Bernheim JC, Brouet M, Daniel MT, Flandrin G: t(8;14) Translocation in a Burkitt's type of lymphoblastic leukaemia (L3). *Br J Haematol* 43:87–90, 1979.
9. Berger R, Bernheim A, Flandrin G: Absence d'anomalie chromosomique et leuce'mie aigue: Relations avec les cellules me'dullaires normales. *CR Acad Sci Paris* 290:1557–1559, 1980.
10. Bernard P, Reiffers J, Lacombe F, Dachary D, David B, Boisseau MR, Broustet A: Prognostic value of age and bone marrow karyotype in 78 adults with acute myelogenous leukemia. *Cancer Genet Cytogenet* 7:153–163, 1982.
11. Bernstein R, Mendelow B, Pinto MR, Morcom G, Bezwoda W: Complex translocations involving chromosomes 15 and 17 in acute promyelocytic leukaemia. *Br J Haematol* 46:311–314, 1980.
12. Bernstein R, Morcom G, Pinto MR, Mendelow B, Dukes I, Penfold G, Bezwoda W: Cytogenetic findings in chronic myeloid leukemia (CML); evaluation of karyotype, blast morphology, and survival in the acute phase. *Cancer Genet Cytogenet* 2:23–37, 1980.

13. Bernstein R, Pinto MR, Behr A, Mendelow B: Chromosome 3 abnormalities in acute nonlymphocytic leukemia (ANLL) with abnormal thrombopoiesis: Report of three patients with a "new" inversion anomaly and a further case of homologous translocation. *Blood.* 60:613–617, 1982.
14. Bernstein R, Pinto MR, Morcom G, Macdougall LG, Bezwoda W, Dukes I, Penfold G, Mendelow B: Karyotype analysis in acute nonlymphocytic leukemia (ANLL): Comparison with ethnic group, age, morphology, and survival. *Cancer Genet Cytogenet* 6:187–199, 182.
15. Bloomfield CD, Lindquist LL, Brunning RD, Yunis JJ, Coccia PF: The Philadelphia chromosome in acute leukemia. *Virchows Archiv (Cell Pathol)* 29:81–91, 1978.
16. Brodeur GM, Williams DL, Kalwinsky DK, Willaim KJ, Dahl GV: Cytogenetic features of acute nonlymphoblastic leukemia in 73 children and adolescents. *Cancer Genet Cytogenet* 8:93–105, 1983.
17. Canellos GP, DeVita VT, Whang-Pheng J, Chabner BA, Schein PS, Young RC: Chemotherapy of the blastic phase of chronic granulocytic leukemia: Hypodiploidy and response to therapy. *Blood* 47:1003–1009, 1976.
18. Chessells JM, Janossy G, Lawler SD, Secker Walker LM: The Ph[1] chromosome in childhood leukaemia. *Br J Haematol* 41:25–41, 1979.
19. Cimino MC, Rowley JD, Kinnealey A, Variakojis D, Golomb HM: Banding studies of chromosomal abnormalities in patients with acute lymphocytic leukemia. *Cancer Res* 39:227–238, 1979.
20. Cunningham, I, Gee T, Dowling M, Chaganti R, Bailey R, Hopfan S, Bowden L, Turnbull A, Knapper W, Clarkson B: Results of treatment of Ph[1]+ chronic myelogenous leukemia with an intensive treatment regimen (L-5 protocol). *Blood* 53:375–395, 1979.
21. Dalla-Favera R, Bregni M, Erikson J, Patterson D, Gallo RC, Croce CM: Human c-myc onc gene is located on the region of chromosome 8 that is translocated in Burkitt lymphoma cells. *Proc Natl Acad Sci USA* 79:7824–7827, 1982.
22. de Klein A, van Kessel AG, Grosveld G, Bartram CR, Hagemeijer A, Bootsma D, Spurr NK, Heisterkamp N, Groffen J, Stephenson JR: A cellular oncogene is translocated to the Philadelphia chromosome in chronic myelocytic leukaemia. *Nature* 300:765–767, 1982.
23. Dewald GW, Morrison-DeLap SJ, Schuchard KA, Spurbeck JL, Pierre RV: A possible specific chromosome marker for monocytic leukemia: Three more patients with t(9;11)(p22;q24) and another with t(11;17)(q24;q21), each with acute monoblastic leukemia. *Cancer Genet Cytogenet* 8:203–212, 1983.
24. First International Workshop on Chromosomes in Leukaemia, 1977: Chromosomes in Ph[1]-positive chronic granulocytic leukaemia. *Br J Haematol* 33:305–310, 1978; Chromosomes in acute non-lymphocytic leukaemia. *Br J Haematol* 39:311–316, 1978.
25. Fitzgerald PH, Morris CM, Fraser GJ, Giles LM, Hamer JW, Heaton DC, Beard MEJ: Nonrandom cytogenetic changes in New Zealand patients with acute myeloid leukemia. *Cancer Genet Cytogenet* 8:51–66, 1983.
26. Fitzgerald PH, Morris CM, Giles LM: Direct versus cultured preparation of bone marrow cells from 22 patients with acute myeloid leukemia. *Hum Genet* 60:281–283, 1982.
27. Golomb HM, Rowley JD, Vardiman JW, Testa JR, Butler A: "Microgranular" acute promyelocytic leukemia: A distinct clinical, ultrastructural, and cytogenetic entity. *Blood* 55:253–259, 1980.
28. Golomb HM, Vardiman J, Rowley JD: Acute nonlymphocytic leukemia in adults: Correlations with Q-banded chromosomes. *Blood* 48:9–21, 1976.
29. Golomb HM, Vardiman JW, Rowley JD, Testa JR, Mintz U: Correlation of clinical findings with quinacrine-banded chromosomes in 90 adults with acute nonlympho-

cytic leukemia. An eight-year study (1970–1977). *N Engl J Med* 299:613–619, 1978.

30. Hagemeijer A, Hahlen K, Abels J: Cytogenetic follow-up of patients with nonlymphocytic leukemia. II. Acute nonlymphocytic leukemias. *Cancer Genet Cytogenet* 3:109–124, 1981.
31. Hagemeijer A, Hahlen K, Sizoo W, Abels J: Translocation (9;11)(p21;q23) in three cases of acute monoblastic leukemia. *Cancer Genet Cytogenet* 5:95–105, 1982.
32. Hagemaijer A, Lowenberg B, Abels J: Analysis of the breakpoints in translocation (15;17) observed in 4 patients with acute promyelocytic leukemia. *Hum Genet* 61:223–227, 1982.
33. Hagemeijer A, Smit EME, Bootsma D: Improved identification of chromosomes of leukemic cells in methotrexate-treated cultures. *Cytogenet Cell Genet* 23:208–212, 1979.
34. Hagemeijer A, van Zanen GE, Smit EME, Hahlen K: Bone marrow karyotypes of children with nonlymphocytic leukemia. *Pediatr Res* 13:1247–1254, 1979.
35. Harper ME, Franchini G, Love J, Simon MI, Gallo RC, Wong-Staal F: Chromosomal sublocalization of human c-*myb* and c-*fes* cellular *onc* genes. *Nature* 304:169–171, 1983.
36. Hossfeld DK, Faltermeier M-T, Wendehorst E: Beziehungen zwischen Chromosomenbefund und Prognose bei akuter nichtlymphoblastischer Leukaemie. *Blut* 38:377–382, 1979.
37. ISCN (1978): An International System for Human Cytogenetic Nomenclature (1978). Birth Defects: Original Article Series, Vol 14, No 8, White Plains, NY, National Foundation, 1978; also in *Cytogenet Cell Genet* 21:309–404, 1978.
38. ISCN (1981): An International System for Human Cytogenetic Nomenclature—High-Resolution Banding (1981). Birth Defects: Original Article Series, Vol 17, No 5, New York, March of Dimes Birth Defects Foundation, 1981; also *Cytogenet Cell Genet* 31:1–23, 1981.
39. Kaiser-McCaw B, Epstein AL, Kaplan HS, Hecht F: Chromosome 14 translocation in African and North American Burkitt's lymphoma. *Int J Cancer* 19:482–486, 1977.
40. Kamada N, Okada K, Ito T, Nakatsui T, Uchino H: Chromosomes 21–22 and neutrophil alkaline phosphatase in leukaemia. *Lancet* 1:364, 1968.
41. Kamada N, Okada K, Oguma N, Tanaka R, Mikami M, Uchino H: C-G translocation in acute myelocytic leukemia with low neutrophil alkaline phosphatase activity. *Cancer* 37:2380–2387, 1976.
42. Kaneko Y, Rowley JD, Check I, Variakojis D, Moohr JW: The 14q+ chromosome in pre-B-ALL. *Blood* 56:782–785, 1980.
43. Kirsch IR, Morton CC, Nakahara K, Leder P: Human immunoglobulin heavy chain genes map to a region of translocations in malignant B lymphocytes. *Science* 261:301–303, 1982.
44. Kondo K, Larson RA, Vardiman JW, Golomb HM, Rowley JD: Further evidence for the specificity of t(15;17) in acute promyelocytic leukemia. *Am J Hum Genet* (Abstr) 34:72A, 1982.
45. Kondo K, Sasaki M, Mikuni C: A complex translocation involving chromosomes 1, 8, and 21 in acute myeloblastic leukemia. *Proc Jpn Acad* 54:21–24, 1978.
46. Larson RA, Le Beau MM, Vardiman JW, Testa JR, Golomb HM, Rowley JD: The predictive value of initial cytogenetic studies in 148 adults with acute nonlymphocytic leukemia: A 12 year study (1970–1982). *Cancer Genet Cytogenet* 10:219–236, 1983.
47. Lawler SD, Summersgill B, Clink HM, McElwain TJ: Cytogenetic follow-up study of acute non-lymphocytic leukaemia. *Br J Haematol* 44:395–405, 1980.

48. Le Beau MM, Larson RA, Bitter MA, Vardiman JW, Golomb HM, Rowley JD: Association of an inversion of chromosome 16 with abnormal marrow eosinophils in acute myelomonocytic leukemia. *N Engl J Med* 309:630–636, 1983.
49. Lindgren V, Rowley JD: Comparable complex rearrangements involving 8;21 and 9;22 translocations in leukaemia. *Nature* 266:744–745, 1977.
50. Misawa S, Testa JR: Cytogenetic abnormalities in patients with acute leukemia studied with banding techniques, including the high-resolution banding method. *In* Cancer: Etiology and Prevention (Crispen RG, ed) New York, Elsevier Science Publishing Company, 1983, 29–42.
51. Mitelman F, Brandt L, Nilsson PG: Relation among occupational exposure to potential mutagenic/carcinogenic agents, clinical findings, and bone marrow chromosomes in acute nonlymphocytic leukemia. *Blood* 52:1229–1237, 1978.
52. Mitelman F, Nilsson PG, Levan G, Brandt L: Non-random chromosome changes in acute myeloid leukemia. Chromosome banding examination of 30 cases at diagnosis. *Int J Cancer* 18:31–38, 1976.
53. Mitelman F, Andersson-Anvret M, Brandt L, Catovsky D, Klein G, Manolov G, Manolova Y, Mark-Vendel E, Nilsson PG: Reciprocal 8;14 translocation in EBV-negative B-cell acute lymphocytic leukemia with Burkitt-type cells. *Int J Cancer* 24:27–33, 1979.
54. Morse H, Hays T, Peakman D, Rose B, Robinson A: Acute nonlymphoblastic leukemia in childhood. High incidence of clonal abnormalities and nonrandom changes. *Cancer* 44:164–170, 1979.
55. Morse HG, Heideman R, Hays T, Robinson A: 4;11 translocation in acute lymphoblastic leukemia: A specific syndrome. *Cancer Genet Cytogenet* 7:165–172, 1982.
56. Nagasaka M, Maeda S, Maeda H, Chen H-L, Kita K, Mabuchi O, Misu H, Matsuo T, Sugiyama T: Four cases of t(4;11) acute leukemia and its myelomonocytic nature in infants. *Blood* 61:1174–1181, 1983.
57. Neel BG, Jhanwar SC, Chaganti RSK, Hayward WS: Two human c-onc genes are located on the long arm of chromosome 8. *Proc Natl Acad Sci USA*, 79:7842–7846, 1982.
58. Nilsson PG, Brandt L, Mitelman F: Prognostic implications of chromosome analysis in acute non-lymphocytic leukemia. *Leukemia Res* 1:31–34, 1977.
59. Nowell PC, Hungerford DA: A minute chromosome in human chronic granulocytic leukemia. *Science* 132:1497, 1960.
60. Oguma N, Misawa S, Testa JR: A variant 8;21 translocation in acute myeloblastic leukemia. *Am J Hematol* 15:391–396, 1983.
61. Oshimura M, Freeman AI, Sandberg AA: Chromosomes and causation of human cancer and leukemia. XXVI. Banding studies in acute lymphoblastic leukemia (ALL). *Cancer* 40:1161–1172, 1977.
62. Oshimura M, Hayata L, Kakati S, Sandberg AA: Chromosomes and causation of human cancer and leukemia. XVII. Banding studies in acute myeloblastic leukemia (AML). *Cancer* 38:748–761, 1976.
63. Paris Conference, 1971: Standardization in Human Cytogenetics. Birth Defects: Original Article Series, Vol 8, No 7, White Plains, NY, National Foundation, 1972.
64. Parkin JL, Arthur DC, Abramson CS, McKenna RW, Kersey JH, Heideman RL, Brunning RD: Acute leukemia associated with the t(4;11) chromosome rearrangement: Ultrastructural and immunologic characteristics. *Blood* 60:1321–1331, 1982.
65. Pasquali F, Casalone R: Rearrangement of three chromosomes (Nos. 2, 8, and 21) in acute myeloblastic leukemia. Evidence for more than one specific event. *Cancer Genet Cytogenet* 3:335–339, 1981.

66. Pearson MG, Vardiman JW, Le Beau MM, Rowley JD: T(6;9): a new cytogenetic subset in acute non-lymphocytic leukemia (ANLL) associated with bone marrow basophilia. *Blood* 64:180a, 1983.
67. Philip P, Jensen MK, Killmann SA, Drivsholm A, Hansen NE: Chromosomal banding patterns in 88 cases of acute nonlymphocytic leukemia. *Leukemia Res* 2:201–212, 1978.
68. Potter AM, Watmore AE, Cooke P, Lilleyman JS, Sokal RJ: Significance of non-standard Philadelphia chromosomes in chronic granulocytic leukaemia. *Br J Cancer* 44:51–54, 1981.
69. Prigogina EL, Fleischman EW: Certain patterns of karyotype evolution in chronic myelogenous leukaemia. Chromosome abnormalities in CML. *Humangenetik* 30:113–119, 1975.
70. Prigogina EL, Fleischman EW, Puchkova GP, Kulagina OE, Majokova SA, Balakirev SA, Frenkel MA, Khvatova NV, Peterson IS: Chromosomes in acute leukemia. *Hum Genet* 53:5–16, 1979.
71. Rowley JD: A new consistent chromosomal abnormality in chronic myelogenous leukaemia identified by quinacrine fluorescence and Giemsa staining. *Nature* 243:290–293, 1973.
72. Rowley JD: Identification of a translocation with quinacrine fluorescence in a patient with acute leukaemia. *Ann Genet* 16:109–112, 1973.
73. Rowley JD: The cytogenetics of acute leukaemia. *Clin Haematol* 7:385–406, 1978.
74. Rowley JD: Ph[1]-positive leukaemia, including chronic myelogenous leukaemia. *Clin Haematol* 9:55–86, 1980.
75. Rowley JD: Do all leukemic cells have an abnormal karyotype? *N Engl J Med* 305:164–166, 1981.
76. Rowley JD, Fukuhara S: Chromosome studies in non-Hodgkin's lymphomas. *Semin Oncol* 7:255–266, 1980.
77. Rowley JD, Golomb HM, Dougherty C: 15/17 translocation, a consistent chromosomal change in acute promyelocytic leukaemia. *Lancet* 1:549–550, 1977.
78. Rowley JD, Golomb HM, Vardiman JW: Nonrandom chromosome abnormalities in acute leukemia and dysmyelopoietic syndromes in patients with previously treated malignant disease. *Blood* 58:759–767, 1981.
79. Rowley JD, Potter D: Chromosomal banding patterns in acute nonlymphocytic leukemia. *Blood* 47:705–721, 1976.
80. Rowley JD, Testa JR: Chromosome abnormalities in malignant hematologic diseases. *Adv Cancer Res* 36:103–148, 1982.
81. Sakurai M, Hayata I, Sandberg AA: Chromosomes and causation of human cancer and leukemia: XV. Prognostic value of chromosomal findings in Ph[1]-positive CML. *Cancer Res* 36:313–318, 1976.
82. Sakurai M, Sandberg AA: Prognosis of acute myeloblastic leukemia: Chromosomal correlation. *Blood* 41:93–104, 1973.
83. Sakurai M, Sandberg AA: Chromosomes and causation of human cancer and leukemia. XI. Correlation of karyotypes with clinical feature of myeloblastic leukemia. *Cancer* 37:285–299, 1976.
84. Sakurai M, Sasaki M, Kamanda N, Okada M, Oshimura M, Ishihara T, Shiraishi Y: A summary of cytogenetic, morphologic, and clinical data on t(8q−;21q+) and t(15q+;17q−) translocation leukemias in Japan. *Cancer Genet Cytogenet* 7:59–65, 1982.
85. Sandberg AA: Chromosomes and causation of human cancer and leukemia: XL. The Ph[1] and other translocations in CML. *Cancer* 46:2221–2226, 1980.
86. Sandberg AA: The Chromosomes in Human Cancer and Leukemia. New York, Elsevier-North Holland, 1980.

87. Sandberg AA: The cytogenetics of chronic myelocytic leukemia (CML): Chronic phase and blastic crisis. *Cancer Genet Cytogenet* 1:217–228, 1980.
88. Sandberg AA, Morgan R, McCallister JA, Kaiser-McCaw B, Hecht F: Acute myeloblastic leukemia (AML) with t(6;9)(p23;q34): A specific subgroup of AML. *Cancer Genet Cytogenet* 10:139–142, 1983.
89. Schwartz S, Jiji R, Kerman S, Meekins J, Cohen MM: Translocation (6;9) (p23;q34) in acute nonlymphocytic leukemia. *Cancer Genet Cytogenet* 10:133–138, 1983.
90. Sheer D, Hiorns LR, Stanley KF, Goodfellow PN, Swallow DM, Povey S, Heisterkamp N, Groffen J, Stephenson JR, Solomon E: Genetic analysis of the 15;17 chromosome translocation associated with acute promyelocytic leukemia. *Proc Natl Acad Sci USA* 80:5007–5011, 1983.
91. The Second International Workshop on Chromosomes in Leukemia, Leuven, Belgium, October 2–6, 1979. *Cancer Genet Cytogenet* 2:89–113, 1980.
92. Shiraishi Y, Taguchi H, Niiya K, Shiomi F, Kikukawa K, Kubonishi S, Ohmura T, Hamawaki M, Ueda N: Diagnostic and prognostic significance of chromosome abnormalities in marrow and mitogen response of lymphocytes of acute nonlymphocytic leukemia. *Cancer Genet Cytogenet* 5:1–24, 1982.
93. Sonta S, Sandberg AA: Chromosomes and causation of human cancer: XXIX. Further studies on karyotypic progression in CML. *Cancer* 41:153–163, 1978.
94. Takeuchi J, Ohshima T, Amaki I: Cytogenetic studies in adult acute leukemias. *Cancer Genet Cytogenet* 4:293–302, 1981.
95. Taub R, Kirsch I, Morton C, Lenior G, Swan D, Tronick S, Aaronson S, Leder P: Translocation of the c-myc gene into the immunoglobulin heavy chain locus in human Burkitt lymphoma and murine plasmacytoma cells. *Proc Natl Acad Sci USA* 79:7837–7841, 1982.
96. Teerenhovi L, Borgstrom GH, Mitelman F, Brandt L, Vuopio P, Timonen T, Almqvist A, de la Chapelle A: Uneven geographical distribution of 15;17-translocation in acute promyelocytic leukaemia. *Lancet* 2:797, 1978.
97. Testa JR, Golomb HM, Rowley JD, Vardiman JW, Sweet DL: Hypergranular promyelocytic leukemia (APL): Cytogenetic and ultrastructural specificity. *Blood* 52:272–280, 1978.
98. Testa JR, Hogge DE, Misawa S, Zandparsa N: Chromosome 16 rearrangement in acute myelomonocytic leukemia with abnormal eosinophils. *N Engl J Med* 310:468, 1984.
99. Testa JR, Mintz U, Rowley JD, Vardiman JW, Golomb HM: Evolution of karyotypes in acute nonlymphocytic leukemia. *Cancer Res* 39:3619–3627, 1979.
100. Testa JR, Oguma N, Misawa S, Wiernik PH: Chromosome abnormalities in acute leukemia: A higher incidence than previously assumed. *Cancer Genet Cytogenet* 9:305–306, 1983.
101. Testa JR, Oguma N, Pollak A, Wiernik PH: Near-tetraploid clones in acute leukemia. *Blood* 61:71–78, 1983.
102. Testa JR, Rowley JD: Chromosomal banding patterns in patients with acute nonlymphocytic leukemia. *Cancer Genet Cytogenet* 1:239–247, 1980.
103. Third International Workshop on Chromosomes in Leukemia. Lund, Sweden, July 21–25, 1980. *Cancer Genet Cytogenet* 4:95–142, 1981.
104. Third International Workshop on Chromosomes in Leukemia. Chromosomal abnormalities and their clinical significance in acute lymphoblastic leukemia. *Cancer Res* 43:868–873, 1983.
105. Tijo JH, Whang J: Chromosome preparation of bone marrow cells without prior *in vitro* culture or *in vivo* colchicine administration. *Stain Technol* 37:17–20, 1962.
106. Trujillo JM, Cork A, Ahearn MJ, Youness EL, McCredie KB: Hematologic and cytologic characterization of 8/21 translocation acute granulocytic leukemia. *Blood* 53:695–706, 1979.

107. Van den Berghe H, Louwagie A, Broeckaert-Van Orshover A, David G, Verwilghen R, Michaux JL, Sokal G: Chromosome abnormalities in acute promyelocytic leukemia (APL). *Cancer* 43:558–562, 1979.
108. Vermaelen K, Michaux J-L, Louwagie A, Van den Berghe H: Reciprocal translocation t(6;9) (p21;q33): A new characteristic chromosome anomaly in myeloid leukemias. *Cancer Genet Cytogenet* 10:125–131, 1983.
109. Whang-Peng J, Canellos GP, Carbone PP, Tijo JH: Clinical implications of cytogenetic variants in chronic myelocytic leukemia (CML). *Blood* 32:755–766, 1968.
110. Whang-Peng J, Kuntsen T, Ziegler J, Leventhal B: Cytogenetic studies in acute lymphocytic leukemia: Special emphasis in long-term survival. *Med Pediatr Oncol* 2:333–351, 1976.
111. Yamada K, Sugimoto E, Amano M, Imamura Y, Kubota T, Matsumoto M: Two cases of acute promyelocytic leukemia with variant translocations: The importance of chromosome No. 17 abnormality. *Cancer Genet Cytogenet* 9:93–99, 1983.
112. Yunis JJ: Comparative analysis of high resolution chromosome techniques for leukemic bone marrows. *Cancer Genet Cytogenet* 7:43–50, 1982.
113. Yunis JJ: New chromosome techniques in the study of human neoplasia. *Hum Pathol* 12:540–549, 1981.
114. Yunis JJ: The chromosomal basis of human neoplasia. *Science* 221:227–236, 1983.
115. Yunis JJ, Bloomfield CD, Ensrud BS: All patients with acute nonlymphocytic leukemia may have a chromosomal defect. *N Engl J Med* 305:135–139, 1981.
116. Yunis JJ, Brunning RD, Howe RB, Lobell M: High-resolution chromosomes as an independent prognostic indicator in adult acute nonlymphocytic leukemia. *N Engl J Med* 311:812–818, 1984.
117. Zech L, Haglund U, Nilsson K, Klein G: Characteristic chromosomal abnormalities in biopsies and lymphoid-cell lines from patients with Burkitt and non-Burkitt lymphomas. *Int J Cancer* 17:47–56, 1976.

12.

Chromosome changes in lymphoma and solid tumors

AVERY A. SANDBERG

Major advances have been made, particularly during the last decade, in the cytogenetics of human neoplasia, including the lymphomas and solid tumors [63, 86, 94, 95]. Much of this has been a result of improvements in chromosomal techniques, the introduction of banding methods, and greater standardization in the classification of lymphomas. Much more needs to be done in the methodologies of solid tumor and lymphoma chromosome examinations, particularly in the ability to obtain more cells in mitosis, so that the percentage of tumors and lymphomas in which cytogenetic findings can be obtained will be improved. This is an area in which examination of established cell lines may be of value, but one must always keep in mind that the cells in such lines do not necessarily represent the situation in vivo. This may be the best we can do, however, until an agent, akin to those for B and T lymphocytes (see below), capable of stimulating tumor cells into mitosis is found.

General background to chromosome analysis in lymphoma and solid tumors

Results of cytogenetic and related studies have been and undoubtedly will be of value not only in the diagnosis [40, 77, 79] and therapy of lymphomas but also in the understanding of some of these diseases including phenotypic and molecular events such as surface immunoglobulin production in Burkitt's lymphoma (BL) [5, 26, 48, 50, 55, 61]. Advances on the chromosome changes in human neoplasia will no doubt contribute further to the application of these findings to the areas of carcinogenesis, molecular biology and genetics, and the clinical aspects of these disorders.

The introduction of mitogens capable of stimulating normal and abnormal lymphoid cells (Table 12–1), particularly B cells, has led to

Table 12–1 Mitogenic stimulators of blood lymphocytes.

Mitogen	Cell type stimulated	Concentrations used
Phytohemagglutinin (PHA)	T lymphocytes	25–100 μg/ml
Pokeweed mitogen (PWM)	B and T lymphocytes	Up to 150 μg/ml
Concanavalin A (Con A)	T lymphocytes (mouse)	1–5 μ/ml
Calcium ionophore-A23187	B & T lymphocytes	$5 \times 10^{-7} - 10^{6}$ M (0.5–2.0 μg/ml)
Sodium metaperiodate ($NaIO_4$)	T lymphocytes	$2\text{–}4 \times 10^{-3}$ M
Epstein-Barr virus (EBV)	B lymphocytes	10–20% supernatant; 1:9 v/v of culture
Lipopolysaccharide W (*E. coli* 055:B5)	B lymphocytes	100 μg/ml
Staphylococcus bacteria strain Cowan I protein	B lymphocytes	100 μg/ml
Conditioned medium from PHA-stimulated T cells (PHA induced soluble factors)	B lymphocytes	1:2 dilution
Protein A	B lymphocytes	20–100 μm/ml
T-cell growth factor (Interleukin-2)	T lymphocytes	10%

the observation of nonrandom karyotypic changes in B-cell chronic lymphocytic leukemia (CLL) [4, 31–34, 69, 82, 88–90, 97] and in some of the lymphomas [51, 109]. We hope the presently available mitogens and growth factors will be used to extend the cytogenetic findings in other lymphomas and tumors and that mitogens with specificity for malignant lymphoma cells, including subgroups within the B- and T-cell varieties, will be developed.

Nearly all lymphomas and tumors adequately studied cytogenetically have chromosome changes that, in my opinion, can be interpreted as *primary* and *secondary* (Table 12–2 and Figure 12–1) [96]. Thus, the

Table 12–2 Significance of chromosome changes in human cancer.

Primary changes

Specific karyotypic anomalies—possibly related to etiology

Secondary changes

May be cytogenetic epiphenomena and/or play an important role in the heterogeneity and variable biology of cancer cells, e.g., invasiveness, metastatic spread, sensitivity or resistance to therapy, etc.

May be the selective proliferation of cells already existing in the earliest phases of the cancer or may develop de novo

The relative heterogeneity of the secondary karyotypic changes may be a reflection of the hosts' genotypes and/or extraneous factors

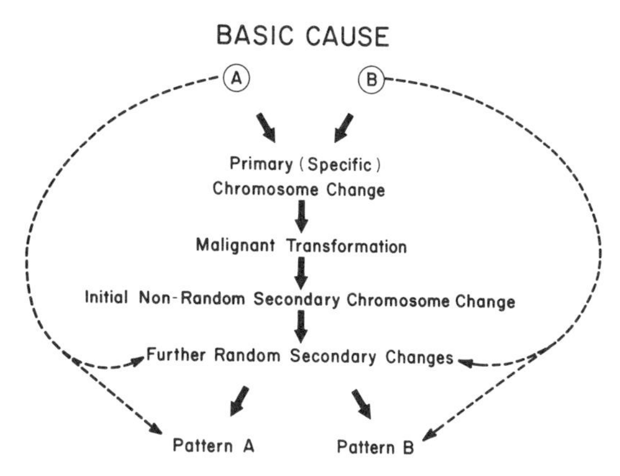

Figure 12–1 Schematic presentation of a hypothesis of chromosomal events in human cancer. Different *basic causes* of the neoplasia, shown as examples Ⓐ and Ⓑ may lead to the same primary (specific) chromosome change in a particular cell, such as the Ph[1]-translocation in CML. Let us assume that basic cause Ⓐ is radiation and basic cause Ⓑ is drug related, respectively, with each being capable of inducing the Ph[1] chromosome in the appropriate marrow cell. This chromosome change then leads to malignant transformation of the cell, in the case of CML, of a marrow precursor cell. When secondary chromosome changes develop, they may be initially nonrandom, for example, +Ph[1],+8, and an i(17q) in CML. Although events in the transformed cell are essentially under the control of the primary karyotypic change, the basic cause may play an auxiliary role (indicated by broken arrows) in the development of secondary chromosome changes. Thus, it could be theorized that Ⓐ may be associated with +8, whereas Ⓑ is associated with +Ph[1], as the preferred initial (nonrandom) secondary change in CML. A similar effect may exist in the development of further and apparently random secondary cytogenetic changes in various cancers, leukemias, and lymphomas.

chromosome changes in lymphoma and solid tumors will be presented with the guiding principle that each condition is characterized by a *primary karyotypic change*, directly related to the causation of the neoplasia, and *secondary karyotypic changes* that are responsible for the subsequent biologic characteristics and behavior of the tumor, such as invasion, metastatic spread, and response to therapy. Since the secondary karyotypic changes vary considerably from lymphoma to lymphoma and from tumor to tumor, it is not surprising that these neoplastic conditions show such variability in their clinical aspects.

Karyotypic findings in Burkitt's lymphoma

Discussion of the chromosome changes in lymphoma will be divided into two sections: those of BL and those of non-BL. Such a division is cogent because BL is a well-defined disease not only histologically and clinically but also cytogenetically [57, 58, 101] and phenotypically. On the other hand, the status in non-BL is not as clear as that in BL.

Even though to date the majority of BL tumors and leukemias have been shown to have a reciprocal translocation between chromosomes 8 and 14 [9, 24, 41, 54, 64–66, 102, 112], more and more cases with variant translocations involving chromosomes 8 and 2 or 22 are being described (Table 12–3) [30, 87, 103, 106]. The consensus exists now that the crucial event, that is, the primary karyotypic change, in the genesis of BL is deletion of the long arm of chromosome 8 at band q24 (q24.13), and that the chromosome to which the material is translocated can be chromosome 14 (band q32), 2 (band p12), or 22 (band q12) (Figures 12–2 through 12–4). The histologic and clinical nature of the disease does not seem to differ significantly in relation to the translocation, but some of the cellular phenotypic manifestations do appear to be related, particularly the expression of the surface light chains [1, 50]. Translocations involving chromosomes 2 and 8 express exclusively kappa light chains and those involving chromosomes 8 and 22 lambda express light chains exclusively. Such observations (Table 12–4) are important because they pinpoint the location of genes for surface immunoglobulin production [5, 26, 48, 55, 61] and also serve to define precisely where the type of changes affecting such tumors both genetically [21, 27, 71, 104] and biologically occur.

Very few BL tumors or leukemias are characterized cytogenetically by these translocations alone (Figures 12–2 through 12–4). Secondary karyotypic changes are present in a majority and perhaps in all cases of this disease [1, 9, 15]. These secondary changes very possibly play a role in the biologic behavior of the tumors, particularly in differences in the clinical pictures of endemic and nonendemic BL, tumor spread, response to therapy, and ultimate survival of the patients [96].

In a substantial number of BL cases, particularly of the nonendemic variety, even though typical karyoptyic changes were present, no evidence of Epstein-Barr virus (EBV) infection was demonstrated [9, 50], possibly indicating that the virus per se may not be responsible for the karyotypic change and that mechanisms other than infection with EBV may be responsible for the development of the disease. This appears to be particularly true of those BL cases that are not of endemic origin and might indicate that at least in nonendemic areas the EBV plays a secondary role in the causation of BL. Once the genesis of the tumor occurs, chromosomal changes appear to be the same in endemic as in nonendemic BL. These changes may, in fact, be closely associated with the development of the tumors. Thus, the key and primary event in the

Table 12–3 Clinical and karyotypic findings of published Burkitt's lymphoma or leukemia (ALL-L3) cases with variant translocations; (2;8) or (8;22).*

Reference	Patient's country	Age and sex	Survival (months)	EBV	Source	Karyotype
			Cases with t(2;8)			
Miyoshi et al. [68]	Japan	29,M	8.0	+	BM	46,XY,t(2;8)(p12;q24)
Van Den Berghe et al. [106]	Belgium	4,F	5.5	NT	LN	46,XX,t(2;8)(p12;q23)
Bornkhamm et al. [18]	West Germany	34,F	3.0	+	Cell line	46,XX,t(2;8)(p11;q24)
Rowley et al. [87]	United States	20,M	2.0	NT	PB	46,XY,t(2;8)(p13;q24)
Abe et al. [2]	Japan	45,F	6.0	NT	PE,PB BM	45,X,−X,t(2;8)(p11;q24) 45,X,−X,−2,t(2;8)(p11;q24),+de r(2)t(2der(8) t(2;8)(p11;q24)(p13;q21)
Slater et al. [100]	Holland	7,F	2.0	−	BM,LN	46,X,Xp+,t(2;8)(p11;q24),1q−,7p+,17p−
Berger & Bernheim [9]	France	22,M	7.0+	−	BM	46,XY,t(2;8)(p12;q24)
Bernheim et al. [14]	Kenya	13,M	−	+	Cell line	46,XY,t(2;8)(p12;q24),der(1)triplication(q22-q32)del(q42-qter)
Bernheim et al. [14]	Uganda	7,F	−	+	Cell line	46,XX,t(2;8)(p12;q24)
Philip et al. [69]	Algeria	8,M	4.0	+	BM	46,XY,t(2;8;9)(p11;q23;q31)
			Cases with t(8;22)			
Berger et al. [11]	Turkey-France	65,M	1.0	−	BM	46,XY,t(8;22)(q23;q12),t(1;6)(q21;qter)
	France	7,M	6.0	−	BM	46,XY,t(8;22)(q23;q12)/46,XY,t(8;22)(q23;q12) −6,t(1;6)(q23;q26)
	France	19,M	6.0	+	LN,PB	46,XY,t(8;22)(q23;q11),14q+
	France (trip to Cameroon prior to Dx)	54,M		Probably negative	PE	45,X,−Y,t(1;14)(q22;q32),t(8;22)(q24;q11), 3q+,6q−,6q+,7q+,9p−,14q−,18q−
Miyoshi et al. [67]	Japan	29,F	5.0	−	Ascites	46,XX,t(8;22)(q24;q13)
Berger & Bernheim [9]	France	70,F	15.0	NT	LN	47,XX,t(8;22)(q24;q21),+8,t(11;21)(q14;q22),4q+6q+ ,others
	France	31,M	6.0	−	PB	46,XX,t(8;22)(q24;q11)
	France	13,M	6.0	NT	BM	47,XY,t(8;22)(q24;q11),t?22/47,XY,t(8;22) (q24;q11),+?22,dup(1)(q23q24)
Abe et al. [1]	United States	6,M	0.5	NT	BM	46,XY,t(8;22)(q24;q12)
Bernheim et al. [14]	Uganda	M	−	+	Cell line	46,XY,t(8;22)(q24;q11)
	Kenya	M	−	+	Cell line	46,XY,t(8;22)(q24;q11),t(4;5)(q22;q13), del(11)(q24)
	France	35,M	−	−	Cell line	44,X,t(8;22),−13,−17,11q+,t(1;17;?Xqter;pter;?)

*Key to abbreviations: EBV = Epstein-Barr virus; NT = not tested; BM = bone marrow, LN = lymph node; PE = pleural effusion; PB = peripheral blood; M = male; F = female; Dx = diagnosis.

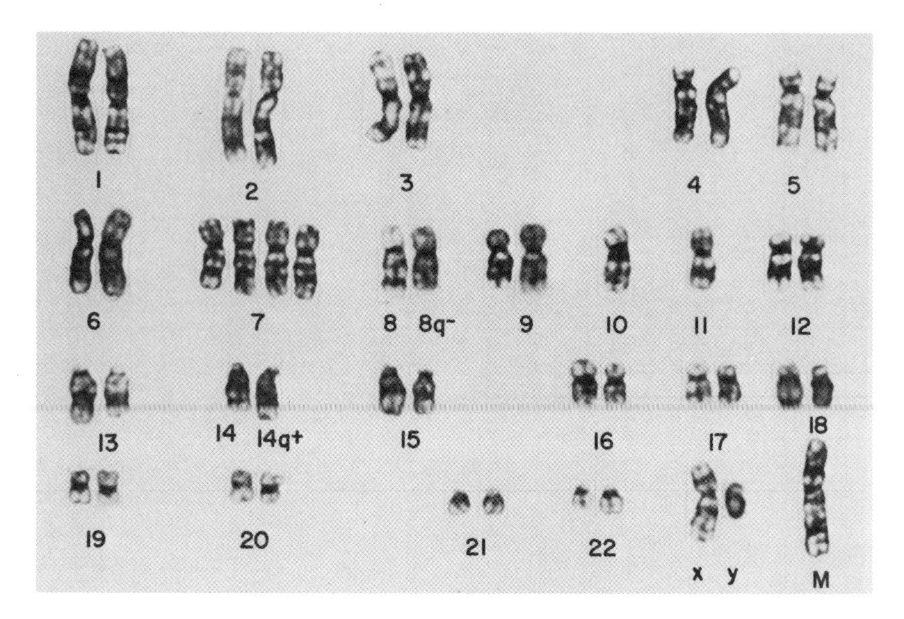

Figure 12–2 G-banded karyotype of a cell from a tumor of a patient with nonendemic Burkitt's lymphoma (BL). The cell contains 47 chomosomes and includes a specific (primary) karyotypic change seen in BL, that is, t(8;14)(q23;q32), shown as 8q− and 14q+. In addition, other (secondary) chromosomal abnormalities (+7, +7, −10, −11, +M) are present and may play a role in the biology and behavior of the tumor and disease. Such secondary changes are common in BL but vary from case to case. The primary karyotypic event, that is, deletion at band q23 of chromosome 8, probably characterizes all BL cases. This case was described by Kakati et al. [41].

karyotypic changes in BL appears to be deletion of the long arm of chromosome 8, which may be the locus for BL development and thus the key cytogenetic event related to the genesis of this disease [96]. The secondary karyotypic changes would then bear directly on the clinical characteristics and biology of BL tumors and leukemias. Failure to observe a translocation or involvement of chromosomes 8 and 14 in a rare case of BL has been reported [23].

Recently published findings, however, raise a number of questions regarding presently held views in BL. Thus, it has been reported [12] that BL (nonendemic) may be associated with a 6q- anomaly without evidence of involvement of chromosomes 8, 14, 2, or 22. The authors raise the question about the definition of BL, at least in cytogenetic terms, and of the involvement of various DNA sequences as a mechanism in the genesis of BL. In another report [53], cell lines derived from a homosexual patient with probable acquired immunodeficiency syndrome (AIDS) and BL and with a consistent t(8;22) produced kappa

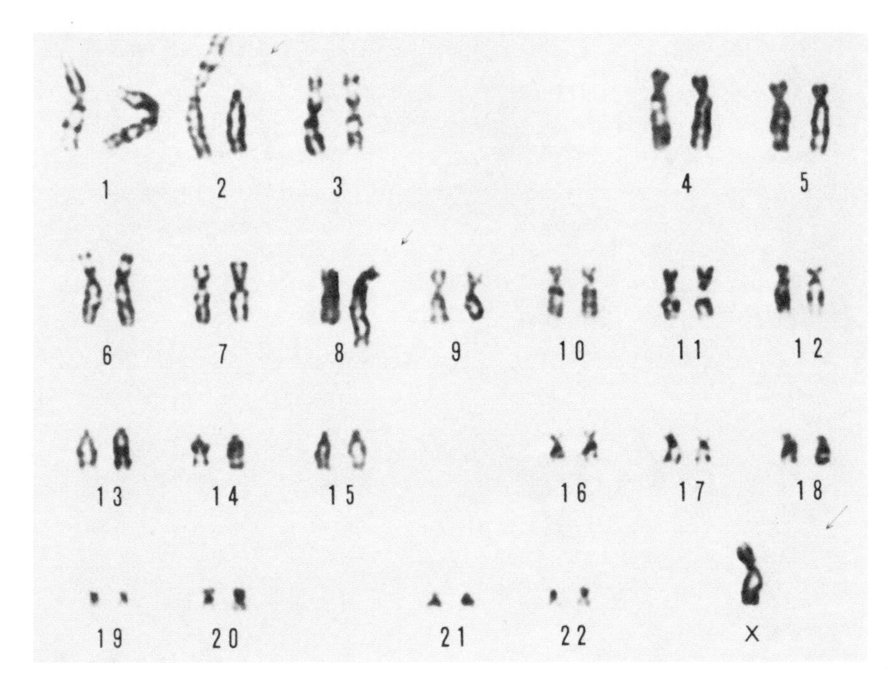

Figure 12–3 Karyotype of a Burkitt's lymphoma cell from a pleural effusion in a 45-year-old Japanese woman with a variant translocation between the short arm of chromosome 2 and the long arm of 8, that is, t(2;8)(p11;q24) *(short arrows)*. The longer arrow points to a missing X chromosome. This case was described by Abe et al. [2].

light chains rather than the expected lambda chains. The findings indicated to the authors that the translocation in BL may occur as a separate event from immunoglobulin gene rearrangement or that the proposed hierarchial sequence of immunoglobulin gene rearrangement is not always adhered to. Furthermore, the authors [53] indicated that cells containing a translocation between the long arm of chromosome 8 and a chromosome bearing an immunoglobulin gene and activation and expression of the cellular *myc* oncogene may occur regardless of the immunoglobulin gene that is expressed.

The frequent involvement of chromosome 1 in the karyotypic changes of Burkitt's lymphoma, as it is in other lymphomas, leukemias, and cancers, was related to EBV status. All nine BL cell lines not associated with EBV were shown to contain an abnormality of the long arm (bands 1q23–q24) of chromosome 1. The authors hypothesized that genetic information resembling that contained within the viral genome was present on the long arm of chromosome 1 and may bear on the relationship between BL cell proliferation and EBV [10].

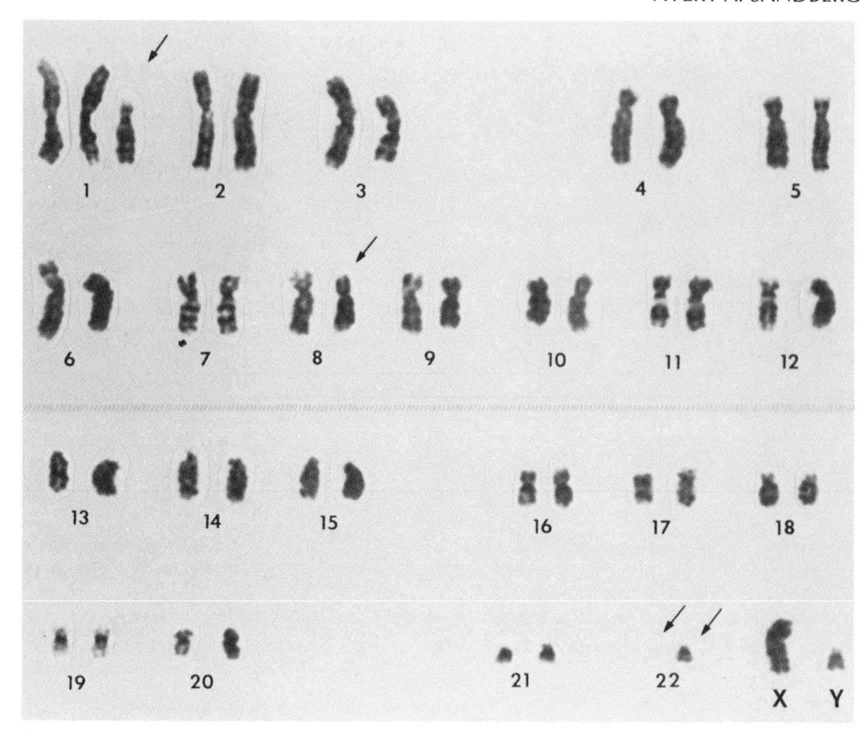

Figure 12–4 Karyotype of a marrow cell from a 6-year-old, white, North American boy with Burkitt-type acute lymphoblastic leukemia (ALL) with a translocation t(8;22)(q24;q12) *(arrows),* a variant of the usual t(8;14) translocation seen in BL. This case was described by Abe et al. [1].

Karyotypic findings in non-Burkitt's, non-Hodgkin's lymphomas

None of the karyotypic findings in lymphomas other than BL has been established with the precision of those in BL. The major reason for this is the failure of pathologists to arrive at a uniform system for classifying the lymphomas, thus making correlations with karyotypic findings difficult, if not impossible. However, recent attempts at bringing the classification of non-Burkitt's lymphomas into a reasonable homogeneous system [70] have already led to definite correlations (Table 12–5) [51, 109], with the possibility that ultimately the chromosome findings may be more specific in classifying a lymphoma than most of the phenotypic characteristics. Furthermore, some correlations between the karyotypic changes, the type of lymphoma, and some of the phenotypic characteristics are already appearing [51]; there is little doubt that ultimately detailed correlations will reveal specificity for the cytogenetic findings in lymphoma, akin to those in various acute and chronic leukemias.

Table 12–4 Surface immunoglobulins in Burkitt's lymphoma cells with variant translocation t(2;8) or t(8;22).

Authors	Type of immunoglobulin chain	
	Heavy	Light
Translocation t(2;8)		
Miyoshi et al. [68]	μ	—
Bornkamm et al. [18]	μ and δ	κ
Abe et al. [2]	μ	—
Slater et al. [100]	α	—
Bernheim et al. [14]	μ	—
Bernheim et al. [14]	μ	κ
Philip et al. [78]	μ	κ
Translocation t(8;22)		
Berger et al. [11]	μ	—
Berger et al. [11]	—	λ
Berger et al. [11, 13, 50]	δ or γ	λ
Berger et al. [9]	μ	λ
Miyoshi et al. [67]	γ	λ
Bernheim et al. [14]	μ	λ
Bernheim et al. [14]	μ	λ
Lenoir et al. [50]	μ	λ
Lenoir et al. [50]	μ	λ
Abe et al. [1]	α	λ
Berger and Bernheim [9]	μ	λ

Results to date indicate that a common change in lymphoma is the presence of the 14q+ anomaly [31, 36, 42, 43, 59, 62, 80, 81], with the band involved being very similar in most cases. However, the donor chromosome, when it can be identified, has shown considerable variability, and it is this variability that must be sorted out as the classification of lymphoma becomes more uniform. Thus, it already appears that certain lymphomas more often than not involve a translocation between chromosomes 8 and 14, others between 14 and 18, and still others between chromosomes 11 or 12 and 14, whereas still other lymphomas may have different types of translocations not involving chromosome 14 or the other chromosomes mentioned. Generally, t(8;14) is common in the diffuse type of lymphoma and t(14;18) in the follicular variety (Figure 12–5). Why chromosome 14 remains such a common recipient chromosome is difficult to state with certainty, but little doubt exists about the vulnerability of band q32 on chromosome 14 to breakage and union with material involved in reciprocal and nonreciprocal translo-

Table 12–5 Common (primary?) chromosome changes in human lymphoma as related to the new international formulation and Rappaport classifications.

Chromosome change	International formulation of malignant lymphomas	Rappaport classification of malignant lymphomas
+12,14q+ or t(11;14)	*Small lymphocytic*	Well-differentiated lymphocytic, diffuse
	Consistent with CLL	
	Plasmacytoid	
t(14;18)	*Follicular, predominantly small cleaved cell*	Poorly differentiated lymphocytic, nodular
	Diffuse areas	Poorly differentiated lymphocytic, nodular, diffuse
	Sclerosis	
t(14;18)	*Follicular, mixed, small cleaved, and large cell*	Mixed, nodular
	Diffuse areas	Mixed, nodular, diffuse
	Sclerosis	
	Intermediate Grade	
t(14;18)	*Follicular, predominantly large cell*	Histiocytic, nodular
	Diffuse areas	Histiocytic, nodular, diffuse
	Sclerosis	
?	*Diffuse, small cleaved cell*	Poorly differentiated lymphocytic, diffuse
	Sclerosis	
?t(12;14)	*Diffuse, mixed, small, and large cell*	Mixed, diffuse
	Sclerosis	
	Epithelioid cell component	Non-Hodgkin's lymphoma with epithelioid reaction
t(8;14)? or t(11;14)	*Diffuse, large cell*	Histiocytic, diffuse
	Cleaved cell	
	Noncleaved cell	
	Sclerosis	
	High Grade	
t(8;14),6q−	*Large cell, immunoblastic*	
	Plasmacytoid	Histiocytic, diffuse
	Clear cell	
	Polymorphous	Mixed, diffuse
	Epithelioid cell component	Non-Hodgkin's lymphoma with epithelioid reaction
?(1q+,6q−,9q+)	*Lymphoblastic*	Malignant lymphoma, lymphoblastic
	Convoluted cell	Convoluted nuclei
	Nonconvoluted cell	Nonconvoluted nuclei
t(8;24) or	*Small noncleaved cell*	Undifferentiated
t(2;8) or t(8;22)	Burkitt's	Undifferentiated, Burkitt's type
	Follicular areas	

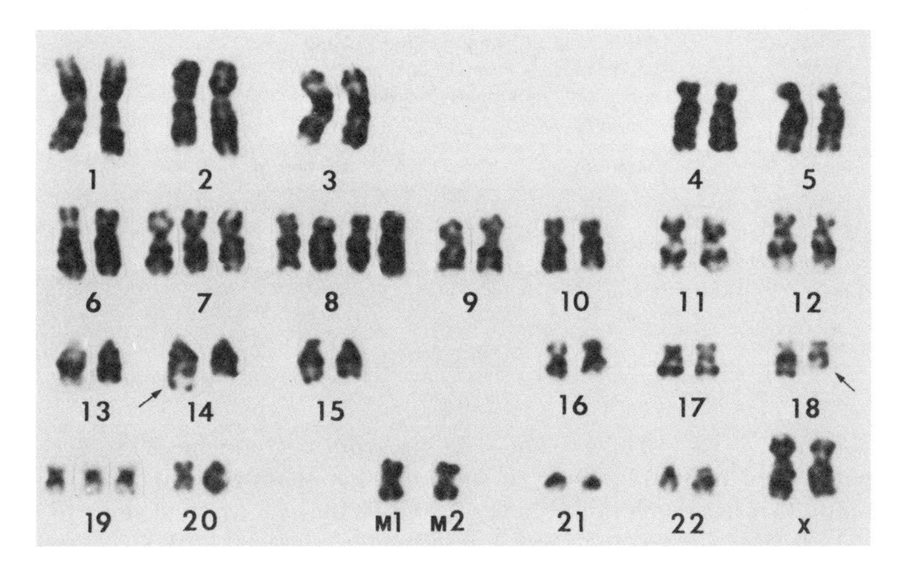

Figure 12–5 Karyotype of a follicular lymphoma cell showing the translocation t(14;18) *(arrows)* common in this type of lymphoma and probably constituting the specific (primary) cytogenetic event in this disease. Secondary changes consist of +7, +8, +8, +19, and two markers (M1, M2). This case was described by Kakati et al. [42].

cations. It is possible that the behavior of the genetic material translocated to and now contiguous with chromosome 14 may assume a major role in the genesis of lymphoma [39], with the secondary karyotypic changes, as pointed out previously, possibly playing a major role in the biologic behavior of the tumors.

A rather common change in lymphoma is deletion of the long arm of chromosome 6 (6q−), originally described by us [76] in acute lymphoblastic leukemia (ALL). It appears to be one of the more common changes in lymphoma, although its biologic and clinical significance is yet to be determined. In the case of ALL, it appears to carry with it a relatively good prognosis [105], and it will be interesting to ascertain whether this applies similarly to lymphomas in which the 6q- anomaly is found, after a sufficient number of cases have been studied and analyzed. Other common chromosome changes in lymphoma are shown in Table 12–6, although their exact significance remains to be determined. However, it is my opinion that these changes play an important role in the biologic behavior of the tumor, if they are not involved in the genesis of these tumors, and contribute significantly to the spread, response to therapy, and ultimately the prognosis of a particular disease.

In a more recent report on a large series of lymphomas [17], it was shown that among the numerical chromosome changes the most fre-

Table 12–6 Common numerical and morphologic karyotypic changes in malignant lymphomas (in order of their frequency).

Morphologic	Numerical
14q	+12
18q	+18
6q	+7
1p	+21
8q	

quent are +12, +18, +7, and +21 (in order of frequency), the +12 being seen most frequently in small-cell lymphocytic lymphoma, an entity that has much in common with CLL. In the latter leukemia, the +12 is the most frequent karyotypic change observed [32]. Structural abnormalities most frequently involved 14q, 18q, 6q, 1p, and 8q (in order of frequency). The long arm of chromosome 14, which is often involved in translocations, was involved in more than 70% of the lymphomas. The most frequent translocation was t(14;18)(q32;q21), followed by t(8;14)(q24;q32) and t(1;14)(q42;q32). Deletions most frequently involved chromosome 6 at band q21 or q23. Correlations have also been shown to exist between the karyotype and the histology and some of the immunologic aspects of the lymphomas [17, 45].

Chromosome changes in T-cell lymphoma

Human T-cell lymphomas have not been satisfactorily classified and, in fact, appear to be a more heterogeneous group of diseases than those of B-cell origin. This applies also to a cytogenetic definition of T-cell lymphomas, which is based primarily on cutaneous forms of these lymphomas and T-cell CLL [25, 46, 52, 72, 73]. However, a recent attempt to establish nonrandom changes in lymphoblastic lymphoma, thought to be a T-cell disease, although it revealed some karyotypic changes (e.g., 9q+, 6q−, and changes of chromosome 1), failed to reveal such changes, with the tumors not containing a 14q+ or definite translocations involving chromosomes 8 or 14 [46]. Although 14q+ has been described in a few cases of T-cell lymphoma [74], the karyotypic changes in these diseases are generally complex, with no specific anomaly characterizing any one entity [44, 46, 74]. In the cases with the 14q+ anomaly, it has been difficult to establish with certainty the origin of the extra material. In any case, the 14q+ anomaly seems to characterize a wide range of cells in the lymphoid system, even though they may differ phenotypically.

Interestingly, in a recently published report [74] the 6q− anomaly was also found in several cases with cutaneous T-cell lymphoma as well as in two cases of lymphoblastic lymphoma [74], again pointing to a sharing of karyotypic similarities by lymphoid malignancies of different cellular origin.

Recent studies indicate that T-cell malignancies are not infrequently associated with a break at band 14q11. The deleted segment of this chromosome may or may not be involved in a translocation. Thus, the q11 band of chromosome 14 appears to contain genetic material intimately involved in the genesis of T-cell lymphomas.

Chromosome changes in Hodgkin's disease

Although Hodgkin's disease (HD) is a relatively common disorder, the karyotypic characteristics of this disease have not been well defined [74, 81]. A 14q+ anomaly has been found in about a third of the cases; the source of the extra material on chromosome 14 has been difficult to ascertain in most cases and when due to a translocation, the chromosomes involved (other than 14) have varied. In addition, technical difficulties and the cellular nature of HD tissues have presented obstacles in obtaining a large volume of reliable cytogenetic data in this disease. However, the karyotypic data unequivocally indicate that HD is a malignancy of lymphoid origin [83]. In fact, the presence of such chromosome changes has been used as a key diagnostic tool in differentiating pleural effusions due to HD from nonmalignant conditions when other diagnostic approaches were equivocal [40].

Karyotypic changes in chronic lymphocytic leukemia

Chronic lymphocytic leukemia (CLL) bridges the gap between leukemia and lymphoma and has been shown interesting cytogenetic characteristics. In the United States and in the Western world CLL is preponderantly a B-cell disease and, as indicated earlier, the introduction of mitogens for B cells (see Table 12–1) has led to the establishment of chromosomal changes in this particular leukemia. To date, it appears that trisomy 12 is the most common anomaly in this disease, followed by 14q+ and some other changes [32, 88–91]. In fact, involvement of chromosome 12 has been said to play a key role in lymphoproliferative disease development [33]. That +12 is not an uncommon anomaly in

Table 12–7 Nonrandom translocations characterizing human cancers.

Translocation	Condition
t(3;8)(p25;q21)	Mixed tumor of parotid
t(6;14)(q21;q24)	Serous cystadenocarcinoma of ovary
t(11;22)(q24;q12)	Ewing's sarcoma
	Peripheral neuroepithelioma (neuroblastoma)

some forms of lymphoma [109] may indicate that CLL has aspects akin to those of lymphomas, a situation that has been appreciated clinically and histologically for many years. Recently, prognostic aspects, including survival in CLL, have been correlated with trisomy 12 in CLL [82], although undoubtedly many more cases will have to be examined before this is established with certainty. The phenotypic manifestations, particularly at the cellular level, of trisomy 12 are also being investigated, and already some interesting leads have been obtained

Table 12–8 Nonrandom morphologic and numerical chromosome changes in human cancer.

Chromosome	Condition
1p−(p34)	Neuroblastoma
1p−(p11p22)	Malignant melanoma
1p−(p31p36)	Neuroblastoma
1p−,i(1q)	Endometrial cancer
1q+,1p−(qter-p21)	Breast cancer
1q+(q24 q44?)	Large bowel cancer
1q+	Breast cancer
3p−(p14p23)	Small cell cancer of lung
3p−	Renal cancer
3q+	Nasopharyngeal cancer
i(5p)	Bladder cancer
5q−,+7,−9	Cervical cancer
6q−(q11q31),i(6p)	Malignant melanoma
i(6p)	Retinoblastoma
6q−(q15)	Ovarian cancer
8q−(q11q22)	Salivary gland tumors
11p−(p13)	Wilms' tumor
11q−	Cervical cancer
i(12p)	Seminoma, teratoma
12q−(q22q24)	Large bowel cancer
12q−(q13q15)	Salivary gland tumors
13q−(q14),−13	Retinoblastoma
22q−(q11),−22	Meningioma

[38]. In addition, the possibility has been raised that segment q13 to q22 on chromosome 12 carries important genes that are duplicated in those lymphoproliferative disorders characterized by trisomy of this chromosome [82].

Chromosome changes in cancers

Tables 12–7 and 12–8 show that the number of solid tumors characterized by nonrandom or specific karyotypic changes is impressive and ever increasing, as more and more conditions are being examined cytogenetically. Besides serving as indicators of malignancy and behavior of

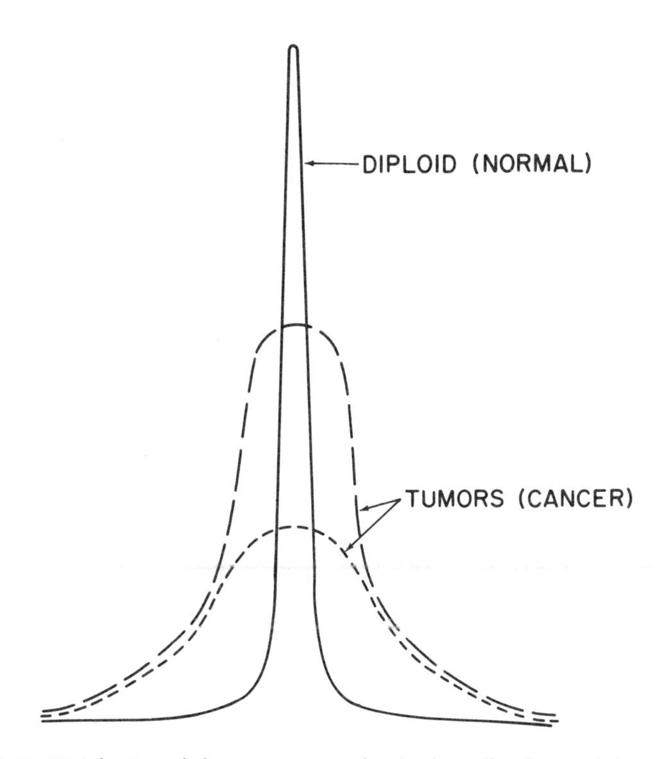

Figure 12–6 Distribution of chromosome number in the cells of normal tissues (diploid) and in primary and metastatic tumors (*long hatches,* primary; *short hatches,* metastatic). As can be seen, in normal tissue the mode of 46 is rather sharp, with only a few cells having missing or extra chromosomes. On the other hand, in cancerous tissue there is a wide distribution of the chromosome numbers with a considerable proportion of cells having chromosome numbers over a wide range below and above the modal number. In metastatic tumors, the cells with the chromosome modal number may constitute a small percentage of the total cell population. Obviously, the wide range in chromosome numbers is reflected in a similar heterogeneity of the karyotypes among the cancer cells.

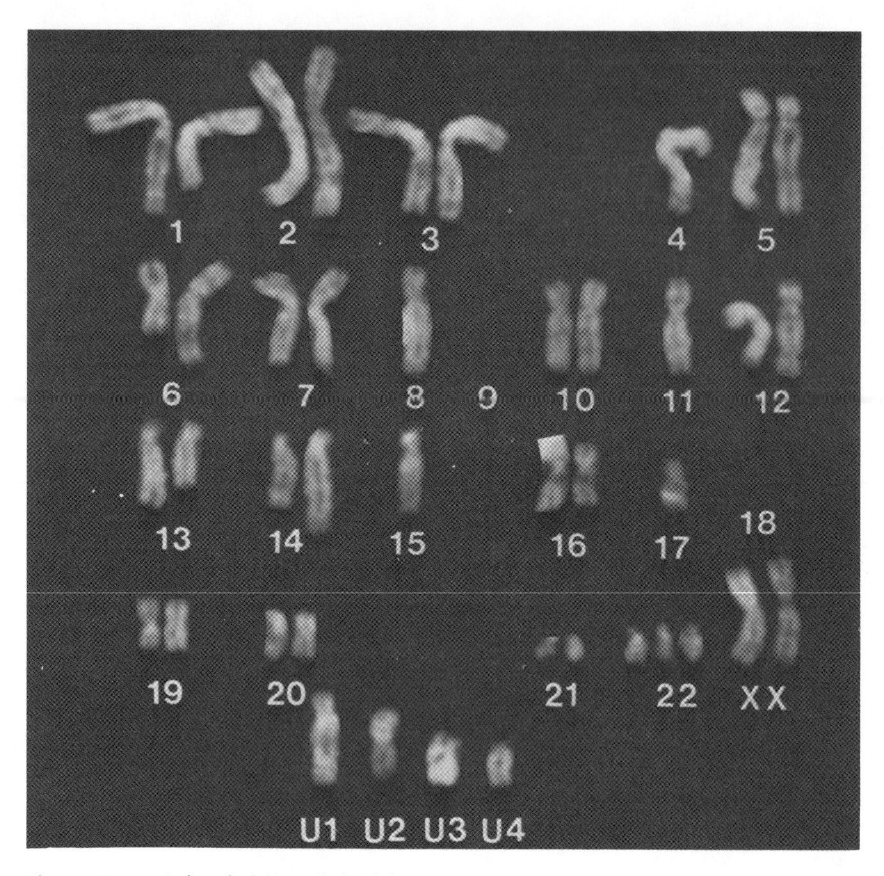

Figure 12–7 Q-banded hypodiploid karyotype consisting of 43 chromosomes of a cell from a serous cystadenocarcinoma of the ovary containing the primary (specific) chromosomal change, that is, t(6;14), and secondary changes [−4, −8, −9, −9, −11, 13q−, −17, −18, −18, +22, and four markers (U1–U4)]. These markers probably contain some of the missing chromosomes, but their exact identity could not be established. The primary karyotypic change is probably related to the genesis of the tumor and the secondary changes to the biology and behavior of the cancer, for example, metastatic spread, resistance or sensitivity to therapy, and invasiveness. This case was described by Wake et al. [107].

tumors, the chromosome changes in solid tumors will ultimately, in my opinion, be shown to carry a significance akin to that of the karyotypic changes in leukemia and lymphoma [96], that is, a primary karyotypic change characterizes each tumor, such as t(6;14) in serous cystadenocarcinoma of the ovary [3, 107], −22 in meningioma [110], 3p− in small cell tumors of the lung [108], and t(3;8) in mixed parotid tumors [60], and that this change plays a crucial role in the genesis of such tumors. The secondary karyotypic changes that are often numerous and complex in solid tumors (Figures 12–6 through 12–9), and may vary

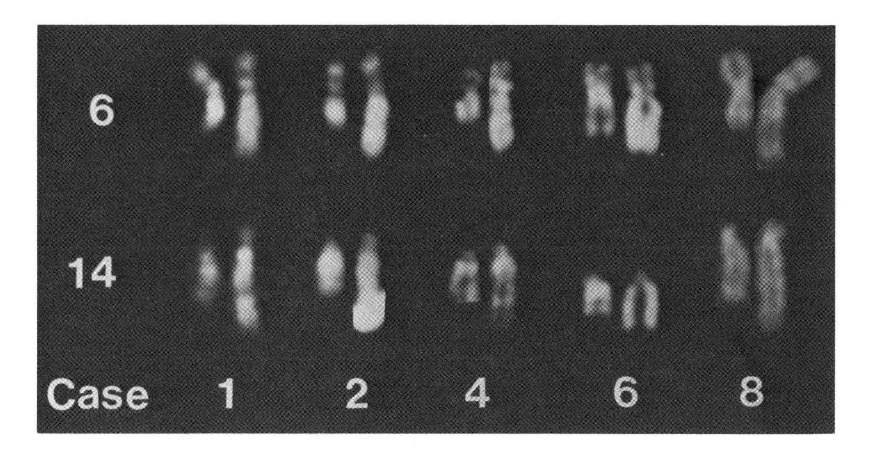

Figure 12–8 Characteristic primary karyotypic change of t(6;14) in five different serous cystadenocarcinomas of the ovary [107].

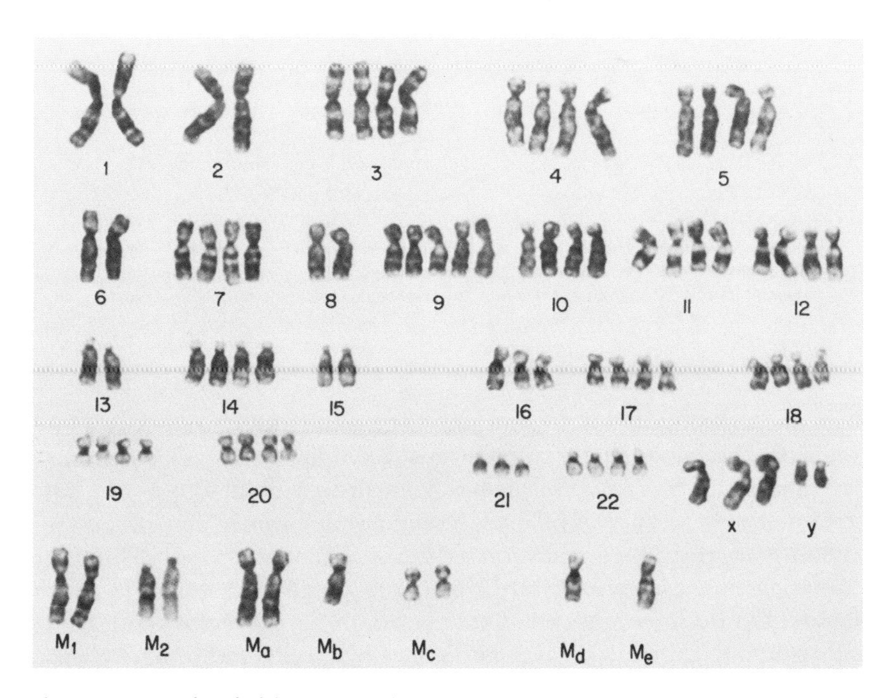

Figure 12–9 G-banded karyotype of a cancer cell with a high chromosome number, including a group of markers. Some of the latter could be identified as to their origin (M_1 and M_2); others (M_{a-e}) could not. Note the large number of extra chromosomes in most groups, including the sex chromosomes.

Table 12–9 Nonrandom translocations characterizing human leukemias and lymphomas.*

Translocation	Condition
t(1;3)(p36;q21)	Myelodyspoietic disorders
t(1;7)(p11;p11)	Dysmyelopoietic disorders (induced?)
t(1;14)(q42;q32)	Lymphoma
t(1;19)(q23;q13)	ALL (L1)
t(1;19)(q23;p13.3)	Pre-B-cell ALL
t(2;8)(p11−13;q24)	BL, ALL (L3)
t(2;11)(p21;q23)	Dysmyelopoietic preleukemia
t(3;5)(q21;q31)	ANLL (M2)
t(3;17)(q26;q22)	Acute disease in myeloproliferative disorders (?)
t(4;11)(q21;q23)	ALL, acute myelomonocytic leukemia
t(6;9)(p23;q34)	ANLL (M2), myeloproliferative diseases
t(6;12)(q15;p13)	Prolymphocytic leukemia
t(8;14)(q32;q32)	BL, ALL (L3), malignant lymphoma (diffuse)
t(8;21)(q22;q22)	AML (M2 with Auer bodies)
t(8;22)(q24;q11)	BL, ALL (L3)
t(9;11)(p21;q23)	AMoL (M5)
t(9;22)(q34;q11)	CML, acute leukemias
t(11;14)(q13;q32)	ALL, CLL, and lymphoma (small cell, diffuse)
t(11;19)(q23;q12 or p12)	ANLL (M5)
t(11;21)(q22;q21)	ANLL, myeloproliferative disorders
t(12;14)(q13;q32)	Diffuse, mixed T-cell lymphoma
t(14;18)(q32;q21)	Malignant lymphoma (follicular)
t(15;17)(q22;q12)	APL (M3)

*Key to abbreviations: ALL = acute lymphoblastic leukemia; AML = acute myeloblastic leukemia; AMoL = acute monocytic leukemia; ANLL = acute nonlymphocytic leukemia; APL = acute promyelocytic leukemia; BL = Burkitt's lymphoma; CLL = chronic lymphocytic leukemia; CML = chronic myelocytic leukemia.

from one cell population to another in the same tumor, may play a key role in the biologic aspects of the tumor, including its response to radiation and chemotherapy [96]. As has been already pointed out, this heterogeneity of the karyotype even within a single tumor may play a key role in drug resistance and metastatic spread of the tumor. The indications from these findings are that, to be totally successful in these tumors, the therapy must take into consideration the cytogenetic variability among the cells, which undoubtedly is reflected in phenotypic variation, including sensitivity or resistance to different forms of therapy. Even though the clinical application of chromosome changes in solid tumors has not been as common or successful as in leukemia, examples have been published, for example, higher invasiveness and recurrence of bladder tumors with marker chromosomes or hyperploidy or both [28, 37, 93], the possibility of differentiating large bowel cancers on the basis of their karyotypes [6, 75], and the application of similar approaches in other cancers [7, 96]. The application of cytogenetic changes in cancer in the future holds much promise, although obviously much more information is needed on the karyotypic changes in many tumors that have not yet been studied successfully or adequately.

Oncogenes and karyotypic changes

Recent publications [29, 35, 101] have indicated that the development of some tumors, such as retinoblastoma and Wilms' tumor, although associated with a constitutional chromosome defect, 13q14 and 11p13, respectively, in a small percentage of cases may be related to expression of recessive alleles, resulting through several mechanisms (not all necessarily karyotypic). From published descriptions, it is difficult to ascertain whether detailed karyotypes were established in each case in order to rule out homozygosity of the tumor locus, possibly resulting from an excessive number of the chromosomes mentioned previously.

Table 12-10 Nonrandom morphologic chromosome changes in human leukemia and lymphoma.

Chromosome	Condition
3p−(p13)	Prolymphocytic (B-cell) leukemia
3p−,3q−	Secondary leukemia
5q−(q22q23)	Refractory anemia, secondary leukemia
6q−	Lymphoma, ALL
7q−(q33q36)	Secondary leukemia
9p−	T-cell ALL
11q−(q23)*	ANLL (M2,M4 and M5), lymphoma
12p−(p12)	ALL
12p−	Secondary leukemia
12q−	ANLL
inv(14)(q11q32)	T-cell CLL
14q+(q32)	Lymphoma, ALL, CLL, adult T-cell acute leukemia
inv(16)(p13q22) or 16q−(q22)	ANLL with eosinophilia
i(17q)	Blastic phase of CML
20q−	Polycythemia vera
21q−	Preleukemia, ANLL

*May be involved in translocations with any of the following chromosomes: 6, 9, 10, 17, and 19.

Table 12–11 Nonrandom numerical chromosome changes in human leukemia, lymphoma, and cancer.

Chromosome	Condition
−5	Secondary leukemia
−7	Secondary leukemia, preleukemia with infection
+7	Cancer of large bowel, bladder cancer, lymphoma
+8	ANLL, blastic phase of CML, polyps of colon, preleukemia
−9	Bladder cancer
+12	CLL, seminoma, lymphocytic lymphoma
+19	Blastic phase of CML
+21	ALL
−22	Meningioma, sarcoma
+22	ANLL

The possibility that retinoblastoma may be of multifocal origin has been provided by recent observations in which abnormalities of 1q, i(6p), and 13q− were each seen in diverse tumors, including tumors at different sites of the same patient [8, 19].

A contribution of the establishment of the primary cytogenetic events in human neoplasias (Tables 12–7 through 12–11) has been the possibility that oncogenes (20, 22), at various levels of expression and function, may be located or affected through these cytogenetic events [47, 84, 92]. Although the exact identity and nature of these oncogenes remain essentially unknown, the strong possibility exists that the primary karyotypic changes in each disease entity affect or involve oncogenetic material (85), which through these changes assumes functions expressed in neoplastic lesions (49, 56, 97, 98). Thus, careful and detailed definition of the primary cytogenetic events in each human leukemia, lymphoma, and cancer may be of crucial help in locating the possible sites of oncogenes or material related to their expression and function (16).

Acknowledgment

Some of the data shown in this article are the result of investigations supported by grants from the National Cancer Institute (CA-14555 and CA-28853).

Literature cited

1. Abe R, Tebbi CK, Yasuda H, Sandberg AA: North American Burkitt-type ALL with variant translocation t(8;22). *Cancer Genet Cytogenet* 7:185–195, 1982.

2. Abe R, Hayashi Y, Sampi K, Sakurai M: Burkitt's lymphoma with 2/8 translocation: A case report with special reference to the clinical features. *Cancer Genet Cytogenet* 6:135–142, 1982.
3. Atkin NB, Baker MC: Specific chromosome change in ovarian cancer. *Cancer Genet Cytogenet* 3:275–276, 1981.
4. Autio K. Turunen O, Pentillä P, Eräma E, de la Chapelle A: Human chronic lymphocytic leukemia: Karyotypes in different lymphocyte populations. *Cancer Genet Cytogenet* 1:147–155, 1979.
5. Balazs I, Purrello M, Rubinstein P, Alhadeff B, Siniscalco M: Highly polymorphic DNA site D14S1 maps to the region of Burkitt lymphoma translocation and is closely linked to the heavy chain $\gamma 1$ immunoglobulin locus. *Proc Natl Acad Sci USA* 79:7395–7399, 1982.
6. Becher R, Gibas Z, Sandberg AA: Involvement of chromosomes 7 and 12 in large bowel cancer: Trisomy 7 and 12q−. *Cancer Genet Cytogenet,* 9:329–332,1983.
7. Becher R, Gibas Z, Karakousis C, Sandberg AA: Non-random chromosome changes in malignant melanoma. *Cancer Res,* 43:5010–5016,1983.
8. Benedict WF, Banerjee A, Mark C, Murphree AL: Nonrandom chromosomal changes in untreated retinoblastomas. *Cancer Genet Cytogenet* 10:311–333, 1983.
9. Berger R, Bernheim A: Cytogenetic studies on Burkitt's lymphoma-leukemia. *Cancer Genet Cytogenet* 7:231–244, 1982.
10. Berger R, Bernheim A: Existe-t-il une équivalence fonctionnelle entre anomalies du bras long du chromosome 1 et présence du virus d'Epstein-Barr dans les lignées continues de lymphome de Burkitt? *CR Acad Sci* (Paris) 298:143–145, 1984.
11. Berger R, Bernheim A, Bertrand S, Fraisse J, Frocrain C, Tanzer J, Lenoir G: Variant chromosomal t(8;22) translocation in four French cases with Burkitt lymphoma-leukemia. *Nouv Rev Fr Hematol* 23:39–41, 1981.
12. Berger R, Bernheim A, Siguax F, Valensi F, Daniel M-T, Flandrin G: Two Burkitt lymphomas with chromosome 6 long arm deletions. *Cancer Genet Cytogenet* 15:159–167,1984.
13. Berger R, Bernheim A, Weh H-J, Flandrin G, Daniel M-T, Brouet J-C, Colbert N: A new translocation in Burkitt's tumor cells. *Hum Genet* 53:111–112, 1979.
14. Bernheim A, Berger R, Lenoir G: Cytogenetic studies on African Burkitt's lymphoma cell lines: t(8;14), t(2;8) and t8;22) translocations. *Cancer Genet Cytogenet* 3:307–315, 1981.
15. Biggar RJ, Lee EC, Nkrumah FK, Whang-Peng J: Direct cytogenetic studies by needle aspiration of Burkitt's lymphoma in Ghana, West Africa. *JNCI* 67:769–776, 1981.
16. Bishop JM: Oncogenes and proto-oncogenes. *Hosp Practice* 18:67–74, 1983.
17. Bloomfield CD, Arthur DC, Frizzera G, Levine EG, Peterson BA, Gajl-Peczalska KJ: Nonrandom chromosome abnormalities in lymphoma. *Cancer Res* 43:2975–2984, 1983.
18. Bornkamm GW, Kaduk B, Kachel G, Schneider U, Fresen KO, Schwanitz G, Hermanek P: Epstein-Barr virus-positive Burkitt's lymphoma in a German woman during pregnancy. *Blut* 40:167–177, 1980.
19. Chaum E, Ellsworth RM, Abramson DH, Haik BG, Kitchin FD, Chaganti, RSK: Cytogenetic analysis of retinoblastoma: Evidence for multifocal origin and *in vivo* gene amplification. *Cytogenet Cell Genet* 38:82–91, 1984.
20. Dalla-Favera R, Bregni M, Erikson J, Patterson D, Gallo RC, Croce CM: Human c-*myc onc* gene is located on the region of chromosome 8 that is translocated in Burkitt lymphoma cells. *Proc Natl Acad Sci USA* 79:7824–7827, 1982.
21. Dalla-Favera R, Martinotti S, Gallo RC: Translocation and rearrangements of c-myc oncogene locus in human undifferentiated B-cell lymphomas. *Science* 219:963–967, 1983.

22. Davis M, Malcolm S, Rabbitts TH: Chromosome translocation can occur on either side of the c-*myc* oncogene in Burkitt lymphoma cells. *Nature* 308:286–288, 1984.
23. Douglass EC, Magrath IT, Terebelo H: Burkitt cell leukemia without abnormalities of chromosomes No. 8 and 14. *Cancer Genet Cytogenet* 5:181–185, 1982.
24. Douglass EC, Magrath IT, Lee EC, Whang-Peng J: Cytogenetic studies in non-African Burkitt lymphoma. *Blood* 55:148–155, 1980.
25. Edelson EL, Berger CL, Raafat J, Warburton D: Karyotype studies of cutaneous T cell lymphoma: Evidence for clonal origin. *J Invest Dermatol* 73:548–550, 1979.
26. Erikson J, Martinis J, Croce CM: Assignment of the genes for human γ immunoglobulin chains to chromosome 22. *Nature* 294:173–175, 1981.
27. Erikson J, ar-Rushdi A, Drwinga HL, Nowell PC, Croce CM: Transcriptional activation of the translocated c-myc oncogene in Burkitt lymphoma. *Proc Natl Acad Sci USA* 80:820–824, 1983.
28. Falor WH, Ward RM: Prognosis in early carcinoma of the bladder based on chromosomal analysis. *J Urol* 119:44–48, 1978.
29. Fearon ER, Volgelstein B, Feinberg AP: Somatic deletion and duplication of genes on chromosome 11 in Wilms' tumor. *Nature* 309:176–178, 1984.
30. Fraisse J, Lenoir G, Vasselon C, Jaubert J, Brizard CP: Variant translocation in Burkitt's lymphoma: 8;22 translocation in a French patient with an Epstein-Barr virus-associated tumor. *Cancer Genet Cytogenet* 3:149–153, 1981.
31. Fukuhara S, Ueshima Y, Shirakawa S, Uchino H, Morikawa S: 14q translocations, having a break point at 14q13, in lymphoid malignancy. *Int J Cancer* 23:739–743, 1979.
32. Gahrton G, Robèrt K-H: Chromosomal aberrations in chronic B-cell lymphocytic leukemia. *Cancer Genet Cytogenet* 6:171–181, 1982.
33. Gahrton G, Robèrt K-H, Friberg K, Juliusson G, Biberfeld P, Zech L: Cytogenetic mapping of the duplicated segment of chromosome 12 in lymphoproliferative disorders. *Nature* 297:513–514, 1982.
34. Gahrton G, Robèrt K-H, Friberg K, Zech L, Bird AG: Nonrandom chromosomal aberrations in chronic lymphocytic leukemia revealed by polyclonal B-cell mitogen stimulation. *Blood* 56:640–647, 1980.
35. Gilbert F: Retinoblastoma and recessive alleles in tumorigenesis. *Nature* 305:761–762, 1983.
36. Gödde-Salz R, Schwarze E-W, Lennert K, Grote W: Letter-to-the Editor: t(Y;14); A new type of 14q+ marker chromosome. *Cancer Genet Cytogenet* 3:89–90, 1981.
37. Granberg-Öhman I, Tribukait B, Wijkström H: Cytogenetic analysis of 62 transitional ALL bladder carcinomas. *Cancer Genet Cytogenet,* 11:69–86,1983.
38. Han T, Ozer H, Sadamori N, Gajera R, Gomez G, Henderson ES, Minowada J, Sandberg AA: Clinical significance of clonal chromosome changes in leukemic lymphocytes of chronic lymphocytic leukemia. *ASCO Abstr* 1:186, 1982.
39. Hect F, Kaiser-McCaw B, Sandberg AA: Chromosome translocations in cancer. *N Engl J Med* 304:1493, 1981.
40. Hossfeld DK, Schmidt CG: Chromosome findings in effusions from patients with Hodgkin's disease. *Int J Cancer* 21:147–156, 1978.
41. Kakati S, Barcos M, Sandberg AA: Chromosomes and causation of human cancer and leukemia. XXXVI. The 14q+ anomaly in an American Burkitt lymphoma and its value in the definition of lymphoproliferative disorders. *Med Pediatr Oncol* 6:121–129, 1979.
42. Kakati S, Barcos M, Sandberg AA: Chromosomes and causation of human cancer and leukemia. XLI. Cytogenetic experience with non-Hodgkin, non-Burkitt lymphomas. *Cancer Genet Cytogenet* 2:199–220, 1980.
43. Kaneko Y, Abe R, Sampi K, Sakurai M: An analysis of chromosome findings in non-Hodgkin's lymphomas. *Cancer Genet Cytogenet* 5:107–121, 1982.

44. Kaneko Y, Larson RA, Variakojis D, Haren JM, Rowley JD: Nonrandom chromosome abnormalities in angioimmunoblastic lymphadenopathy. *Blood* 60:877–887, 1982.
45. Kaneko Y, Rowley JD, Variakojis D, Haren JM, Ueshima Y, Daly K, Kluskens LF: Prognostic implications of karyotype and morphology in patients with non-Hodgkin's lymphoma. *Int J Cancer* 32:683–692, 1983.
46. Kaneko Y, Variakojis D, Kluskens L, Rowley JD: Lymphoblastic lymphoma: Cytogenetic, pathologic and immunologic studies. *Int J Cancer* 30:273–279, 1982.
47. deKlein A, van Kessel AG, Grosveld G, Bartram CR, Hagemeijer A, Bootsma D, Spurr NK, Heisterkamp N, Groffen J, Stephenson JR: A cellular oncogene is translocated to the Philadelphia chromosome in chronic myelocytic leukaemia. *Nature* 300:765–767, 1982.
48. Kirsch IR, Morton CC, Nakahara K, Leder P: Human immunoglobulin heavy chain genes map to a region of translocations in malignant B lymphocytes. *Science* 216:301:303, 1982.
49. Leder P, Battey J, Lenoir G, Moulding C, Murphy W, Potter H, Stewart T, Taub R: Translocations among antibody genes in human cancer. *Science* 222:765–771, 1983.
50. Lenoir GM, Preud'Homme JL, Bernheim A, Berger R: Correlation between immunoglobulin light chain expression and variant translocation in Burkitt's lymphoma. *Nature* 298:474–476, 1982.
51. Levine EG, Arthur DC, Frizzera G, Gajl-Peczalska KJ, Lindquist L, Peterson BA, Hurd DD, Bloomfield CD: Cytogenetic analysis of 55 cases of malignant lymphoma (ML): Correlation with histologic and immunologic phenotype. *AACR Abstr* 23:116, 1982.
52. Liang JC, Gaulden ME, Herndon JH: Chromosome markers and evidence for clone formation in lymphocytes of a patient with Sézary syndrome. *Cancer Res* 40:3426–3429, 1980.
53. Magrath I, Erikson J, Whang-Peng J, Sieverts H, Armstrong G, Benjamin D, Triche T, Alabaster O, Croce CM: Synthesis of kappa light chains by cell lines containing 8;22 chromosomal translocation derived from a male homosexual with Burkitt's lymphoma. *Science* 222:1094–1098, 1983.
54. Magrath IT, Pizzo PA, Whang-Peng J, Douglass EC, Alabaster O, Gerber P, Freeman CG: Characterization of lymphoma-derived cell lines: Comparison of cell lines positive and negative for Epstein-Barr virus nuclear antigen. I. Physical, cytogenetic, and growth characteristics. *JNCI* 64:465–476, 1980.
55. Malcolm S, Barton P, Bentley DL, Ferguson-Smith M, Murphy C, Rabbitts T: Assignment of a V Kappa locus for immunoglobulin light chain to the short arm of chromosome 2 (2 cen to 2 p13). In Human Gene Mapping Conference. VI. White Plains, NY, National Foundation, 1982.
56. Malcolm S, Barton P, Murphy C, Ferguson-Smith MA, Bentley DL, Rabbitts TH: Localization of human immunoglobulin k light chain variable region genes to the short arm of chromosome 2 by in situ hybridization. *Proc Natl Acad Sci USA* 79:4957–4961, 1982.
57. Manolov G, Manolova Y: Marker band in one chromosome 14 from Burkitt lymphomas. *Nature* 237:33–34, 1972.
58. Manolova Y, Manolov G, Kieler J, Levan A, Klein G: Genesis of the 14q+ marker in Burkitt's lymphoma. *Hereditas* 90:5–10, 1979.
59. Mark J, Dahlenfors R, Ekedahl C: Recurrent chromosomal aberrations in non-Hodgkin and non-Burkitt lymphomas. *Cancer Genet Cytogenet* 1:39–56, 1979.
60. Mark J, Dahlenfors R, Ekedahl C, Steuman G: The mixed salivary gland tumor—A normally benign human neoplasm frequently showing specific chromosomal abnormalities. *Cancer Genet Cytogenet* 2:231–241, 1980.

61. McIntosh RV, Cohen BB, Steel CM, Read H, Moxley M, Evans HJ: Evidence for involvement of the immunoglobulin heavy-chain gene locus in the 8:14 translocation of human B lymphomas. *Int J Cancer* 31:275–279 1983.
62. Mitelman F: Marker chromosome 14q+ in human cancer and leukemia. *Adv Cancer Res* 34:141–170, 1981.
63. Mitelman F, Levan G: Clustering of aberrations to specific chromosomes in human neoplasia. IV. A survey of 1,871 cases. *Hereditas* 95:79–139, 1981.
64. Mitelman F, Andersson-Anvret M, Brandt L, Catovsky D, Klein G, Manolov G, Manolova Y, Mark-Vendel E, Nilsson PG: Reciprocal 8;14 translocation in EBV-negative B-cell acute lymphocytic leukemia with Burkitt-type cells. *Int J Cancer* 24:27–33, 1979.
65. Miyamoto K, Miyano K, Miyoshi I, Hamasaki K, Nishihara R, Terao S, Kimura I, Maeda K, Matsumura K, Nishijima K, Tanaka T: Chromosome 14q+ in a Japanese patient with Burkitt's lymphoma. *Acta Med Okayama* 34:61–65, 1980.
66. Miyamoto K, Sato J, Miyoshi I, Nishihara R, Terao S, Hara M, Kimura I: 8;14 translocation in a Japanese Burkitt's lymphoma. *Acta Med Okayama* 34:139–142, 1980.
67. Miyoshi I, Hamazaki K, Kubonishi I, Yoshimoto S, Kitajima K,Kimura I, Miyamoto K, Sato J, Yorimitsu S, Tao S, Ishibashi K, Tokuda M: A variant translocation (8;22) in a Japanese patient with Burkitt lymphoma. *Gann* 72:176–177, 1981.
68. Miyoshi I, Hiraki S, Kimura I, Miyamoto K, Sato J: 2/8 translocation in a Japanese Burkitt's lymphoma. *Experientia* 35:742–743, 1978.
69. Morita M, Minowada J, Sandberg AA: Chromosomes and causation of human cancer and leukemia. XLV. Chromosome patterns in stimulated lymphocytes of chronic lymphocytic leukemia. *Cancer Genet Cytogenet* 3:293–306, 1981.
70. National Cancer Institute sponsored study of classifications of non-Hodgkin's lymphomas. Summary and description of a working formulation for clinical usage. The Non-Hodgkin's Lymphoma Pathologic Classification Project. *Cancer* 46:2112–2135, 1982.
71. Neel BG, Jhanwar SC, Chaganti RSK, Hayward WS: Two human c-onc genes are located on the long arm of chromosome 8. *Proc Natl Acad Sci USA* 79:7842–7846, 1982.
72. Nowell P, Daniele R, Rowlands D Jr, Finan J: Cytogenetics of chronic B-cell and T-cell leukemia. *Cancer Genet Cytogenet* 1:273–280, 1980.
73. Nowell P, Finan J, Glover D, Guerry D: Cytogenetic evidence for the clonal nature of Richter's syndrome. *Blood* 58:183–186, 1981.
74. Nowell PC, Finan JB, Vonderheid EC: Clonal characteristics of T-cell lymphomas: Cytogenetic evidence from blood, lymph nodes and skin. *J Invest Dermatol* 78:69–75, 1982.
75. Ochi H, Takeuchi J, Holyoke D, Sandberg AA: Possible specific chromosome changes in large bowel cancer. *Cancer Genet Cytogenet* 10:121–122, 1983.
76. Oshimura M, Sandberg AA: Chromosomal 6q- anomaly in acute lymphoblastic leukaemia. *Lancet* 2:1045–1046, 1976.
77. Pearson J, Ilgren EB, Spriggs AI: Lymphoma cells in cerebrospinal fluid confirmed by chromosome analysis. *J Clin Pathol* 35:1307–1311, 1982.
78. Philip T, Lenoir GM, Fraisse J, Philip I, Bertoglio J, Ladjaj S, Bertrand S, Brunat-Mentigny M: EBV-positive Burkitts lymphoma from Algeria, with a three-way rearrangement involving chromosomes 2,8 and 9. *Int J Cancer* 28:417–420, 1981.
79. Pierre RV, Dewald GW, Banks PM: Cytogenetic studies in malignant lymphoma: Possible role in staging studies. *Cancer Genet Cytogenet* 1:257–261, 1980.
80. Prieto F, Badia L, Herranz C, Redón J, Caballero M, Artiges E, Báguena J: Cromosoma marcador (14q+) y linfomas. *Rev Esp Oncologia* 25:399–408, 1978.
81. Reeves BR, Pickup VL: The chromosome changes in non-Burkitt lymphomas. *Hum Genet* 53:349–355, 1980.

82. Robert K-H, Gahrton G, Friberg K, Zech L, Nilsson B: Extra chromosome 12 and prognosis in chronic lymphocytic leukaemia. *Scand J Haematol* 28:163–168, 1982.
83. Rowley JD: Chromosomes in Hodgkin's disease. *Cancer Treat Rep* 66:59–63, 1982.
84. Rowley JD: Human oncogene locations and chromosome aberrations. *Nature* 301:290–291, 1983.
85. Rowley JD: Biological implications of consistent chromosome rearrangements in leukemia and lymphoma. *Cancer Res* 44:3159–3168, 1984.
86. Rowley JD, Fukuhara S: Chromosome studies in non-Hodgkin's lymphomas. *Semin Oncol* 7:255–266, 1980.
87. Rowley JD, Variakojis D, Kaneko Y, Cimino M: A Burkitt-lymphoma variant translocation (2p−;8q+) in a patient with ALL, L3 (Burkitt type). *Hum Genet* 58:166–167, 1982.
88. Sadamori N, Han T, Minowada J, Henderson ES, Ozer H, Sandberg AA: Clonal chromosome changes in stimulated lymphocytes of CLL. *Blood* 58 (Suppl 1): 151a, 1981.
89. Sadamori N, Han T, Minowada J, Sandberg AA: Clinical significance of cytogenetic findings in untreated patients with B-cell chronic lymphocytic leukemia. *Cancer Genet Cytogenet* 11:45–52, 1984.
90. Sadamori N, Matsui SI, Han T, Sandberg AA: Comparative results with various polyclonal B-cell activators in aneuploid chronic lymphocytic leukemia. *Cancer Genet Cytogenet* 11:25–30, 1984.
91. Sadamori N, Takeuchi J, Abe R, Sandberg AA: Kappa and lambda immunoglobulin expression associated wtih abnormalities of chromosomes 2 and 22 in lymphoma and leukemia. *Cancer Genet Cytogenet* 10:209–212, 1983.
92. Sakaguchi AY, Naylor SL, Shows TB, Toole JJ, McCoy M, Weinberg RA: Human c-Ki-ras2 proto-oncogene on chromosome 12. *Science* 219:1081–1083, 1982.
93. Sandberg AA: Chromosome markers and progression in bladder cancer. *Cancer Res* 37:322–329, 1977.
94. Sandberg AA: The Chromosomes in Human Cancer and Leukemia, New York, Elsevier-North Holland, 1980, 776.
95. Sandberg AA: Chromosome changes in the lymphomas. *Hum Pathol* 12:531–540, 1981.
96. Sandberg AA: Chromosomal changes in human cancer: Specificity and heterogeneity. *In* Tumor Cell Heterogeneity: Origins and Implications (Ownes AH Jr, Coffey PS, Baylin SB, eds) vol 4, New York, Academic Press, 1982.
97. Sandberg AA: A chromosomal hypothesis of oncogenesis. *Cancer Genet Cytogenet* 8:277–285, 1983.
98. Sandberg AA: Chromosomal changes and cancer causation. Chromatin's re-awakening. *In Accomplishments in Cancer Research* (Fortner JC, Rhoads JE, eds), Philadelphia, JB Lippincott Co, 1984, 157–169.
99. Schröder J, Vuopio P, Autio K: Chromosome changes in human chronic lymphocytic leukemia. *Cancer Genet Cytogenet* 4:11–21, 1981.
100. Slater RM, Behrendt H, van Heerde P: Cytogenetic studies on four cases of non-endemic Burkitt lymphoma. *Med Pediatr Oncol* 10:71–84, 1982.
101. Solomon E: Recessive mutation in aetiology of Wilms' tumor. *Nature* 309:111–112, 1984.
102. Stewart SE, Lovelace E, Whang JJ, Ngu VA: Burkitt tumor: Tissue culture, cytogenetic and virus studies. *J Natl Cancer Inst* 34:319–327, 1965.
103. Tanzer J, Frocrain C, Alcalay D, Desmarest MC: Pleuropericardite a cellules de Burkitt chez un poitevin. Chromosome 14q+ par translocation (1;14), association a une translocation (8;22), a la perte de l'y et a de nombreuses autres anomalies. *Nouv Rev Fr Hematol* (Suppl, Abstr) 22:107, 1980.

104. Taub R, Kirsch I, Morton C, Lenoir G, Swan D, Tronick S, Aaronson S, Leder P: Translocation of the c-myc gene into the immunoglobulin heavy chain locus in human Burkitt lymphoma and murine plasmacytoma cells. *Genetics* 79:7837–7841, 1982.
105. Third International Workshop on Chromosomes in Leukemia, *Cancer Genet Cytogenet* 4:94–142, 1981.
106. Van Den Berghe H, Parloir C, Gosseye S, Englebienne V, Cornu G, Sokal G: Variant translocation in Burkitt lymphoma. *Cancer Genet Cytogenet* 1:9–14, 1979.
107. Wake N, Hreshchyshyn MM, Piver SM, Matsui SI, Sandberg AA: Chromosomes and causation of human cancer and leukemia. XLII. Specific cytogenetic changes in ovarian cancer. *Cancer Res* 40:4512–4518, 1980.
108. Whang-Peng J, Bunn PA Jr, Kao-Shan CS, Lee EC, Carney DN, Gazdur A, Minna JD: A non-random chromosomal abnormality, del 3p(14-23), in human small cell lung cancer (SCLC). *Cancer Genet Cytogenet* 6:119–134, 1982.
109. Yunis JJ, Oken MM, Kaplan ME, Ensrud KM, Howe RR, Theologides A: Distinctive chromosomal abnormalities in histologic subtypes of non-Hodgkin's lymphoma. *N Engl J Med* 307:1231–1236, 1982.
110. Zang KD, Singer H: Chromosomal constitution of meningiomas. *Nature* 216:84–85, 1967.
111. Zech L, Haglund U, Nilsson K, Klein G: Characteristic chromosomal abnormalities in biopsies and lymphoid-cell lines from patients with Burkitt and non-Burkitt lymphomas. *Int J Cancer* 17:47–56, 1976.
112. Zhang S, Zech L, Klein G: High-resolution analysis of chromosome markers in Burkitt lymphoma cell lines. *Int J Cancer* 29:153–157, 1982.

13.

The significance of identifying a cancer-predisposed person: Lessons from the chromosome-breakage syndromes

JAMES GERMAN

Chromosome instability and cancer-proneness are shared by four rare, recessively transmitted disorders sometimes collectively called the chromosome-breakage syndromes: Bloom's syndrome (BS), ataxia-telangiectasia (AT), Fanconi's anemia (FA), and xeroderma pigmentosum (XP). Even though these four entities are unrelated in both pathogenesis and clinical presentation, grouping them can be illuminating for the clinician, because, of all the recessively inherited disorders known, these four feature the most striking predisposition to cancer. In a broad sense, the clinical feature that the syndromes share is the probability that any of a striking array of serious complications will develop. It is the complication-proneness of the disorders that I shall emphasize in this paper rather than either their basic scientific aspects or their clinical spectra, which are covered extensively elsewhere. In discussing the chromosome-breakage syndromes, I shall first present concise clinical summaries of each and then tell how the clinician can inform and thereafter care for the patient who at diagnosis is identified as having enormous risk of serious complications.

Clinical features of the chromosome-breakage syndromes

In column A of Table 13–1 are listed the major clinical features of each of the four syndromes; Table 13–2 lists additional clinical features and certain laboratory findings. (Comprehensive clinical descriptions of each of these conditions appear in *Chromosome Mutation and Neoplasia* [1–3, 5].)

Bloom's syndrome

The major clinical features of BS are the following: abnormally small size both pre- and postnatally, with approximately normal body pro-

Table 13–1 Clinical aspects of the chromosome-breakage syndromes.

Entity	A: Clinical features of major diagnostic value	B: Some possible clinical complications or features of delayed onset
BS	Growth deficiency, intrauterine (fetus & placenta) and postnatal	Unusual facial configuration
	Characteristic facies	Disfiguring facial skin lesion
	Facial erythema & telangiectasia, sun-sensitive	Respiratory tract infection, recurrent
	Immunodeficiency	Short adult stature
		Psychological problems secondary to ostracism because of short stature and facial disfigurement
		Infertility
		Diabetes
		Cancer†
AT	Cerebellar ataxia	Neurologic deterioration
	Telangiectasia, conjunctival & dermal	Endocrine insufficiencies
	Immunodeficiency	Respiratory tract infection, recurrent
		Cancer†
FA	Anatomic malformations (e.g., thumb, kidney)	Bone marrow failure (anemia, hemorrhage, infection)
	Bone marrow failure	Complications of multiple transfusions
	Growth deficiency (variable)	Complications of prolonged androgen therapy
	Dermal hyperpigmentation	Low intelligence
		Cancer†
XP	Freckling, keratosis, atrophy of the skin*	Facial disfigurement from sunlight damage to the skin or surgical removal of cancers
	Corneal scarring*	Blindness
	±Neurologic deficiency	Neurological deterioration
		Cancer†

*In response to sunlight.

†The distinctive distribution of types of neoplasia is tabulated and commented on elsewhere [4].

portions; dolichocephaly, slight relative microcephaly, and a characteristic facies (narrow, with nasal prominence associated with hypoplastic malar and mandibular areas); and, a sun-sensitive, telangiectatic erythema that affects the face almost exclusively, particularly the butterfly area, the lower eyelids, and the lower lip. Abnormally low concentrations of one or more plasma immunoglobulins are usually demonstrable, along with other manifestations of a moderate to severe immunodeficiency. The incidence of minor developmental defects appears to be increased, and mild mental deficiency is sometimes present. A

Table 13–2 Additional clinical features and some laboratory findings of value in diagnosis or characterization of the chromosome-breakage syndromes*

Entity	Cytogenetic disturbances	Hypersensitivity to environmental agents: Clinical	Hypersensitivity to environmental agents: Laboratory	Additional abnormalities
BS	↑SCE	Sunlight	EMS	↓Replicon-fork-progression rate
	↑Homologous interchanges	Cancer chemotherapeutic agents	MC	↑Mutation rate in vitro and in vivo
	↑Gaps, breaks		UV, 313 nm	↑Recombination in adenovirus infecting BS cells
AT	↑Clones with translocations	Roentgentherapy	Ionizing radiation	↓Lymphocyte response to PHA
	↑Gaps, breaks, nonhomologous interchanges		MNNG	↓Inhibition of replicon chain initiation following γ-irradiation
			Bleomycin	↓Repair of γ-endonuclease-sensitive sites
				↑α-fetoprotein
FA	↑Gaps, breaks, nonhomologous interchanges	Chemotherapeutic agents used for immunosuppression	MC	↓Removal of DNA interstrand cross-links
	↑Clones with translocations		DEB	
			Trenimon	
			CP	
			INH	
			PUVA	
XP-A to -I	↑Gaps, breaks, SCE after UV, 254 nm (variable)	Sunlight	UV, 254 nm	↓Dimer-excision repair
			4NQO	
			AAAF	
XP-V		Sunlight		↓Daughter-strand-gap repair

*Key to abbreviations: AAAF = 8-acetoxy-*N*-acetylaminofluorene; CP = cyclophosphamide; DEB = diepoxybutane; EMS = ethyl methanesulfonate; 4NQO = 4-nitroquinoline oxide; INH = isonicotinic acid hydrazide; MC = mitomycin C; MNNG = *N*-methyl-*N'*-nitro- *N*-nitrosoguanidine; PHA = phytohemagglutinin; PUVA = psoralen plus UV irradiation; UV = ultraviolet radiation; XP-A to -I = complementation groups A to I of excision-repair-defective XP; XP-V = "variant" XP.

manyfold increase in the number of exchanges between both sister and nonsister-but-homologous chromatids is a consistent and apparently unique finding in cultured cells, making it of great diagnostic value. The several known DNA-repair pathways appear to function normally.

Ataxia-telangiectasia

Progressive ataxia and other signs of central nervous system dysfunction beginning early in infancy in combination with telangiectasia of the bulbar conjuctivae and skin are the clinical features responsible for this syndrome's name. Defective immune function associated with hypoplasia of various lymphoid organs also is characteristic but not invariably present. Various endocrinopathies are characteristic but as yet have not been studied adequately. Increased numbers of chromatid rearrangements are present in blood lymphocytes and skin fibroblasts in culture, and mutant clones bearing rearranged chromosome Nos. 7 and 14 are a characteristic finding in blood T lymphocytes. Hypersensitivity to X-rays has been observed clinically, and AT cells in culture respond abnormally to damage experimentally induced by ionizing radiation.

Fanconi's anemia

FA classically consists of the onset of pancytopenia in a preadolescent child with anatomical malformations. Common features include small stature; microcephaly; diminutive (infantile) facial features; aplasia, hypoplasia, or anomalous formation of the thumb, with or without malformation of the radius and other bones of the hand; kidney anomaly; mild mental deficiency; and skin hyperpigmentation. Considerable variation exists in both the severity and pattern of the developmental anomalies and the age of onset of the pancytopenia, so that the diagnosis of FA is often difficult. Bone marrow failure sometimes is the only clinical manifestation, anatomic malformations being absent; on the other hand, an affected person occasionally fails to develop clinical anemia. Increased numbers of chromatid aberrations are present in cultured cells; however, not every blood sample from an affected individual exhibits chromosome instability, a fact that contributes further to the difficulties in diagnosing FA. Cells in culture are unusually sensitive to DNA cross-linking agents (e.g., mitomycin C, psoralen plus ultraviolet [UV]-irradiation).

Xeroderma pigmentosum

Hypersentivitiy of the skin and cornea to sunlight is the major clinical feature of XP. Most persons with XP appear normal at birth, but without heroic measures to avoid sun exposure an inexorable destruction of the skin sets in. Freckles and solar keratoses usually have formed by late infancy. In a minor proportion of affected persons, neurological abnormalities including mental deficiency are an integral part of the clinical syndrome. The cells of the affected homozygote are defective in repairing the damage produced by UV-irradiation and UV-like chemicals. In XP, irradiated cells die after exposure to an array of DNA-damaging agents, and of those surviving, many presumably contain mutations, the consequence of defective repair of lesions introduced into the DNA. Most XP patients ("classic" XP) are deficient at excising the lesions introduced into DNA by ultraviolet light. Cell-hybridization studies in excision-deficient XP have demonstrated a remarkable degree of genetic heterogeneity [2], remarkable because of the great rarity of the clinical entity itself. The remainder of patients with clinical XP ("variant" XP) manifest a disturbance in semiconservative DNA replication following such irradiation. XP cells exhibit no increased chromosome instability unless challenged with UV light or UV-like chemicals, in which case those of at least some of the complementation groups (Table 6-3) respond with an abnormally great number of sister-chromatid exchanges and chromosome breaks.

Clinical significance of identifying a person with a chromosome-breakage syndrome

The exceedingly rare disorders just described have special implications for affected families. Diagnosis of any one of them identifies the patient as an heir later in life to any of several important complications (Table 13–1, column B), ranging from unpleasant and partially controllable to life threatening. In most cases these consequences could not be suspected by the uninitiated parents, not only because the conditions themselves are not well known but also because the relatively normal phenotype at birth belies the seriousness of the diagnosis. Only accurate diagnosis and an adequate understanding of the natural course of the disorder by the physician in charge will permit the appropriate care of the patient and counseling of the family. Even then, the seriousness and distressing nature of many of the complications make the physician's task difficult. The affected person will experience a lifetime of medical problems. Doctors' visits, hospitalization, periodic tests (some research-related), and unusual and prolonged treatment regimens doom the patient to a special and in many respects unpleasant life. The many medical problems repeatedly interrupt schooling and may deprive the child of a good education and normal social adjustment, which compounds the problem. In fact, in most cases the family will be required to provide lifetime care of the person. But of the myriad serious complications possible in each of the chromosome-breakage syndromes, cancer is the sword of Damocles that hangs most threateningly over those affected.

Complications

In BS, infancy is commonly hectic for the parents because of a combination of recurrent gastrointestinal and respiratory tract problems. The child exhibits little interest in food. Repeated episodes of severe but unexplained diarrhea are common, sometimes requiring hospitalization for the resulting dehydration. Upper and lower respiratory tract infections occur repeatedly throughout infancy and childhood in the majority of affected persons, often with progression to purulent otitis media or pneumonia. These infections, which can reach life-threatening proportions quickly, respond well to antibiotics. As is the case in most other genetically determined immunodeficiencies, these repeated illnesses and hospital visits very possibly will interfere with the child's education and social adjustment. As childhood progresses, new problems are encountered because of the unusual appearance of the person with BS: when the red facial lesion is severe, other children will deride the child, and the striking degree of shortness can be a further cause of ostracism. Shortness of stature in adult life is the rule, and short adults, especially men, can expect to be discriminated against in many occupations. Smallness during infancy and childhood also can have another undesirable psychological consequence: emotional maturation may be retarded if these children are "babied" by parents, teachers, and medical attendants, that is, treated as though they were younger than their chronological age. Infertility in men with BS appears to be the rule, and women conceive only rarely. Diabetes, sometimes insulin-dependent, is much more common than in the general population. Of the known recessively transmitted disorders, BS features perhaps the greatest risk of cancer, and the risk is for any of many types [4]. Leukemia is the main neoplasm during childhood, and in adulthood other types of cancers occur at much earlier ages than expected for the general population. Superimposed on this last-named complication is the probability of hypersensitivity to the cytotoxic chemotherapeutic agents that very well may be required to treat the cancer.

The future of the young child *with AT* is also gloomy. Sooner or later, serious to overwhelming infections, particularly of the lower respiratory tract, will arise. Relentless neurological deterioration will set in:

an expressionless face and a nasal, expressionless voice will develop, and life in a wheelchair and bed is a common outcome; intelligence sometimes deteriorates late in the course of the disorder. Secondary sexual characteristics may develop incompletely. The lymphoid tissue, defective in AT with respect to immune function, is often the site of neoplasia. Neoplasia in nonlymphoid tissues also occurs with increased frequency, at exceptionally early ages [4]. If death from pneumonia or sepsis is escaped in childhood or adolescence, neoplasia of some type very possibly will be the cause of an untimely death.

In FA, bone marrow failure is almost inevitable; therefore, early death is to be expected with or without treatment. The age at which manifestations of marrow failure appear is variable, and in exceptional cases it is delayed until early adulthood. Theoretically, bone marrow transplantation should be a useful therapeutic measure, but hypersensitivity to the immunosuppressive drugs required after transplantation produces new complications that in themselves usually prove lethal. The current management of persons with FA has other serious complications as well, including the consequences of repeated blood transfusions and the effects of long-term administration of anabolic steroid hormones (androgens). As to neoplasia, several types of cancers have been documented in FA [4], with acute nonlymphocytic leukemia and liver tumors of various types predominating. Actually, the magnitude of the cancer predisposition is unknown. It may be very high, but it is curious that cancer was not recognized as a feature of FA for four decades after 1927 when Fanconi described the condition. Death in childhood from anemia, hemorrhage, or infection obviously would preclude neoplasia to some extent, but elsewhere [4] I have discussed the possibility that prolonged androgen therapy (introduced in 1959) contributes to what unquestionably is a very high incidence of leukemia and liver neoplasia in FA; if the therapy rather than the genetic constitution is responsible, FA itself might be less cancer-predisposing than is generally believed.

In XP, skin cancers of standard types begin to appear in late childhood. Without surveillance and appropriate excision these will destroy neighboring tissues and may cause death if cranial structures or major blood vessels become involved. Melanoma is much more common than in the general population, often with a lethal outcome. Nonintegumentary neoplasia appears not to be increased in frequency [4]. Many persons with XP suffer serious corneal damage, the consequence being varying degrees of blindness; it is unknown why some are spared this complication. Also for unknown reasons, some but by no means all persons with XP are mentally deficient, some show other neurological deficits, and some have stunted growth.

Handling the complications

The proper handling of any developmentally defective child and that child's family requires the physician in charge to be well prepared and equipped with standard medical knowledge. The physician also often will do well to learn whether the condition—often a rare one—is under investigation at some medical center, and whether new diagnostic and therapeutic approaches might be available through contact with the investigating group. However, in the complication-prone syndromes under consideration here, even more is required of the patient's own personal physician than in most rare conditions, and now I speak of the art as well as the science of medicine. Usually the course is a long and difficult one for patient and parents—and for the physician. The properly prepared physician can hope to help the family cope not only with the complications of the disorder but also with the fear of them.

Complications other than cancer-proneness can be handled by any competent medical group. But, as stated earlier, the diagnosis of any of the chromosome-breakage syndromes identifies a cancer-prone person, and this is a special problem the solution to which has not often been discussed. Cancer is still a greatly feared disease because of the possibility of much suffering and a lethal outcome even with the best of therapy. An obvious objective for both the physician and the affected family would be to convert the unpleasant awareness of cancer predisposition into useful information. If a given individual is identified as being at increased risk of cancer, it might be assumed that such knowledge automatically would be of considerable value in that it would facilitate appropriate preventive measures. But, in fact, what are the "appropriate preventive measures"? Unfortunately, not a great deal can be done currently, and more general recognition is needed of this unpleasant fact, particularly among practicing physicians.

Today, there is widespread interest in identifying cancer-predisposed persons and populations. Methods for detecting damaged genetic material in somatic cells are being developed. As the countless environmental carcinogens are identified and as understanding of the interplay between them and the genetic material of somatic cells increases, the accuracy with which epidemiologists and geneticists can identify persons at increased risk of cancer will improve; increasing numbers of people at an exceptionally high risk will be identified. But what is to be done once such an identification is made? How and when should the affected family be informed? What preventive measures can be taken by those predisposed? As I have dealt during the past 25 years with parents of children with the four rare disorders addressed in this paper, and with affected adults, I have become increasingly aware of the paucity of available preventive measures. Few important new measures have been developed during this period of time. Presumably, means of very early detection as well as effective antitransformation measures and agents that will prevent tumor progression will become available in the future. (In due time, of course, the treatment of cancer will become less formidable and more often curative, thus lessening the seriousness of this complication in genetically predisposed persons.)

The course of action obviously will differ depending on the type of tumor that is likely to emerge; some situations are easier to handle than others. If the predisposition pertains to a single tissue or anatomical region, a reasonable surveillance program is easier to devise than when the predisposition is to cancer generally, which is the case with BS and AT. Thus, if colonic polyps are anticipated, as in the dominantly transmitted polyposis coli syndromes, a plan of surveillance of just the colonic mucosa can readily be worked out; a very high risk of breast cancer raises the possiblity of prophylactic mastectomy. But what is to be done if retinoblastoma in infancy has been successfully treated? Osteosarcoma then can be anticipated before age 35! And what if leukemia is the neoplasia anticipated? The answer at present to the question regularly posed by parents who learn that their child is leukemia-prone—"How often should blood counts be made on our child?"—unfortunately is, "No more often than for other children." A potentially useful measure is the aspiration and storage of the patient's own bone marrow before leukemia has developed. Marrow storage for potential reimplantation conceivably could have therapeutic value also in relation to neoplasms other than leukemia, in that it would permit more aggressive therapy aimed at totally ablating malignant cells, during which therapy the normal marrow and lymphoid stem cells probably also would be destroyed.

Cancer-prone persons may need to be reminded that the occurrence of one clinical neoplasm does not relieve them of their predisposition. Even with the successful care or control of one neoplasm, the increased risk persists, possibly even greater if therapeutic agents that themselves are carcinogenic have been used. (In the 110 known persons with BS, 31 have developed clinical cancer, at a mean age of 25.5 years (range 4–44); but 6 of these 31 have developed two cancers, the second unrelated to the first.)

To be sure, even now the knowledge that a person is likely to develop cancer at an earlier-than-normal age sometimes can be valuable. It can even be life-saving for that person because, with the physician's help, that person can take special measures to detect surgically curable cancers as early as possible—intensifying his or her personal cancer-surveillance program. An appropriate program would include a complete evaluation by an internist annually once adulthood is reached, with prompt consultation by the patient with that physician at the appearance, even at an early age, of such premonitory symptoms as hoarseness, cough, indigestion, change in bowel habit, blood in the stool, a lump in the breast, or unexplained vaginal bleeding. Women with one of the syndromes might begin uterine cervical surveillance and self-examination for breast tumor considerably earlier than is generally advised. Also, affected persons can take more than the usual precautions in avoiding exposure to unnecessary oncogenic radiation and chemicals, including tobacco.

The physician can hope to open the patient's eyes to the special danger in ways that will be useful, even life-saving, but simultaneously in as emotionally nondisruptive a way as possible. For the conscientious physician, the disclosure to a patient that cancer predisposition exists is a matter of considerable moment. In making such a disclosure, the physician will strive not to violate the first law of therapeutics, *primum non nocere.* (This expression implies that the physician, whether the patient can be benefited or not, aims at making matters no worse.) The internist dealing with a person with organic heart disease is well aware of the obligation not to produce additional disease by causing undue fear, the so-called cardiac neurosis. Now, the oncologist-physician, epidemiologist-physician, and geneticist-physician find themselves in an analogous position when they are called on to deal with the problem of cancer proneness, whether it be on an environmental or a genetic basis. Their aim is to circumvent or contain the neoplasia but not to superimpose inordinate fear—a cancer neurosis. Our recently acquired knowledge of cancer etiology, along with the current emphasis by health agencies and the general public on the delineation of cancer-prone populations, makes this a new and serious matter. Lessons of general value can be learned from those few physicians who have the opportunity to care for persons and families affected with one of the rare, cancer-predisposing syndromes. Therefore, it seems desirable to have the clinical experiences of such physicians and clinics reported in medical journals—not just basic scientific observations.

It is my conviction that the presentation of a detailed description and explanation of the clinical disorder to the parents of persons with most rare genetic disorders, or to the patients themselves if they are grown, is desirable soon after a diagnosis is made. Ideally, this would be made by a knowledgeable, sensitive, and compassionate physician who also is capable of communicating in appropriate and comprehensible terms. In the case of the so-called chromosome-breakage syndromes this explanation includes a presentation of the complications that very possibly will occur, in terms that are useful to the family and that will facilitate long-term and effective cooperation with the physician. By such a plan, chronic disability can best be avoided or postponed, and the best chance of early cure of cancers will be secured. An understanding of the natural course of the disorder with which affected persons and their families are to live during the coming years is a major source of support to those families—if the information is presented properly. Furthermore, such knowledge spares the affected person from undue medical "handling," by which I mean unnecessary consultation with one doctor or clinic after another; unnecessary testing and hospitalizations in themselves can be psychologically damaging to a child, and financially disastrous to a family.

Perhaps most important of all for the affected patient and family, however, is just to be in the hands of someone who understands the nature of the rare disorder afflicting them. Probably the most desirable situation for a family affected with a rare and little-known disorder, particularly one for which a long and stormy course is predicated, is long-term contact with a physician or a medical group who not only identifies the disorder accurately but understands and has a special academic interest in it and counsels freely with them during the years. Together they can cope in the best way possible with each complication as it arises.

Literature cited

1. Alter BP, Potter NU: Long-term outcome in Fanconi's anemia: Description of 26 cases and review of the literature. *In* Chromosomes Mutation and Neoplasia (German J, ed), New York, Alan R Liss, 1983, 43–62.
2. Andrews AD: Xeroderma pigmentosum. *In* Chromosomes Mutation and Neoplasia (German J, ed), New York, Alan R Liss, 1983, 63–83.
3. Gatti RA, Hall K: Ataxia-telangiectasia: Search for a central hypothesis. *In* Chromosomes Mutation and Neoplasia (German J, ed), New York, Alan R Liss, 1983, 23–41.
4. German J: Patterns of neoplasia associated with the chromosome-breakage syndromes. *In* Chromsomes Mutation and Neoplasia (German J, ed), New York, Alan R Liss, 1983, 97–134.
5. Passarge E: Bloom's syndrome. *In* Chromosome Mutation and Neoplasia (German J, ed), New York, Alan R Liss, 1983, 11–21.

14.

Managing families genetically predisposed to cancer: The "cancer-family syndrome" as a model

CHRISTOPHER J. WILLIAMS

The clinical management of patients and their families affected with certain well-recognized genetic conditions that predispose to cancer is becoming reasonably well understood, for instance, dominantly transmitted familial polyposis of the colon. Others such as Bloom's syndrome [7] are being studied intensively. However, in the case of less well known conditions, the care of persons who actually develop a cancer, the advice to their relatives, and the long-term management of the family is far from standardized. The cancer-family syndrome (CFS) is an example of a less well known condition; here, management continues to evolve, and the various problems and considerations it raises make it a useful model for approaching the clinical problem generally of families genetically predisposed to cancer.

First, the CFS itself will be described briefly. Then, a patient with CFS will be described. Finally, several aspects of the management of persons with cancer in an affected family and of the family as a unit will be addressed.

The cancer-family syndrome

In 1913 Aldred Warthin [39] described a cancer-prone family—"Family G." Since then, many similar families have been reported [3, 5, 10, 16, 18, 19, 23, 27, 29, 33, 37], including the one I shall describe below [41]. The features of the CFS are listed by Lynch as follow:

- An unusually large number of members of the family are affected with adenocarcinoma of the colon [26], endometrium [26], and probably ovary [21]. Adenocarcinomas of the breast [25], stomach [25], and other tissues [41], and possibly lymphomas [16], also may affect members of the family.

- The colon cancers tend to occur in its proximal portion of the colon, that is, the right side [22].
- The cancers appear at an unusually early age, the mean being about 45 years [24], in comparison with the mean of 65 in the general population.
- Affected family members tend to have multiple primary cancers [30, 34, 41].
- This distribution of affected persons in the family is compatible with autosomal dominant inheritance [22].
- In some cases, features of the Muir-Torre syndrome occur (sebaceous adenomas, epitheliomas, carcinomas).
- Affected persons possibly have an unusually prolonged survival [28].

Evidence for autosomal dominant inheritance

Several observations support the hypothesis of autosomal dominant transmission of the CFS. Supporting dominant inheritance is a high lifetime incidence of colon and endometrial carcinoma among the offspring of matings that include an affected person [26]. In "Family G" the lifetime risk of colon or endometrial carcinoma for 581 of the members is 55%, compared to about 6% for the general population (Figure 14–1). The risk to the offspring of a couple one of whom is affected with CFS is approximately 50%, which is consistent with the hypothesis that one half of these individuals inherit the deleterious gene that predisposes to cancer.

Lynch [24] has shown by segregation analysis using maximum likelihood techniques that the pattern of inheritance is consistent with autosomal dominance. Gene carriers show a lifetime susceptibility of 89%, with a mean age of onset of 42.5 years.

Further evidence for or against this pattern of transmission should be obtainable if and when consistent genetic markers for CFS become available (see below).

Case report and family history

A 34-year-old woman presented herself in 1975 with a history of adenocarcinoma of the cecum found in 1963, when she was 22 [41]. In 1975, at a second laparotomy, she was found to have two new, separate adenocarcinomas of the colon, with metastatic involvement of the ovaries. Surgery at each operation had been a standard resection of the involved colon with its area of blood supply; at the second operation, a bilateral salpingo-oophorectomy and hysterectomy had been per-

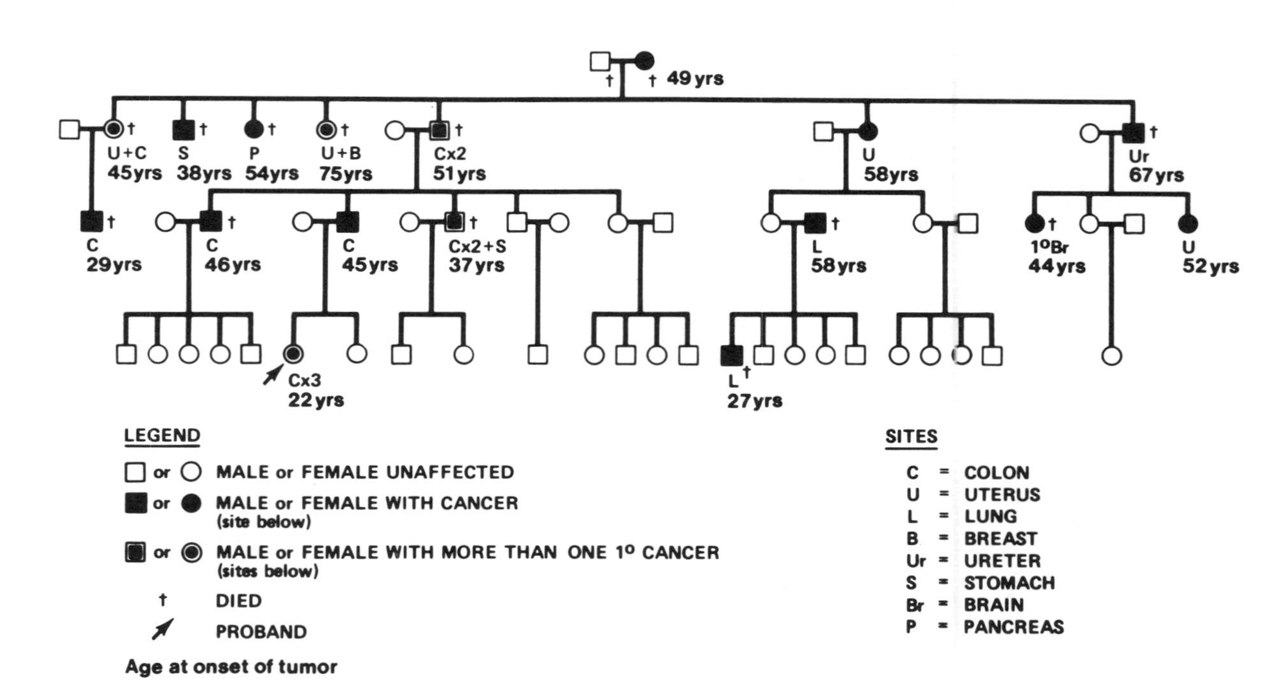

Figure 14–1 Family pedigree showing the increased incidence of various carcinomas and the early age of tumor onset in a family displaying characteristics of the CFS (Williams [41]).

formed. Early in 1977, she was found to have a rising carcinoembryonic antigen (CEA) titer and an abnormal barium enema. A third laparotomy revealed a new carcinoma of the colon; the remaining colon and 30 cm of the terminal ileum were removed, the rectum being preserved.

A family history taken from this young woman revealed a high incidence of carcinoma in several generations; this history stimulated the documentation of more cancers in the family through use of available death certificates and medical records, including pathology reports. Forty-four family members in four generations were traced (Figure 14–2). (For the second generation, the diagnosis was by history only in some cases.) Sixteen (36%) had a cancer, even though most of the 24 individuals in the fourth generation were under the age of 31 years. Five of the 16 persons with cancer had multiple cancers. Of the 20 persons in the first three generations, 14 (70%) had cancer. The ages of onset (range 22–75 years) tended to be earlier than expected

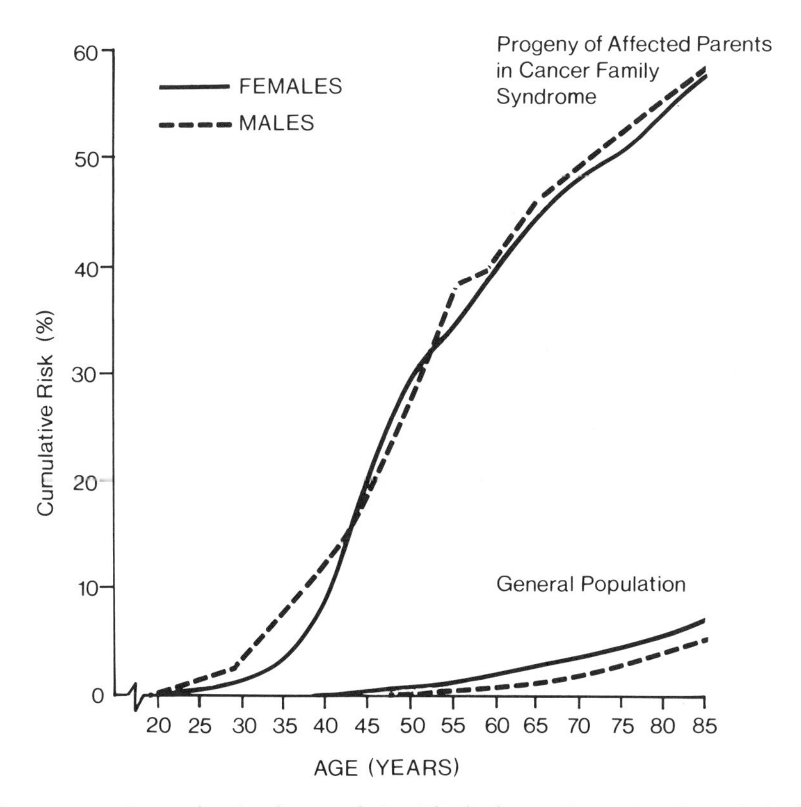

Figure 14–2 Curves showing the cumulative risk of colon carcinoma in males and of colon and endometrial carcinoma in females among the progeny of affected CFS parents. These are compared with the risk of these tumors in the general population based on the third U.S. National Cancer Survey (Lynch et al. [23]).

in the general population. Cancers of several types had occurred in the family (Figure 14–2), colon and uterine cancers being the most common.

Management of cancer and the risk of cancer in families transmitting the CFS

The importance of identifying an affected family

The taking of a family history of disease has become a routine matter when a physician undertakes the care of a new patient. Nevertheless, it is my observation that in the past and even now little attention is paid to a patient's statement that some of his or her close relatives have cancer. With respect to colon cancer specifically, many physicians erroneously appear to believe that familial polyposis of the colon (polyposis coli) is the only cause of colon cancer in families; Table 14–1 shows that there are many others [36]. The finding of a carcinoma of the proximal colon in a young person (only 1 in 8700 white American males develops such a tumor by the age of 44), especially if there have been multiple cancers, should suggest the possibility of the CFS; it then becomes important to obtain an extensive and accurate family history. The family on which I have reported in some detail here illustrates the need for extensive family assessment when the physician encounters a single patient with this history. In that particular family, the proband developed her first proximal colon cancer at the exceptionally early age of 22; and by 36 she had developed three additional ones, without evidence of polyposis. The young woman was able—and quite willing—to supply a detailed family history, and it showed all the features of the CFS (Figure 14–1). Her initial treatment and subsequent management probably would have been different had she been recognized to be a member of a family affected by the CFS when she developed her first carcinoma. Some clinicians still fail to consider the possibility of a genetic syndrome when they encounter cancer presenting in an unusual fashion or setting or in a patient with a family history of an excessive number of cancers.

The early identification of persons who are members of a CFS family is important for two reasons. Such identification can influence the planning of treatment of the person who actually has a cancer; it can also help identify cancer-predisposed individuals who then may benefit from surveillance or prophylactic measures (to be mentioned later). Endoscopic assessment is indicated for the patient and his or her relatives when the possibility of CFS arises in order to rule out the presence of familial polyposis of the colon. (If there is evidence of multiple adenomatous polyps in relatives, even if none is seen in the proband, the syndrome probably should be regarded as familial polyposis coli or a polyposis-like condition rather than CFS.)

Constructing a pedigree

Once the diagnosis of CFS or any other genetic syndrome that predisposes to cancer has been made, it is important to consider the high risk of family members as well as the proband himself or herself (i.e., the person who called the family to the physician's attention in the first place). The doctor, or in some cases genetic counselors or social workers [14], needs to meet key family members in order to construct as complete a pedigree as possible. The pedigree will incorporate dates of birth and death, diagnoses, names and addresses of family members, and names and addresses of family physicians or hospitals.

It is important to ask family members for their cooperation and to explain to them what is known about the syndrome and its possible consequences. Many members in a given family certainly will be entirely well and probably will not be aware of the possibility of an increased risk of cancer in their family. Because of the delicate nature of the situation, it may be easier for the private physicians of the various family members to participate in these discussions. Not surprisingly, family members, even including those who have or have had cancer, may display a negative attitude toward such information, and denial is frequent. Krush [15] found that only 7 of 32 members of a "cancer family" she investigated had engaged in regular medical examinations and that only 2 of 6 of the members with cancer had had regular check-ups despite attempts by the investigators to carry out a cancer-surveillance program.

Surveillance for cancer in the family and consideration of prophylactic surgery for selected at-risk members is contingent on the cooperation of members of a CFS family, and every effort should be made to determine why some particular family member might be uncooperative. Considerable tact is required in arranging the screening for cancer in family members (see below) because some if not most of them will be frightened and will just not want to know of their high cancer risk. If the situation is handled clumsily, there is a real risk of causing cancer phobia, thereby worsening the problem. High-risk family members should receive counseling that emphasizes the possible benefits of surveillance and prophylactic surgery in the prevention of cancer. Care is needed in these initial encounters because frightened family members may refuse screening and deny symptoms of cancer, therefore presenting themselves for treatment only at advanced stages of the disease [20].

Table 14–1 Differential diagnosis in hereditary colon cancer.

Syndrome	Mode of inheritance	Anatomic location and pathology of polyps and/or cancer	Associated clinical features
"Cancer-family syndrome"	Possible autosomal dominant	Colon and endometrium; no polyps	Various adenocarcinomas, particularly colon and endometrium; multiple primary malignant neoplasms; early age at onset of cancer
Familial polyposis coli	Autosomal dominant	Colon; adenomatous polyps	None
Gardner's syndrome	Autosomal dominant	Colon (occasionally small intestine and stomach); adenomatous polyps	Soft tissue (sebaceous cysts, fibromas) and bone lesions (osteomas of mandible, sphenoid, and maxilla); rarely, thyroid carcinoma and retroperitoneal sarcoma, cancer of ampulla of Vater, pancreas, adrenal
Peutz-Jeghers syndrome	Autosomal dominant	Entire gastrointestinal tract except esophagus harbors polyps that may show malignant degeneration (Cancer issue remains controversial in spite of recent documentation.) Hamartomas of muscularis mucosa	Melanin spots of oral and vaginal mucosa and distal portions of fingers; ovarian (granulosa-cell) tumor

Solitary polyps	Possible autosomal dominant	Colon	None
Ulcerative colitis	Possible autosomal dominant in certain families	Colon; pseudopolyposis	Occasionally arthritis, systemic manifestations, and psychological aberrations
Juvenile polyposis coli	Autosomal dominant	Colon; hamartomatous polyps of colon; increased occurrence of adenomatous polyps and adenocarcinoma in relatives; connective tissue abnormality. (One family showed both adenomatous and juvenile polyps in the same patients, and segregating separately in others, with increased frequency of colon cancer.)	None
Turcot's syndrome	Autosomal recessive	Colon and central nervous system; adenomatous polyps	None
Generalized gastrointestinal adenomatous polyposis	Possible autosomal dominant	Stomach, small and large bowel; adenomatous polyps	Desmoid reported
Familial combined breast and colon cancer	Possible autosomal dominant	Colon and breast; no polyps	None

Investigating the family

Once a family is identified as being affected with CFS, or suspected of being affected, the members of the family predisposed to cancer must be identified, as far as is possible. Those who already have cancer identify themselves, of course, but are there other means? One or more biologic "markers" of the CFS are badly needed. These might be distinguishing features of cells or serum from persons carrying the gene (i.e., effects of the gene itself), or they might be traits that would be inherited along with the CFS gene (i.e., a gene linked to the postulated CFS gene on the chromosome). Tight linkage between the gene for such markers and the gene for cancer-proneness would make the testing for those markers in affected families of great practical value. Unfortunately, the results of investigations of this nature have been rather disappointing (reviewed recently by Lynch et al. [26]). Nevertheless, the question of the use of markers in such families arises repeatedly—sometimes raised by the physician and at other times by knowledgeable members of the family—therefore, I shall mention several that may possibly be useful and describe their status.

Carcino-embryonic antigen (CEA)

Guirgis et al. [8, 9] measured plasma CEA levels in 272 members of 6 CFS families together with 191 controls. The mean CEA levels were similar in the CFS family members and controls of the same smoking status. When the CFS family members were classified by their genetic risk (closeness of relationship to a cancer patient or known presence of cancer), there was a significant increase in the mean CEA level of high-risk family members compared to controls or relatives at low risk. However, an unexpected finding in this study was that the spouses of high-risk family members had a correspondingly higher mean CEA level, and those married to low-risk family members had a lower mean CEA! These data are compatible with the existence of a genetic and/or a connubial effect on CEA, possibly as a result of an environmental agent acting in concert with a genetic predisposition to cancer; Guirgis et al. have suggested a hypothetical model for this [8]. In fact, CEA levels are affected by many factors (such as age, smoking habit, chronic lung disease, and chronic inflammatory diseases), and they are of little use in predicting whether an individual CFS member has inherited the CFS gene.

SV40 T antigen

Expression of SV40 T antigen has been reported to be elevated in some CFS members in comparison to normal controls [26]. However, the pattern of elevation was independent of the person's genetic cancer risks, age, sex, family branch, and family generation. The finding of elevated SV40 T antigen is not clearly associated with cancer risk and so far is confined to "family G." Therefore, SV40 T antigen levels are of no general value in predicting susceptibility to cancer, although its presence in a kindred may indicate an increased risk at the family level.

Actin

Kopelovich et al. [13] examined the cytoskeleton of cultured skin fibroblasts from CFS members and patients with familial polyposis of the colon. An altered distribution of actin cables was found in polyposis patients but not in CFS patients, making the test useless as a predictor of cancer risk in CFS.

HLA typing

In "Cancer Family N," 115 members have been investigated by HLA typing [32]. In the cancer-prone branches of the family, 20 of 21 members had one particular HLA haplotype (HLA 2-HLA 12; relative odds = 6.30). Eleven of 12 family members with cancer in the cancer-prone branches had the HLA 2-HLA 12 haplotype (relative odds = 6.06). The single patient with cancer and a different haplotype is the child of a family member who had haplotype HLA 2-HLA 12. The high association of HLA 2-HLA 12 with cancer is suggestive of linkage of a possible cancer susceptibility gene to HLA ($P = 0.025$). However, more studies are needed as the HLA 2–12 haplotype is the most common in Caucasian populations. If linkage of cancer susceptibility with HLA eventually is established, the susceptibility gene could be linked to different haplotypes in other families. HLA typing of a large kindred is time-consuming and expensive; because it is an unproved method for identifying family members at an increased risk of cancer, it remains an area for research.

Mixed leukocyte cultures (MLC)

Berlinger et al. [2] presented preliminary data that suggest that some patients with familial colon cancer of the nonpolyposis type have an immune defect. Eight of 18 patients examined in their study had significantly decreased MLC responses, whereas patients with familial polyposis coli did not exhibit this defect. This possible subclinical immune defect is nonspecific and unproven and cannot be used as a reliable disease marker.

Skin-fibroblast studies

Kopelovich [11] and Danes [4] have studied skin fibroblasts from members of families predisposed to colon cancer. Kopelovich [12], based on results from his studies on polyposis families, has suggested that it may be possible to use repeated skin biopsies to determine the cancer risk of individuals. Using this single-cell type (i.e., the fibroblasts), he was able to search for differences between polyposis patients and normal individuals. Fibroblasts derived from affected members of the polyposis families exhibited abnormal phenotypic properties that segregated as expected for autosomal dominant traits. The abnormalities were present in members of families with polyposis (persons certain to develop cancer), in cancer cells of spontaneously occurring human tumors, and in cells treated in vitro with certain physical and chemical agents. Screening studies, using a panel of abnormal phenotypic features of skin fibroblasts suggested by Kopelovich [11], need to be tested in a variety of different populations, including CFS families. Individuals determined to be at risk on the basis of the skin fibroblast screen would then be followed to see if they developed cancer, since the only confirmation of the accuracy of the screening panel would be the ultimate clinical development of cancer. If cancer-predisposed individuals can be identified by the skin fibroblast screen, the value of prophylactic measures [11] and chemoprevention [6] could then be tested.

Treatment of a cancer patient in a CFS family

Treatment of a cancer in a person who is a member of a CFS family may be different from that given a person in the general population with the same tumor. A modified approach to treatment should at least be considered, for example, in colon cancer. Current experience with CFS suggests that the entire CFS colonic mucosa is a potential source of malignant disease, and patients who undergo conventional ("regional") surgical resection will have an unacceptably high risk of developing further colonic carcinomas [31]. Therefore, total or subtotal colectomy with preservation of a rectal stump should be considered for colon cancer patients with CFS.

The case reported by Ruma et al. [36] shows the utility of this approach: an additional occult distal carcinoma was found in the colectomy specimen of a patient who had presented himself with a tumor in the proximal colon and a family history of the CFS. Further support is provided by "family X" [31]: six of eight family members treated conservatively, with a hemicolectomy, have died of recurrent cancer; the only person who remained alive and disease-free 12 years later was the one who had had a total colectomy.

Subtotal colectomy removes the majority of the colonic mucosa at risk and permits proctoscopic assessment of the remaining colon. As most carcinomas in the CFS are in the proximal colon (Table 14–2), this

Table 14–2 Proximal colonic cancer incidence among 19 cancer-family-syndrome families [26]

Family No.	Number of cases of colonic cancer* (initial diagnosis only)	Number first occurring at proximal colon*	Percentage of cases affecting proximal colon
1	7	7	100
33	27	11	41
51	6	3	50
120	7	5	71
196	11	4	36
200	7	6	86
30	4	4	100
35	3	2	67
115	7	6	86
113	5	4	80
164	6	4	67
199	4	4	100
250	6	2	33
007	2	2	100
198	26	21	81
069	4	4	100
194	3	1	33
010	2	1	50
203	2	1	50
Total	139	92	66

*Only tumors that have been verified as to subsite have been included: all the families include additional cases in which colonic cancers could not be verified or could not be verified as to subsite.

may be justified, but it remains to be proven that such an approach is adequate and that the later development of rectal carcinoma is not a problem. This type of prophylactic surgery is an established procedure in long-standing ulcerative colitis and familial polyposis of the colon and also should be considered in patients who have a very high risk of recurrent tumor. Most patients tolerate this type of procedure well and have only two or three bowel movements a day, although some older patients have more severe diarrhea [17]. (In the affected family I described earlier, when the patient presented herself at age 22 with a strong family history of cancer, more extensive surgery could have been considered. Also, when she developed two new primary cancers 12 years later, total colectomy with or without preservation of the rectum would have been reasonable, in that curative surgery at that time was considered possible.)

Because of the risk of endometrial and ovarian carcinomas in these patients, total abdominal hysterectomy and bilateral salpingo-oophorectomy also may be considered as an additional prophylactic measure in selected patients, particularly those who have completed their family. However, it should be remembered that extensive surgery should be done only when the operation is potentially curative; advanced or metastatic disease should be treated conventionally. Extensive surgery has been used little in the CFS and is unproven. This approach to the prevention of further cancer, however, has been successful in familial polyposis of the colon, and so should be considered in other syndromes in which the complete removal of an organ can prevent cancer from developing.

Prophylactic surgery

Total or subtotal colectomy with ileorectal anastomosis, although not truly prophylactic, is recommended in selected members of CFS families who develop a curable colonic carcinoma (as just stated). In female family members who have shown themselves to be cancer-prone by developing a colonic lesion, prophylactic removal of the uterus and ovaries in addition to the colon may be considered in selected patients, such as the following: those found to have endometrial atypia on routine endometrial biopsy, members of families with a high incidence of ovarian carcinoma, and those likely to be unable to comply with rigorous screening. This rather radical surgical approach remains experimental, and long-term follow-up of patients who elect prophylactic surgery is required before its value will be known.

Also remaining to be investigated in high-risk patients (but not yet with overt cancer) is the effect of the prophylactic subtotal removal of the colon and, in women who have completed their child-bearing, the removal of the uterus and ovaries. Such measures may be considered in highly selected cases when it is remembered that the lifetime-risk of cancer is as high as 55% for all family members and penetrance of the trait (i.e., developing a clinical cancer) is 89% for gene carriers. However, it must not be assumed that this approach is guaranteed to work, and a paper has already reported on the possible failure of prophylactic surgery for carcinoma in 28 members of families at high-risk of ovarian carcinoma [38]. Three of the women included in the report developed tumors that were indistinguishable from ovarian primaries (papillary pattern and psammoma bodies) despite prophylactic, bilateral salpingo-oophorectomy and total abdominal hysterectomy, and the authors postulate that these patients may be at risk of developing cancers from coelomic tissues.

Prophylactic surgery of this type has been used in women at very high risk of carcinoma of the breast (subcutaneous bilateral mastectomies and concurrent implantation), and a place may exist for such an approach in the CFS, particularly if a biologic marker for the susceptibility gene is found (as discussed previously); but, again, long-term studies are required before we will know. Although prophylactic measures and screening for asymptomatic cancers (to be discussed below) may be beneficial, they are second best to finding ways of preventing tumors from developing in genetically predisposed individuals.

Survival of persons with the CFS who develop cancer

Anecdotal reports of long-term survival in patients with advanced tumors have led to studies of survival patterns in members of CFS families with colonic carcinoma [18, 28, 36, 41]. One hundred seventeen patients from 18 families were examined [26], and their survivals by tumor location in the colon were compared with data from tumor registries. Survival at 5 years for all locations was 52.8% for CFS, compared with 35.3% in the control group from tumor registries. Although, because of its nature, this collection of data cannot be considered conclusive, it seems likely that the natural history of colon cancer is modified in CFS. Such information, if substantiated, may be helpful to the physician dealing with members of an affected family.

Avoidance of carcinogens

If we knew what to tell patients, advice concerning the avoidance of specific carcinogens would be timely. Current theories on the development of colon cancer may not be applicable to members of CFS families because most of the colonic tumors are in the proximal colon, as already mentioned. Despite the lack of definitive information about carcinogens, it would seem useful to reinforce the current empirical recommendations for cancer prevention when discussing the syndrome with family members [40].

Cancer surveillance (screening)

In large studies of the general population, screening for cancer has been disappointing on the whole. However, it is probably worthwhile in CFS families, because the stage at the time of diagnosis of both colon and endometrial carcinoma is crucial if a cure is to be achieved. In persons who have been cured of a cancer, careful lifelong follow-up is essential. In the other high-risk persons, clinical screening for asymptomatic cancer(s) should begin at age 20 because one of the features of this syndrome is an unusually early onset of cancer.

Testing for occult fecal blood (e.g., the "hemoccult" technique) is simple and effective because many colorectal tumors bleed. Such testing can be done at home by the persons at risk, provided they comply with the necessary dietary restrictions. Winawer and Sherlock [42] detected 119 colon cancers beyond the reach of the sigmoidoscope with only one false negative; 85% of these tumors were localized to the bowel wall. Further investigation depends on the findings of the test for occult fecal blood, positive or negative. A schema I have already proposed [41] for subsequent investigation is shown in Figure 14–3. Air-contrast barium study of the colon or colonoscopy, or both, should be made in persons with occult fecal blood. Some authors [26, 31] recommend colonoscopy and proctosigmoidoscopy in asymptomatic, occult-blood-negative persons at high risk; however, these procedures may be of uncertain value when used at the recommended intervals (2–

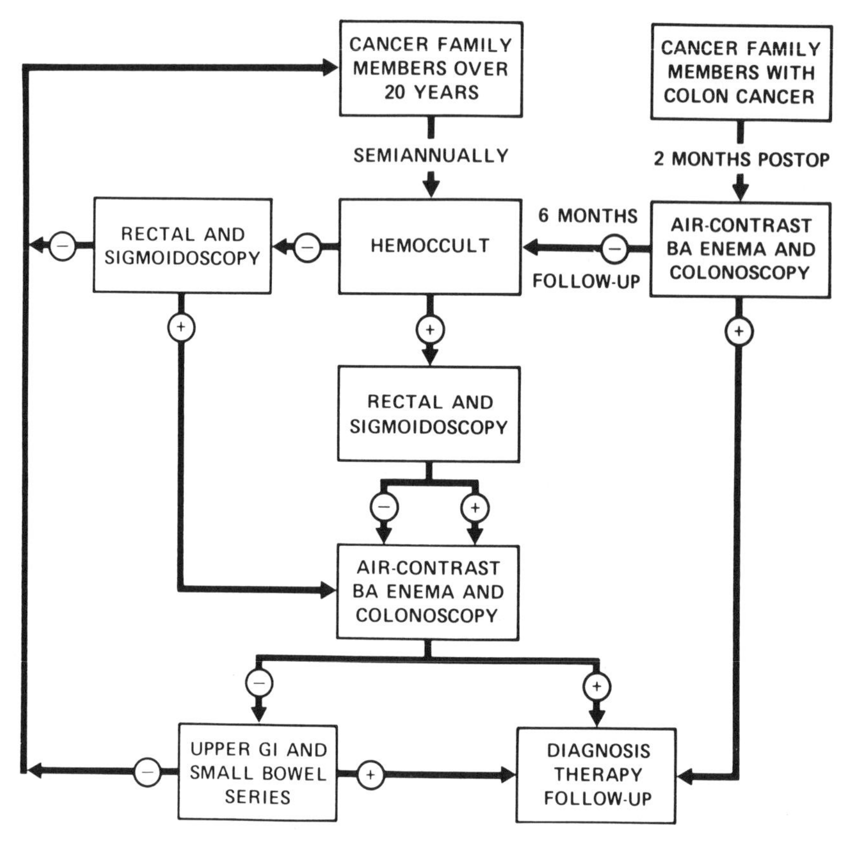

Figure 14–3 Schema for screening investigations for colon cancer that could be used in CFS members at high risk of developing these tumors (Williams [41]).

3 years), and they are instrusive on persons' life-styles. Also, because most colonic tumors in the CFS are in the proximal (right) colon, proctosigmoidoscopy is largely unhelpful. If the facilities are readily available, frequent screening colonoscopy (1–2 yearly) might be the method of choice in all CFS members, although critical assessment of the value of such screening is needed.

Annual pelvic examination and cervical smear is indicated in female family members, and endometrial biopsy or jet-washings have been recommended every 3 years, although at present there are no positive data to support their use. Women should be advised that any symptoms, especially abnormal vaginal bleeding, should be investigated immediately using fractional dilitation and curettage. Ovarian carcinoma is difficult to detect, and screening programs have not been tried, but yearly ultrasound or computed tomography (CT) scanning and pelvic examination may be useful for members of families that exhibit a high incidence of ovarian carcinoma [21]. Breast self-examination techniques should be taught to all female members in families with a high incidence of breast cancer, and routine breast examination should be arranged together with annual mammography.

Cancer surveillance in families genetically predisposed to cancer makes intuitive sense. However, it will be important in coming years to determine whether the members of such families actually do benefit and whether the benefit offsets the emotional upset caused by the entire surveillance process. It should also be remembered that screening for asymptomatic cancer is useful only when it is known that early diagnosis increases the chance of cure.

Conclusions

1. The cancer-family syndrome (CFS) poses many practical and ethical problems for clinicians and serves as a model for examining some of the clinical problems to be dealt with in other genetic syndromes that predispose families to cancer.

2. Surveillance of an affected family with a formal cancer-screening program for selected members is of unproven value, but it seems to make sense and at least allows the family members to feel that they are being assisted with their special problem. If surveillance is undertaken, this should be coordinated, preferably with groups testing the effectiveness of this approach.

3. Genetic markers for cancer susceptibility are badly needed to help identify members of a family actually at risk.

4. Prophylactic surgery so far has been little used in the CFS; it needs to be assessed critically.

5. One of the major obstacles to progress is the understandable emotional response of family members when learning about the segregation in their family of a gene that causes such a syndrome, specifically, that results in a cancer predisposition.

6. Clinicians should be more aware of the possibility of genetic syndromes that predispose to cancer, so that they can identify potential cancer patients early.

7. Furthermore, the clinical management of genetically cancer-prone persons who actually have a cancer can be adapted to take into account the person's continued risk, that is, of the appearance of additional primaries.

Literature cited

1. Arndt RD, Kositchetc RJ, Boasberg PD: Colon carcinoma and the cancer family syndrome. *JAMA* 237:2847–2848, 1977.
2. Berlinger NT, Lopez C, Lipkin M, Vogel JE, Good RA: Defective recognitive immunity in family aggregates of colon carcinoma. *J Clin Invest* 59:761–769, 1977.
3. Cannon MM, Leavell BS: Multiple cancer types in one family. *Cancer* 19:538–540, 1966.
4. Danes BS: Increased in vitro tetraploidy: Tissue specific within the heritable colorectal syndromes with polyposis coli. *Cancer* 41:2330–2334, 1978.
5. Dunstone GH, Knaggs TW: Familial cancer of the colon and rectum. *J Med Genet* 9:451–456, 1972.
6. Eyers AA, DeCosse JJ: Nutrition. *In* Gastrointestinal Disease (Kurtz RE, ed) New York, Churchill-Livingstone, 1981, 54–69.
7. German J: Bloom's syndrome 1. Genetical and clinical observations in the first twenty-seven patients. *Am J Hum Genet* 21:196–226, 1969.
8. Guirgis HA, Lynch HT, Harris RE, Vandevoorde JP: Carcinoembryonic antigen (CEA) in the cancer family syndrome. *Cancer* 42:1574–1578, 1978.
9. Guirgis HA, Lynch HT, Harris RE, Vandevoorde JP: Genetic and communicable effects on carcinoembryonic antigen expressivity in cancer family syndrome. *Cancer Res* 8:2523–2528, 1978.
10. Kluge T: Familian cancer of the colon. *Acta Chir Scand* 127:392–398, 1964.
11. Kopelovich L: Genetic predisposition to cancer in man: In vitro studies. *Int Rev Cytol* 77:63–88, 1982.
12. Kopelovich L: Identification of individuals at risk for cancer: Biological aspects and logistics. *In* Membranes and Tumour Growth (Galeotti et al., eds) Amsterdam, Elsevier Biomed Press, 1982, 301–312.
13. Kopelovich L, Lipkin M, Blattner WA, Fraumeni JF Jr, Lynch HT, Pollack RE: Organization of actin-containing cables in cultured skin fibroblasts from individuals at high risk of colon cancer. *Int J Cancer* 26:301–306, 1980.
14. Krush AJ: Social work roles in research studies of families having hereditary cancer and pre-cancer diagnoses. *Soc Work Health Care* 7(2):39–48, 1982.

15. Krush AJ, Lynch HT, Magnuson C: Attitudes towards cancer in a 'cancer family': Implications for detection. *Am J Med Sci* 249:432–438, 1965.
16. Law IP, Herberman RB, Oldham RK, Bouzoukis J, Hanson SM, Rhode MC: Familial occurrence of colon and uterine carcinoma and of lymphoproliferative malignancies; clinical description. *Cancer* 39:1224–1228, 1977.
17. Lillehei RC, Wangensteen OH: Bowel function after colectomy for cancer, polyps and diverticulitis. *JAMA* 159:163–166, 1955.
18. Li FP, Fraumeni JF: Prospective study of a family cancer syndrome. *JAMA* 247:2692–2696, 1982.
19. Lynch HT, Krush AJ: Heredity and adenocarcinoma of the colon. *Gastroenterology* 53:517–527, 1967.
20. Lynch HT, Krush AJ: The cancer family syndrome and cancer control. *Surg Gynecol Obstet* 132:247–251, 1971.
21. Lynch HT, Lynch PM: Tumor variation in the cancer family syndrome: Ovarian Cancer. *Am J Surg* 138:439–442, 1979.
22. Lynch PM, Lynch HT, Harris RE: Hereditary proximal colonic cancer. *Dis Colon Rectum* 20:661–668, 1977.
23. Lynch HT, Lynch PM, Harris RE: Minimal genetic findings and their cancer control implications: A family with the cancer family syndrome. *JAMA* 240:535–538, 1978.
24. Lynch HT, Guirgis HA, Harris RE: Clinical, genetic, and biostatistical progress in the cancer family syndrome. *Front Gastrintest Res* 4:142–150, 1979.
25. Lynch HT, Krush AJ, Guirgis H: Genetic factors in families with combined gastrointestinal and breast cancer. *Am J Gastroenterol* 59:31–40, 1973.
26. Lynch HT, Lynch PM, Albano WA, Lynch JF: The Cancer syndrome: A status report. *Dis Colon Rectum* 24:311–322, 1981.
27. Lynch HT, Lynch PM, Fusaro R, Pester J: The cancer family syndrome: Rare cutaneous phenotypic linkage of Torre's syndrome. *Arch Intern Med* 141:607–611, 1981.
28. Lynch HT, Bardawil WA, Harris RE, Lynch PM, Guirgis HA, Lynch JF: Multiple primary cancers and prolonged survival: Familial colonic and endometrial cancers. *Dis Colon Rectum* 21:165–168, 1978.
29. Lynch HT, Shaw MW, Magnuson CW, Larsen AL, Krush AJ: Hereditary factors in cancer: Study of two large midwestern kindreds. *Arch Intern Med* 117:206–212, 1966.
30. Lynch HT, Harris RE, Lynch PM, Guirgis H, Lynch JF, Bardawil, WA: Role of heredity in multiple primary cancer. *Cancer* 40:1849–1854, 1977.
31. Lynch HT, Harris RE, Organ CH, Guirgis HA, Lynch PM, Lynch T, Nelson EJ: The surgeon, genetics and cancer control. *Ann Surg* 185:435–440, 1977.
32. Lynch HT, Thomas RJ, Terasaki PI, Ting A, Guirgis HA, Kaplan AR, Magee H, Lynch J, Kraft C, Chaperon E: HL-A in cancer family "N." *Cancer* 36:1315–1320, 1975.
33. Metzmaker CO, Sheehan P: Report of a family with cancer family syndrome. *Dis Colon Rectum* 24:523–525, 1981.
34. Moertel CG, Bargen JA, Dockerty MB: Multiple carcinomas of the large intestine. *Gasteroenterology* 34:85–89, 1958.
35. Peltokallio P, Peltokallio V: Relationship of familial factors to carcinoma of the colon. *Dis Colon Rectum* 9:367–370, 1966.
36. Ruma TA, Lynch HT, Albano WA, Sterioff S: Total colectomy and the cancer family syndrome. *Dis Colon Rectum* 25:582–585, 1982.
37. Savage D: A family history of uterine and gastro-intestinal cancer. *Br Med J* 2:341–343, 1956.
38. Tobacman JE, Tucker MA, Kase R, Greene MH, Costa J, Fraumeni JF Jr: Intra abdominal carcinomatosis after prophylactic oophorectomy in ovarian cancer-prone families. *Lancet* ii:795–797, 1982.

39. Warthin AS: Heredity with reference to carcinoma: As shown by the study of the cases examined in the pathological laboratory of the University of Michigan, 1895–1913. *Arch Intern Med* 12:546–555, 1913.
40. Whelan E: Preventing cancer: What You Can Do to Cut Your Risks by Up to 50 Per Cent. London, Sphere Books, 1980.
41. Williams C: Management of malignancy in "cancer families." *Lancet* i:198–199, 1978.
42. Winawer SJ, Sherlock P: Screening for colon cancer. *Gastroenterology* 70:783–790, 1976.

15.

Familial cancer: Implications for healthy relatives

DAVID E. ANDERSON WICK R. WILLIAMS

A family history of breast cancer or colon cancer has long been considered a major risk determinant. On the average, with a parent or sib affected by one of these diseases, an individual's risk for developing these neoplasms is double or triple the risk of individuals from the general population. When these risks are equated to lifetime probabilities, a woman with a family history of breast cancer may have a 15 or 20% probability of developing the disease by age 72; a person with a family history of colon cancer may have a probability of 10 to 15% of developing the same neoplasm. Risks of these magnitudes or the presence of one of these neoplasms in a close relative, such as a parent or sib, or in two such relatives may influence decisions concerning treatment.

Evidence becoming available indicates, however, that not all individuals with a family history of breast or colon cancer are at high risk. We have found that some family histories are associated with very high risks, in the range of 50%, whereas others are little or no different from the 5% risk for colon cancer or 7% risk for breast cancer applicable to the general population. The majority of risk determinations presently available for genetic counseling for breast and colon cancer are *empiric;* that is, they reflect the prevalence of the disease in the sibs of a series of patients in relation to its prevalence in a series of control sibs or some other type of control. In some cases, however, the estimation of an *analytic* recurrence risk will be preferable, because it is a prediction of the probability of affection based on that person's own family history and an underlying genetic model.

Since the determination of risk has important relevance to medical decisions leading to cancer control through early detection or primary or secondary prevention, it is vital that accurate risks be used in counseling. The first part of this chapter will review some of the risks determined from an ongoing family study of breast cancer, the familial and clinical criteria associated with high or low risks for breast as well as colon cancer, inherited types of breast and colon cancer, and finally empiric risks applicable to frequently encountered pedigree situations.

In the second part we will describe our proposed approach to deriving analytic recurrence risks for breast and colon cancer, an approach that ultimately could provide more accurate risks on which to base medical decisions than the empiric risks presently available.

Studies that have provided risk figures for unaffected relatives in families with breast and colon cancer

Familial breast cancer

The material on which much of the present discussion on breast cancer is based has been derived from a pedigree study that was initiated in 1969. A basic assumption of this study was that breast cancer, rather than being a homogeneous disease as was assumed in previous familial studies, is a heterogeneous disease. This assumption was precipitated by epidemiologic data that pointed to differences between patients who developed their disease in the premenopausal period and those who developed it in the postmenopausal period [11], and the long-standing evidence of different pathologic and clinical types of the disease, such as lobular, Paget's, inflammatory, and intraductal, each with its characteristic natural history [13].

Whereas earlier studies had utilized a consecutive series of breast cancer patients, our study was based only on patients with histologic diagnoses of breast cancer who also had family histories of the same disease in a first- or second-degree relative [1, 3, 6, 7]. This selection was made in an attempt to reduce the heterogeneity among patients, but at the same time it was hoped that such selection would also enhance the determinations of any genetic effect [28].

The controls in our study also differed from those used in previous studies in that they were selected from the same patient population and for the same reason as the breast cancer patients. The control patients had histologic diagnoses of cancer other than of the breast and were required to have family histories of the other cancer. Because both the breast cancer and control patients were selected from the same source and for the same reason, they should have been subjected to the same selection biases and admission criteria and should have had access to the same diagnostic and treatment facilities. Furthermore, both groups of patients and their families were studied in the same manner by the same personnel using the same verification procedures, and, at least initially, during the same time interval; therefore, the two groups of patients and their families should have been as comparable as possible.

Age of onset and multiplicity of neoplasms determine the risk

Earlier pedigree studies of certain other cancers had indicated that the familial and inherited forms of neoplasms occurred or were diagnosed at a significantly earlier average patient age and had a higher frequency of multiple primaries than the noninherited forms of these same cancers [17]. These two features also characterize inbred mouse strains that are highly susceptible to a given tumor. The occurrence of the tumor at an early age and tumor multiplicity thus appeared to have biologic meaning. (The significance of these two characteristics has been clarified by a two-step mutation model advanced by Knudson [16] to explain the occurrence of inherited and noninherited cancers.)

The occurrence of inherited forms at an early age plus the epidemiologic findings of different risk factors for premenopausal and postmenopausal patients led to a classification of patients with a family history of breast cancer into two age classes (shown in Table 15–1): one for patients with diagnoses occurring between 20 and 49 years of age, referring primarily to the premenopausal period, and one for patients with diagnoses occurring after age 49, the postmenopausal period. The only criterion available to determine tumor multiplicity was the occurrence of primary disease in both breasts. So, in addition to being classified according to age at diagnosis, the patients were classified according to whether they had unilateral or bilateral disease. (The index

Table 15–1 Risk of breast cancer in first-degree relatives of familial breast cancer patients classified according to age at diagnosis and laterality of disease compared with controls classified according to age at diagnosis.*

Breast cancer patient			
Age at diagnosis	Laterality	Control patients age at diagnosis	Risk (± standard error) of breast cancer occurrence in relatives of patients versus relatives of controls
20–49	NC†	20–49	3.1 ± 1.8
50+	NC	50+	1.5 ± 1.4
NC†	Bilateral	NC	5.4 ± 1.7‡
NC	Unilateral	NC	1.3 ± 1.4
20–49	Bilateral	20–49	8.8 ± 2.0‡
50+	Unilateral	50+	1.2 ± 1.5
20–49	Unilateral	20–49	1.8 ± 1.9
50+	Bilateral	50+	4.0 ± 1.7‡

*Any affected relative who led to the ascertainment of a family, as well as the patient herself, was excluded from all calculations of risks.

†NC = not classified.

‡$P < 0.01$.

patients, through whom the families were ascertained, were used only for classification purposes and were excluded from all calculations of risk.)

This classification of patients served to identify groups of their relatives at varying risks for the disease [1]. As shown in Table 15–1, the sisters, daughters, and mothers of the primary breast cancer patients whose disease developed in the 20- to 49-year age interval had a threefold higher risk than the same relatives of control patients whose cancer had also developed in the 20- to 49-year age interval. The breast cancer risk in the first-degree relatives of patients whose disease developed after age 49 only manifested a 1.5-fold higher risk than the control relatives of control patients whose disease also developed after age 49 years. The risk for relatives of patients with bilateral disease was 5.4-fold higher than that of all relatives of control patients, whereas the risk for relatives of patients with unilateral disease was only 1.3-fold higher. Relatives of patients whose disease developed to the 20- to 49-year age interval and was bilateral exhibited an 8.8-fold higher risk than the relatives of 20- to 49-year-old control patients. The risk for relatives of patients whose disease developed after age 49 and was unilateral was 1.2-fold higher than that of relatives of control patients. Incidentally, the risks to relatives when no classification was made for age at diagnosis or laterality of disease in the patient were the same as those applying to relatives of patients with no family histories of the disease. These results show that an early age at diagnosis and tumor multiplicity in a patient may have genetic significance and should be considered in any assessment of risk in the patients' first-degree relatives.

Variability in risk was also observed when the primary breast cancer patients were classified according to the pattern of breast cancer in their families [7]; that is, the patient had an affected mother (mother pedigrees), an affected sister (sister pedigrees), or an affected second-degree relative (second-degree pedigrees). These affected relatives were responsible for the primary breast cancer patient being selected for study; henceforth, they will be referred to as secondary patients. The secondary patients, as well as primary patients, were used for classification purposes only and were not included in any calculations of lifetime probabilities.

The three pedigree groups resulting from this classification exhibited marked differences in risk [3, 6, 7]. The lifetime probability for the sisters of patients with affected mothers was 32%, 11% for those with affected sisters, and 3% for those with affected second-degree relatives. The controls had a probability of developing breast cancer of 5.2%, not far removed from the 7% probability reported for women in general. The daughters of the patients had probabilities similar to those observed for the sisters [3]. In the pedigrees with affected mothers, the patients' daughters had a lifetime probability of 27%, whereas the daughters of patients with affected sisters had a 14% probability, and daughters of patients with affected second-degree relatives had a 19% probability. The latter group would be expected, on genetic grounds, to have a lower probability than the sister-pedigree group. This discrepancy was probably the result of the classification procedure and the fact that the pedigree groups were not mutually exclusive.

The classification procedure was modified so that the secondary patients as well as the primary patients were classified according to their age at diagnosis and disease laterality [4]. (Again, neither the primary nor secondary patients were involved in any calculations of risk.) Table 15–2 indicates that premenopausal disease in both the primary and secondary patient and bilateral disease in one or the other relative increases the lifetime risk for sisters of the primary patient to 50% in a pedigree with affected mothers and affected sisters. With postmenopausal disease in either patients or affected relatives and no evidence of bilateral disease, the risks for the remaining sisters of patients with affected sisters were little different from the expected 7% in the general population and only slightly higher (10–16%) in pedigrees with affected mothers.

Rare dominantly inherited clinical entities that feature breast cancer

The 50% probabilities for women in pedigrees characterized by premenopausal and bilateral disease point to a dominantly inherited basis for some cases of breast cancer. However, not one but at least three and possibly four different, dominantly inherited clinical entities have

Table 15–2 Lifetime probabilities (± standard errors) of breast cancer developing in sisters of primary patients when the primary and secondary patients were classified according to menopausal status and laterality of disease.*

Age at diagnosis in patients		Laterality of disease			
		Unilateral		Bilateral	
Primary	Secondary	Mother Pedigrees	Sister pedigrees	Mother pedigrees	Sister pedigrees
Premenopausal	Premenopausal	0.33 ± 0.07	0.18 ± 0.07	0.51 ± 0.25	0.50 ± 0.22
Premenopausal	Postmenopausal	0.10 ± 0.04	0.08 ± 0.03	0.23 ± 0.09	0.11 ± 0.07
Postmenopausal	Premenopausal				
Postmenopausal	Postmenopausal	0.16 ± 0.06	0.05 ± 0.03	0.28 ± 0.17	—

*Classified bilateral if either proband or other family member had bilateral disease. The primary and secondary patients were used only for classification purposes and were excluded from all calculations of lifetime probabilities.

now been described with these characteristics. One of these was described by Li and Fraumeni, first in 1969, then in 1975, and most recently in 1982 [19]. This entity, now called the Li-Fraumeni syndrome, is characterized principally by very early onset of breast cancer, the ages ranging from 25 to 35 years, and by a high rate of bilaterality. The distinguishing feature is the associated occurrence of breast cancer, brain tumor, soft tissue sarcoma, and leukemia. Lifetime risk of tumor development in the children of affected parents belonging to such families has been calculated at 46% [22], which is consistent with a dominant inheritance pattern.

Another inherited type of breast cancer characterized by early onset and bilaterality involves a different associated neoplasm, namely, ovarian cancer. Fraumeni et al. [12] and Lynch et al. [21] have described this association. The gene for this entity was reported by King et al. [15] to be dominantly inherited and to be closely linked to the gene for glutamate pyruvate transaminase (GPT), which at that time had been provisionally mapped to chromosome 10 (but now to 16). They estimated a penetrance value of 87% before age 80 years. The entity involving breast cancer, sarcoma, brain tumor, and leukemia was linked neither to the GPT gene locus nor to any other genetic marker locus. Evidence of close linkage between the gene for such an entity and one for a genetic marker would have important utility in identifying a high-risk individual in a given family prior to the onset of a tumor.

Still another inherited, premenopausal, bilateral, and perhaps most frequent type seems to involve only breast cancer, since no consistent associated neoplasms have been observed. This site-specific type has been less well studied than the others, but it appears to be inherited in a dominant pattern and at a reduced penetrance level, so that lifetime risk would be approximately 30 to 35% [7].

Cowden's disease is another dominantly inherited form of breast cancer. It is characterized by hamartomatous lesions of the skin and oral cavity that occur in association with thyroid and breast tumors. More than 90% of female gene carriers develop breast tumors, at an average of 40 years of age. Among those who develop breast cancer, over 40% develop it bilaterally. Also, other inherited forms developing in patients at later ages and pointing further to the heterogeneity of familial breast cancer have been described [4, 5].

It is clear from the foregoing that the risks for breast cancer are highly variable and can be influenced by factors such as the average age at which the disease tends to occur in a family, the tendency of the disease toward multiplicity, the type of family history, and the types of associated neoplasms. These associated neoplasms can occur as multiple primaries in a breast cancer patient and as single primaries in other family members. Because of these features, identical-appearing pedigrees may be associated with highly dissimilar risks; thus, in a pedigree with an unaffected mother and father and breast cancer in two of their four daughters, the risks for their unaffected daughters may range from 7%, to 18%, to 50%, depending on whether the disease was postmenopausal and unilateral in both affected sisters, premenopausal and unilateral in both, or premenopausal and bilateral in both. One important conclusion from the study is that not all women with a family history of breast cancer are at an increased risk for the disease.

Familial colon cancer

Colon cancer is similar to breast cancer in many respects. Patients with this neoplasm and a family history of it have an earlier average age at first diagnosis than colon cancer patients in general. Lovett [20] observed that the percentage of patients with family histories of the disease was highest for those with diagnoses at 40 years of age and younger. In another study [9], patients with family histories of the neoplasm averaged 46.8 years of age at diagnosis, whereas the average age of an unselected series of patients was 60.3 years, a highly significant difference.

Another characteristic of familial colon cancer is the development of multiple primary neoplasms. This is a long-standing observation in patients with familial polyposis coli, Gardner's syndrome, or the other inherited forms of polypoid disease of the gastrointestinal tract, all of which are known to predispose to colon cancer. In addition, however, the occurrence of multiple primaries also characterizes the familial forms of colon cancer that develop independently of any evidence of multiple polyposis [10].

The familial form of colon cancer has another distinguishing feature in that it originates in the cecum and ascending and transverse colon or the right colon much more frequently than in the descending colon and rectosigmoid areas [9]. This is contrary to familial polyposis coli or colon cancer in general, in which at least two thirds of the neoplasms originate in the left colon. In patients with familial disease, only one third develop the disease in this area, and two thirds develop it in the right colon. This difference, in addition to its genetic significance, has clinical implications: examinations for familial colon cancer should utilize methods that reveal the entire colon and not merely the distal portion.

Empiric risk figures

Unlike that of breast cancer, the risk of nonpolypoid colon cancer developing in the first-degree relatives of patients with familial disease has not been estimated according to a classification based on the ages at diagnosis, tumor multiplicity, or site of origin in the colon. Available risks apply only to relatives of a consecutive series of patients. Lovett [20] has recently conducted a study that excluded patients with multiple polyposis. Twenty-six percent of the patients in her series had family histories of the malignancy, and this frequency was highest for those 40 years of age and younger. Of 352 deaths among the first-degree relatives of all patients, 41 were from colon cancer, whereas only 11.7 were expected, yielding a risk for first-degree relatives of 3.5. Similar risks were reported by previous workers [5, 9, 10].

If colon cancer is at all similar to breast cancer—and it certainly seems to be in some respects—then individuals at high risk for developing colon cancer might be identified among the first degree relatives of patients not only with a family history of the disease but also with early occurring disease, multiple primaries, or disease of the right side of the colon. Patients with these features have a high probability of having a genetic basis for their disease.

Rare, dominantly inherited clinical entities that feature nonpolypoid colon cancer

Evidence supporting the proposal that individuals at high risk of developing colon cancer will be identified among the relatives of patients with early, multiple, or right-side disease comes from the different inherited types of large bowel cancer that have been identified to date. Probably the most publicized of these is the so-called cancer-family syndrome [24]. The term *hereditary adenocarcinomatosis* has also been proposed for this disorder, to emphasize its hereditary potential and the occurrence of adenocarcinomas at multiple sites, primarily of the colon and uterus, but also in the stomach, ovaries, and possibly the breast [8]. In addition to their multiplicity, the neoplasms in this disorder are generally first detected in patients about 45 years, and those of the colon are more frequently on the right than on the left side. The disorder is inherited in a dominant fashion and penetrance is about 75% [23]. As such, the risk that a member from such a family will develop a colon or uterine neoplasm or both is about 40%.

In another type of inherited large bowel cancer, referred to as hereditary gastrocolonic cancer, colon and stomach cancer occur either as double primaries or as a combination of single primaries among relatives [5, 9]. This type may be considered a variant or incomplete expression of hereditary adenocarcinomatosis, because the two types have overlapping clinical features, both follow a dominant inheritance pattern, and both are characterized by a right-sided distribution of colon cancers. The types differ, however, in the average age at diagnosis—about 40 years for gastrocolonic cancer and 45 years for adenocarcinomatosis. They differ also in their penetrance values—90% versus 75%, respectively. The high penetrance value indicates that the risk to a first-degree relative from a family with gastrocolonic cancer developing one or both of the neoplasms is about 45%.

Another inherited disorder with a more limited range of phenotypic expressivity is "hereditary colon cancer," in which cancer of the large bowel is the only malignancy that occurs at a high frequency in an affected family. This type is also characterized by a right-sided distribution, an average patient age at diagnosis of approximately 40 years, and a penetrance value of about 85% [5, 10] or a risk of occurrence of about 40% to first-degree relatives of an affected person.

Torre-Muir syndrome is another inherited type of large bowel cancer, and perhaps the least recognized [2]. This disorder differs both quantitatively and qualitatively from the other types. It can be identified by diagnosis at an average patient age of 50 years and the occurrence of skin tumors, multiple adenocarcinomas of the large and small intestine and uterus, squamous cell carcinomas of the mucous membranes, transitional cell carcinoma of the urinary system, sporadic polyps of the intestinal tract and bladder, diverticulosis, and, surprisingly, a low degree of malignancy as reflected by a relatively long survival following the occurrence of an internal malignancy. This syndrome follows a dominant inheritance pattern with a high penetrance value of about 80 to 85% [2, 5]. The risk of a first-degree relative of an affected member developing this disorder is about 50%.

In summary, the risks associated with a family history of colon cancer appear to vary. The familial form of the disease has earlier onset than colon cancer in general, tends to develop in multiple sites, develops more frequently on the right than on the left side of the colon, and may involve associated neoplasms. Furthermore, familial colon cancer may be a component of an inherited disorder. So, as with breast cancer, identical-appearing pedigree diagrams may be associated with dissimilar risks. The specific risks applying to a given individual will require careful analysis of both the family and clinical data.

Genetic epidemiologic methods for analytic risk in families affected with cancer

As stated earlier, an individual's risk for developing breast or colon cancer may be the basis of an important medical decision; that is, the risk could determine the type of individualized surveillance program he or she will follow, whether or not the individual will undergo biopsy, or, if the risk is deemed to be very high, whether or not the individual should undergo some form of prophylactic surgery. It is important, therefore, that the risks on which these kinds of decisions are based be realistic and accurate.

In practice, however, the risks for a given situation may not always be available, for one of the following reasons: (1) no empiric risk has as yet been estimated, or if available, the risk might be too nonspecific or based on so few individuals as to be unreliable or (2) the family and medical history are not at all suggestive of a known clinicogenetic disorder with dominant inheritance and high penetrance. Fortunately, methods are now becoming available by which analytic risks can be estimated for any given pedigree situation. The pedigree provides the framework on which to base genetic inference. A variety of models can then be fitted to pedigree data, hypotheses and goodness-of-fit tests can be performed, and when possible, a parsimonious solution can be obtained from which an analytic risk can be calculated.

Segregation analysis of some breast cancer pedigrees

To demonstrate this approach, complex segregation analysis was performed on 200 randomly selected breast cancer pedigrees of Danish origin [31]. The pedigree and a description of each and the manner in which each was ascertained can be found in Jacobsen's monograph entitled, "Heredity in Breast Cancer" [14]. (Pedigrees on colon cancer could have been used for the same purpose had a large series of such pedigrees been readily available.) Segregation analysis was performed under the "mixed model" of Morton and MacLean [25] as modified by Lalouel and Morton [18]. Briefly, the mixed model of inheritance considers an individual's phenotype to be the result of a major gene effect, a multifactorial transmissible component, and a residual environmental contribution, each acting independently. The major locus is biallelic, yielding three major genotype classes. The gene frequency of the allele associated with having breast cancer (termed G') is represented by q, and $1 - q$ represents the sum of the frequencies of all other alleles (G) at that locus. The distance between the two homozygous genotype class means is t and is referred to as the displacement. The degree of dominance at the major locus is d, which reflects the position of the heterozygous genotype class in relation to the means of the two homozygous genotype classes. The range of d is from zero to 1, the two limits representing complete recessivity and complete dominance, respectively. Multifactorial inheritance is represented by h, the heritability, which is defined as the fraction of the total phenotypic variance attributable to a large number of genetic and environmental influences that act in an additive fashion.

Table 15–3 Liability indicator for segregation analysis.

Liability class	Sex	Age	Cumulative incidence*
1	F	0–30	0.00013
2	F	30–40	0.00162
3	F	40–50	0.00810
4	F	50–60	0.01951
5	F	60–70	0.03417
6	F	70–80	0.05441
7	F	80+	0.06622†
8	M	0–50	0.00002
9	M	50–70	0.0002
10	M	70+	0.00051†

*To midpoint of interval.

†Cumulative incidence up to age 80 years.

For a dichotomous phenotype (affected versus normal), a correction for differing risk of affection according to age and sex was performed by assignment of individuals to liability classes by sex and age at last observation. The liability values (Table 15–3) and were calculated from Danish age- and sex-specific incidence rates [29]. The complete details of this analysis are reported elsewhere [31]. Segregation analysis was carried out on the 200 pedigrees using the program POINTER written by Lalouel and Yee [26]. Likelihood-ratio tests provide a means of discriminating among possible modes of inheritance for a trait. The difference between -2 ln likelihood for the full model and -2 ln likelihood for the model with q set to zero provides a test of whether a major gene effect is significant; the difference with -2 ln likelihood for the model with h set to zero provides a test of whether a multifactorial contribution is significant. Results of segregation analysis are presented in Table 15–4.

Table 15–4 Results of segregation analysis.

	Parameters*				
Hypothesis	d	t	q	h	−2 ln L
Multifactorial, $q = 0$				0.27	−928.53
Recessive, $d = 0, h = 0$	(0)	1.98	0.124		−929.73
Dominant, $d = 1, h = 0$	(1)	1.68	0.0067		−938.86

*Key to symbols: q = gene frequency, d = degree of dominance, h = heritability, ln L = natural log likelihood, t = displacement between the two homozygous genotype class means.

Table 15-5 Characteristics of the major locus at each liability class.

Sex	Liability class	Age	P(Affection/Genotype)*			P(Genotype/Affection)†		
			GG	$G'G$	$G'G'$	GG	$G'G$	$G'G'$
F	1	0–30	0.0	0.01		0.15	0.85	
F	2	30–40	0.0	0.07		0.43	0.57	
F	3	40–50	0.0	0.20		0.66	0.34	
F	4	50–60	0.02	0.33		0.78	0.22	
F	5	60–70	0.03	0.43		0.83	0.17	
F	6	70–80	0.05	0.52		0.87	0.13	
F	7	80+	0.06	0.56		0.89	0.11	
M	8	0–50	0.0	0.00		0.06	0.94	
M	9	50–70	0.0	0.01		0.18	0.82	
M	10	70+	0.0	0.03		0.27	0.73	

*P(Affection/$G'G'$) ≈ 0 for all classes, due to the rarity of the G' gene.

†P ($G'G'$/Affection) is the same as given for the $G'G$ class because the gene is dominant.

For the general model (i.e., fitting both a multifactorial component and a generalized major locus simultaneously), the multifactorial contribution went to zero, signifying that a major locus alone could account for the observed segregation pattern of breast cancer in these pedigrees. The major gene segregated as a dominant. A recessive major gene demonstrated a significantly poorer fit, as did a simple multifactorial model.

Characteristics of the general solution (major locus) are given in Table 15–5. For both sexes, penetrance of the abnormal gene (G') increased with the carrier's age. By age 80, 56% of female gene carriers ultimately will become affected. This contrasts markedly with the situation for males, in which only 3% of the carriers will ultimately become affected.

The proportion of phenocopies increased with age, and, by age 80 years, only 11% of those affected were carriers of the major gene; the remaining 89% affected had a normal genotype. The proportion of affected women who carried the abnormal gene was higher at premenopausal than at postmenopausal ages. This result is in agreement with previous studies [1, 16, 17]. There were only six affected males in the present study; although the results in males are similar to those seen in females, a larger sample would be necessary to confirm these findings.

Analytic risk analysis in two examples of pedigrees with breast cancer

Analytic recurrence risks were calculated for two extreme pedigree examples (pedigrees 1 and 200) from Jacobsen's survey [14]. The pedigrees are presented in Figure 15–1. Genetic risks were derived under the dominant major locus model obtained from segregation analysis. A description of the probability algebra involved may be found in Williams [30]. The risks were computed using the computer program COUNSEL [26, 30]. Pedigree I exhibits a high frequency of affection, with both the mother and sister of the primary proband having breast cancer. The probability that the proband's 41-year-old daughter, IV-1, would ultimately develop breast cancer was estimated at 21%. This estimate is based both on her family history and on the fact that she

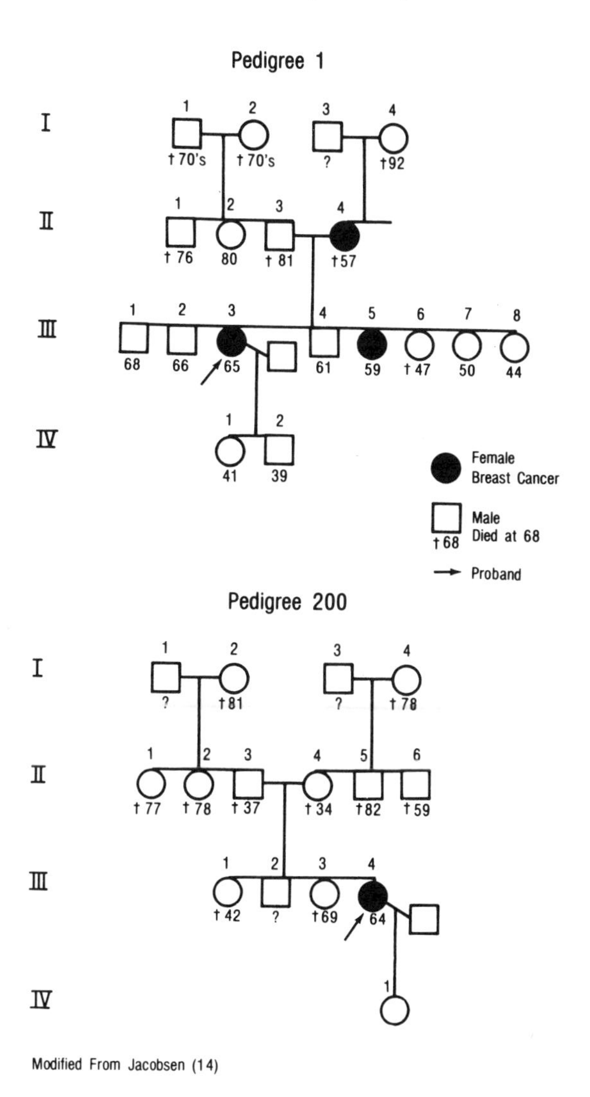

Figure 15–1 Two pedigrees from Jacobsen [14] used in estimating analytic recurrence risks.

does not have breast cancer at age 41. If her age were not considered, the lifetime probability would have been 28%. These estimates are close to those found by different methods used for selected data by Anderson [3, 6]; in these data the risk for a 41-year-old woman with an affected mother and aunt was estimated at 23%, and her lifetime probability of developing breast cancer was estimated at 27%. In the second example, pedigree 200, the only affected person is the proband. For this example, the probability that a daughter of the proband, IV-1, would ultimately develop the disease is 8%, only slightly above the lifetime probability of 7% for women in the general population. The high frequency of affection in pedigree 1 suggests the segregation of an abnormal gene in this family; consequently, the probability that individual IV-1 will ultimately be affected is high. In contrast, the negative family history presented by pedigree 200 suggests that the proband is not a carrier of an abnormal gene; therefore, the risk that her daughters will be affected is similar to that applying to women in the general population.

As a final caveat, it should be mentioned that analytic recurrence risks are not necessarily error free. Aside from mistakes in calculation, misspecification of the genetic model will result in incorrect probability calculations. To minimize this possibility, proper segregation analysis should include tests of heterogeneity whereby the data are partitioned according to mating type, mode of ascertainment, and so on. Partition by objective clinical-epidemiologic features of the disease would also provide better discrimination between genetic and nongenetic cases of disease. Furthermore, the aforementioned analytic risks are population specific; they apply only to the Danish data. Still required are segregation analyses of pedigree data collected in a complete and unbiased manner on other samples of patients, to broaden the applicability of analytic recurrence risks. Until such time, the empiric risks and lifetime probabilities reviewed in this report and elsewhere [27] will have to be used to identify individuals at high or low risk. This should help in providing counseling concerning cancer detection and control measures and in reassuring individuals who may hold exaggerated beliefs about their cancer risks.

Conclusions

In this chapter, we have reviewed empiric risk figures presently available for breast and colon cancer. We have also outlined our proposed approach to generate analytic risks of recurrence for these two diseases.

Acknowledgment

This work was supported in part by National Institutes of Health grant GM 19513 and training grant CA 09299.

Literature cited

1. Anderson DE: A genetic study of human breast cancer. *J Natl Cancer Inst* 48:1029–1034, 1972.
2. Anderson DE: An inherited form of large bowel cancer: Muir's syndrome. *Cancer* 45:1103–1107, 1980.
3. Anderson DE: Breast cancer in families. *Cancer* 40:1855–1860, 1977.
4. Anderson DE: Die familiare und genetische pradisposition bei erkrankungen der brust. *In* Die Erkrankungen der Weiblichen Brustdruse: Epidemiologie, Endokrinologie, Histopathologie, Diagnostik, Therapie, Nachsorge, Psychologie (Frischbier H, ed) Stutgart, Georg Thieme Verlag, 1982, 1–6.
5. Anderson DE: Familial cancer and cancer families. *Semin Oncol* 5:11–16, 1978.
6. Anderson DE: Genetic predisposition to breast cancer. *In* Recent Results in Cancer Research (St Arneault G, Band P, Israel L, eds) vol 57, Berlin, Springer-Verlag, 1976, 10–20.
7. Anderson DE: Genetic study of breast cancer: Identification of a high risk group. *Cancer* 34:1090–1097, 1974.
8. Anderson DE: Genetic varieties of neoplasia. *In* Genetic Concepts and Neoplasia, Baltimore, Williams & Wilkins, 1970, 85–104.
9. Anderson DE: Risk in families of patients with colon cancer. *In* Colorectal Cancer: Prevention, Epidemiology, and Screening (Winawer S, Schottenfeld D, Sherlock P, eds) New York, Raven Press, 1980, 109–115.
10. Anderson, DE, Strong LC: Genetics of gastrointestinal tumors. *In* Proceedings of XIth International Cancer Congress, Amsterdam, Excerpta Medica, Series No 351, 1974, 267–271.
11. DeWaard F, Baanders-van Halewijn EA, Huizinga J: The bimodal age distribution of patients with mammary carcinoma: Evidence for the existence of two types of human breast cancer. *Cancer* 17:141–151, 1964.
12. Fraumeni JF Jr, Grundy GW, Creagan ET, Everson RB: Six families prone to ovarian cancer. *Cancer* 36:361–369, 1975.
13. Haagensen CD: Diseases of the Breast, 2nd ed, Philadelphia, WB Saunders Co, 1971.
14. Jacobsen O: Heredity in Breast Cancer, London, HK Lewis & Co, 1964.
15. King M-C, Go RCP, Elston RC, Lynch HT, Petrakis NL: Allele increasing susceptibility to human breast cancer may be linked to the glutamate-pyruvate transaminase locus. *Science* 208:406–408, 1980.
16. Knudson AG Jr: Mutation and cancer: Statistical study of retinoblastoma. *Proc Natl Acad Sci USA* 68:820–823, 1971.
17. Knudson AG Jr, Strong LC, Anderson DE: Heredity and cancer in man. *In* Progress in Medical Genetics (Steinberg AG, Bearn AG, eds) vol 9, New York, Grune & Stratton, 1973, 113–158.
18. Lalouel JM, Morton NE: Complex segregation analysis with pointers. *Hum Hered* 31:312–321, 1981.
19. Li FP, Fraumeni JF Jr: Prospective study of a family cancer syndrome. *JAMA* 247:2692–2694, 1982.
20. Lovett E: Family studies in cancer of the colon and rectum. *Br J Surg* 63:13–18, 1976.

21. Lynch HT, Guirgis HA, Albert S, Brennan M, Lynch J, Kraft C, Pocekay D, Vaughns C, Kaplan A: Familial association of carcinoma of the breast and ovary. *Surg Gynecol Obstet* 138:717–724, 1974.
22. Lynch HT, Harris RE, Organ CH Jr, Lynch JF: Management of familial breast cancer. *Arch Surg* 113:1053–1058, 1978.
23. Lynch HT, Kaplan AR: Cancer concordance and the hypothesis of autosomal dominant transmission of cancer diathesis in a remarkable kindred. *Oncology* 30:210–216, 1974.
24. Lynch HT, Krush AF, Thomas RJ, et al: Cancer family syndrome. *In* Cancer Genetics (Lynch HT, ed) Springfield, IL, Charles C Thomas, 1976, 355–388.
25. Morton NE, MacLean CJ: Analysis of family resemblance III, Complex segregation of quantitative traits. *Am J Hum Genet* 26:489–503, 1974.
26. Morton NE, Rao DC, Lalouel JM: Methods in Genetic Epidemiology, New York, S Karger, 1983, 1–26.
27. Ottman R, King M-C, Pike MC, Henderson BE: Practical guide for estimating risk for familial breast cancer. *Lancet* 2:556–558, 1983.
28. Schull WJ: Genetics of man: Some of the developments of the last decade. *In* Genetics and Cancer, Austin, University of Texas Press, 1959, 377–390.
29. Waterhouse J, Muir C, Correa P, Powell J: Cancer Incidence in Five Continents, vol III, Lyon, International Agency for Research on Cancer, 1976, 292–295.
30. Williams WR: Computer-assisted genetic counseling. Unpublished doctoral dissertation, University of Hawaii, 1981.
31. Williams WR, Anderson DE: Genetic epidemiology of breast cancer: Segregation analysis of 200 Danish breast cancer pedigrees. *Genet Epidemiol* 1:7–20, 1984.

Contributors

David E. Anderson received his Ph.D. from Iowa State University and then worked as a postdoctoral fellow on the genetics of eye cancer in cattle at Oklahoma State University, continuing this work after his move in 1958 to M. D. Anderson Hospital. He served as a special postdoctoral fellow in the Department of Human Genetics at the University of Michigan from 1963 to 1965. His research interests focus on the genetics of human skin and colon cancer and the genetics of human breast cancer. He is Professor of Genetics at the University of Texas System Cancer Center, M. D. Anderson Hospital and Tumor Institute, Houston.

R. S. K. Chaganti received his Ph.D. from Harvard University, where he conducted research in corn genetics under the direction of Paul C. Mangelsdorf. Since his graduation, he has worked with mammalian systems, first with Charles E. Ford and then with James German, during which periods he became committed to the study of human chromosomes. His major interest is in the role of heredity in the onset of human cancer. Dr. Chaganti is Director of the Genetics Laboratory in the Pathology Department at Memorial Hospital, New York.

James German received his medical education at the University of Texas in Galveston and Southwestern Medical College in Dallas, thereafter specializing in internal medicine. During medical college, he worked in the tissue-culture laboratory of Charles M. Pomerat in Galveston. His research interests are in the field of human genetics, particularly the genetics and cytogenetics of human development and human cancer. He directs the Laboratory of Human Genetics at the New York Blood Center and holds additional appointments at Cornell University, Memorial Hospital, and the Rockefeller University, all in New York.

William S. Hayward received his Ph.D. from the University of California at San Diego, where he studied bacteriophage gene expression. His research interests are centered on RNA tumor viruses and the role of oncogenes in viral and nonviral neoplasia. He is a Member of the Memorial Sloan-Kettering Cancer Center, heading the Laboratory of Molecular Genetics and Oncology. He also holds appointments at the Rockefeller University and Cornell University.

Suresh C. Jhanwar obtained his Ph.D. from the University of Delhi in India. His research interests are in the localization of cellular oncogene positions on human chromosomes and their relationship to nonrandom chromosome changes in cancer. Currently, he is Assistant Attending Geneticist and Cytogeneticist in the Laboratory of Genetics, Department of Pathology at Memorial Hospital, New York.

Mary-Claire King received her Ph.D. in genetics from the University of California at Berkeley. Her dissertation suggested a major role for changes in regulatory genes in human evolution. She is interested in mechanisms of genetic susceptibility in common chronic diseases of multifactorial determination and how genetic susceptibility can be modified by environmental exposure. She is Professor of Epidemiology at the University of California at Berkeley.

John B. Little was graduated in physics from Harvard College and in medicine from Boston University. He continued his clinical training at the Johns Hopkins Hospital and the Massachusetts General Hospital. He was a research fellow at

Harvard University in physiology and cell biology. His research interests are in cellular and molecular mechanisms of carcinogenesis, including genetic susceptibility to cancer. He has been on the faculty of the Harvard School of Public Health since 1965 and currently is Professor in the Department of Cancer Biology and Director of the Kresge Center for Environmental Health.

Shinichi Misawa received his medical training at the Kyoto Prefecture University of Medicine in Japan; his clinical training was in internal medicine. From 1982 to 1984, he was a research fellow in the Section of Cancer Genetics at the University of Maryland Cancer Center. Currently, he is Assistant Professor in the Third Department of Internal Medicine at the Kyoto Prefecture University.

Avery A. Sandberg received his M.D. from Wayne State University in Detroit and subsequent medical training at Mount Sinai Hospital in New York and the University of Utah under M. M. Wintrobe. His research activities have focused on the cytogenetic aspects of various human malignancies, with emphasis being placed in recent years on tumors of the urinary and gastrointestinal tracts. He is Chief of the Department of Genetics and Endocrinology at Roswell Park Memorial Institute in Buffalo.

R. Neil Schimke graduated from Kansas University Medical School. He received his postgraduate training in clinical genetics at Johns Hopkins University. He has a longstanding interest in how single gene defects relate to cancer, having become interested in this area of research through his work on the multiple endocrine neoplasia syndromes. He is Professor of Medicine and Pediatrics at Kansas University and Director of the Division of Endocrinology, Metabolism, and Genetics.

Nancy R. Schneider received her M.D. and Ph.D. from Cornell University. She is interested in the role of genetics and cytogenetics in the developmental pathology of cancer susceptibility. Currently, she is a resident in pathology at the University of Texas Health Science Center at Dallas.

Louise C. Strong received her M.D. from the University of Texas Medical Branch at Galveston. Her postdoctoral training in cancer genetics was in the laboratory of Alfred G. Knudson, Jr., at the University of Texas Graduate School of Biomedical Sciences in Houston. She is interested in the genetic epidemiology of human cancer, especially childhood cancer, and genetic models for cancer. Currently, she is Associate Professor of Experimental Pediatrics/Genetics, and Sue and Radcliffe Killam Professor at M. D. Anderson Hospital and Tumor Institute in Houston.

Joseph R. Testa received his Ph.D. from Fordham University. His postdoctoral training was in the laboratory of Janet Rowley at the University of Chicago, where he conducted cytogenetic studies on hematologic disorders. He is Associate Professor and Head of the Section of Cancer Genetics at the University of Maryland Cancer Center.

Christopher J. Williams graduated from London University. His initial training in oncology was with Professor Hamilton Fairley at St. Bartholomew's Hospital; subsequently, he completed a postdoctoral fellowship at Stanford University, where he worked under Saul Rosenberg. His research interests center on clinical trials in the management of solid tumors. Currently, he is Consultant Medical Oncologist and Senior Lecturer at the University of Southampton Medical School in England.

Wick R. Williams received his Ph.D. from the University of Hawaii, where he was affiliated with the Population Genetics Laboratory (Newton E. Morton, director). His postdoctoral training was in the Department of Genetics at the University of Texas Cancer Center, M. D. Anderson Hospital and Tumor Institute, Houston, where he subsequently was appointed Assistant Professor. He now is on the staff of the Division of Clinical Research of the Fox Chase Cancer Center in Philadelphia. Dr. Williams's training and research interests are in the genetic epidemiology of human cancer.

Index

Abelson murine leukemia virus, 23, 24–25
abl oncogene. *See also* Oncogene
and Abelson murine leukemia virus, 23, 24–25
in chronic myelogenous leukemia, 24–25, 31
map location of, 31, 161
and Philadelphia chromosome, 31, 161
and protein-kinase activity, 28
species of origin, 24–25
ABO blood group, 87, 153
Acetoxy-*N*-acetylaminofluorene, 213
Acid phosphatase, 126
Acoustic neuroma, 84, 107, 112
Acquired immunodeficiency syndrome, 93, 191
Actin, 231
Activation of chemical carcinogen, 19
Activation of oncogenic potential, 22, 22 *n*, 65, 69, 71
Acute leukemia. *See also* Leukemia
and cancer-family syndrome, 111
in Down's syndrome, 67, 85
high-resolution banding analysis of, 175–77
in infancy, 85
risk to relatives of patients, 136
Acute lymphoblastic leukemia
Burkitt-type, 192
chromosome banding patterns in, 171–72, 192, 202, 203
chromosome translocation (4;11) in, 172–72, 202
chromosome translocation 14q+, involving chromosome 8 in, 173, 202
chromosome translocation 14q+, not involving chromosome 8 in, 173–74, 202
chromosome translocation 6q− in, 174, 203
high-resolution banding analysis of, 176–77
hyperdiploidy in, 174–75
incidence of abnormal karyotype in, 159, 171–72, 176
modal chromosome number in, 164, 171–72
and *myc* oncogene, 192
near-haploid type, 174
nonrandom chromosome change in, 202, 203
Philadelphia chromosome positive type, 172, 202
prognostic significance of karyotype in, 172, 175, 177
subgroup recognition by karyotype, 159
Acute monoblastic leukemia, 162, 167, 170, 202, 203
Acute monocytic leukemia. *See* Acute monoblastic leukemia
Acute myeloblastic leukemia
Auer rod-positive cells in, 166, 202
chromosome translocation (8;21) in, 162, 165–66, 202, 203
leukocyte alkaline phosphatase level, 166
median survival in, 166, 169–70
and *mos* oncogene, 165
other chromosome abnormalities in, 166, 202, 203
Acute myelomonocytic leukemia, 162, 167, 170, 203
Acute nonlymphocytic leukemia, de novo, 164–70. *See also* Acute monoblastic leukemia; Acute myeloblastic leukemia; Acute myelomonocytic leukemia; Acute promyelocytic leukemia
basophilia in, 168
chromosome banding patterns in adults with, 69, 164–68, 202, 203
chromosome banding patterns in children with, 169
chromosome gains and losses in, 164–65, 169
chromosome translocation (8;21) in, 162, 165–66, 177, 202
chromosome translocation (15;17) in, 162, 166–67, 202
eosinophilia in, 167–78, 203
in Fanconi's anemia, 217
high-resolution chromosome banding analysis of, 176–77
incidence of abnormal karyotype in, 159, 164, 176, 202, 203
karyotypic evolution in, 165
modal chromosome number in, 164
normal karyotype in, 176
other specific structural rearrangements in, 167–68, 169, 202, 203
prognostic significance of karyotype in, 169–70, 177
sex chromosome abnormalities in, 165, 166, 169
subgroup recognition by karyotype, 159
Acute nonlymphocytic leukemia, secondary, 170–71, 176–77
Acute promyelocytic leukemia, 162, 166–67, 202
Adenocarcinoma
of the breast, 82, 86, 110–11, 222, 235
of the cecum, 141, 223
of the colon, 82, 85–86, 110, 222–23, 228–29
of the endometrium, 85, 110, 222–23, 234, 228–29
of the intestine, 249

of the ovary, 222, 234, 237
of the stomach, 222
of the uterus, 110, 249
Adenomatosis of the colon and rectum syndrome, 9, 14, 18, 153
Adenovirus, 29, 213
Adrenal carcinoma, 115, 228–29
Adrenal cortex tumor, 108, 111
Adrenal-cortical carcinoma, 124
Adrenocorticotrophic hormone, 108
African Burkitt's lymphoma, 173. *See also* Burkitt's lymphoma
Aganglionic megacolon, 105
Albinism, 83, 86, 93, 94
Alpha-fetoprotein, 213
Amplification. *See* Gene amplification
Anatomic malformation, 212, 214
Anchorage independence. *See* Cell growth
Androgen therapy, 212, 217
Anemia, refractory, 203
Aneuploidy. *See* Chromosome number, nonrandom change of
Angioma, 108
Aniridia, 48, 85, 105, 155. *See also* Wilms' tumor
Antipain, 12
A-particle-induced plasmacytoma, 24–25
Arrhenoblastoma, 110
Arthritis, 228–29
Arylhydrocarbon hydroxlase, 19
Ashkenazi Jew, 90
Ataxia, cerebellar, 212, 214
Ataxia-telangectasia
abnormalities associated with, 213, 216–17
cancer predisposition in, 16, 91–92, 211–12, 217
cancer risk for heterozygotes, 85, 137, 151
cancer surveillance in, 218–20
and cellular recessiveness, 95
as a chromosome breakage syndrome, 16, 92, 93, 211
chromosome instability in, 150, 151, 213
clastogenic factor in, 152
clinical complications of, 212, 216–17
clonal expansion of T lymphocytes in, 150, 213, 214
DNA damage in, 94
in a genetic isolate, 99
heterozygote detection, 151
hypersensitivity to environmental agents, 16, 151, 213, 214
as an immunodeficiency disorder, 16, 83, 91, 94, 214
inheritance pattern of, 16
and ionizing radiation, 151, 213, 214
physician handling of, 218–21
prenatal diagnosis in, 152
and radiomimetic drugs, 151
spontaneous chromosome damage, type of, 16, 150, 213, 214
Atypical mole/melanoma. *See* Dysplastic nevus syndrome
Auer rod-positive cell, 166, 202
Autosomal dominant disease. *See* Dominantly inherited clinical disorders predisposing to cancer
Autosomal dominant genes predisposing to cancer. *See* Dominantly inherited cancer predisposition
Autosomal dominant inheritance. *See* Dominant inheritance
Autosomal recessive disease. *See* Recessively inherited clinical disorders predisposing to cancer
Autosomal recessive genes predisposing to cancer. *See* Recessively inherited cancer predisposition
Autosomal recessive inheritance. *See* Recessive inheritance
Avian bursal lymphoma, 24–25
Avian erythroblastosis virus, 23, 24–25
Avian leukosis virus, 22, 23, 24–25, 33
Avian leukosis virus-induced B-cell lymphoma (chicken), 24–25
Avian leukosis virus-induced erythroleukemia (chicken), 24–25, 33
Avian myeloblastosis virus, 23, 24–25
Avian myelocytomatosis virus-29, 23, 24–25, 35

Basal cell carcinoma, 15, 16–17, 52, 108
Basal cell nevus syndrome, 108, 111
Base substitution. *See* Point mutation
Basophil production, regulation of, 168
Basophilia of the bone marrow, 168
B-cell leukemia, 24–25, 173
B-cell lymphoma, 23, 24–25, 29, 33
B-cell lymphoma cell line, 31
B-cell tumor, 65. *See also* Burkitt's lymphoma
bcl-1 sequence, 66
Beckwith-Wiedemann syndrome, 105
Benzopyrene, 13–14
Bladder cancer
in cancer-family syndrome, 111
dominant inheritance of, 114
in Gardner's syndrome, 115
and Harvey *ras* oncogene, 24–25
and nonrandom chromosome change, 198, 203, 204
and Muir-Torre syndrome, 110
Bleomycin, 16, 213
Blindness, 212, 217
Bloom's syndrome
abnormalities associated with, 213, 216

Bloom's syndrome (*continued*)
cancer predisposition in, 16, 91–93, 211, 212, 216
cancer surveillance in, 218–20
and cellular recessiveness, 70, 95
as a chromosome breakage syndrome, 16, 92, 93, 211
chromosome instability in, 70, 150, 213
clastogenic factor in, 152
clinical complications of, 212, 216
clinical features of, 211–14, 216
esophageal cancer in, 115
hypersensitivity to environmental agents in, 16, 213, 216
as an immunodeficiency disorder, 16, 83, 91, 94, 216
inheritance pattern of, 16
physician handling of, 218–21
registry of patients with, 93
sensitivity to chemotherapeutic agents in, 213, 216
sister chromatid exchange in, 70
spontaneous chromsome damage, type of, 16, 150, 213
Blym oncogene, 24–25, 33, 35, 66. *See also* Oncogene
Bone core biopsy, 160
Bone lesion, 228–29
Bone marrow aspirate, 160
Bone marrow failure, 212, 214, 217
Bone marrow transplantation, 217, 219
Boveri, Theodore, 60
Bowel cancer. *See* Colon cancer
Brain tumor, 17, 107, 112, 124, 246. *See also* Meningioma; Neuroblastoma
Breast cancer. *See also* Familial breast cancer
in cancer-family syndrome, 111
in Cowden syndrome, 108
early onset of, 82
and estrogen use, 129
in familial aggregation of cancer, 85, 86, 136, 139–40
family studies of, 97
hyperdiploidy in tumor cells, 150
incidence in families of breast cancer patients, 123–24, 136, 139–40, 245
incidence in families of unaffected individuals, 123–24
in Klinefelter's syndrome, 67, 87
laterality of disease, 245
menopausal status in, 245
nonrandom chromosome change in, 198
pedigrees of affected families, 139–40
prophylactic surgery in, 219
in Muir-Torre syndrome, 110
Breast disease, benign, 129
Breast self-examination, 237
Bronchus tumor, 110
Burkitt's lymphoma
and acute lymphoblastic leukemia, 173, 192
and *Blym* oncogene, 24–25, 33
cell lines, 162
chromosome translocation (8;14), 30, 173, 188, 190, 202
endemic origin of, 188
Epstein-Barr virus and, 29, 188, 191
and immunoglobulin genes, 30, 64–65
and *myc* oncogene, 24–25, 26, 30, 33, 65, 191
nonendemic origin of, 188, 190
primary karyotypic changes in, 30, 61, 63–65, 188, 190, 202
and *ras* oncogene, 33
secondary karyotypic changes in, 63–65, 188, 190, 191, 202
surface immunoglobulin production in, 185, 190–91, 193
variant chromosome translocations in, 188, 189, 190, 191, 192, 193, 202

Cafe-au-lait spot, 112
Calcitonin, 108
Calcium ionophore, 186
Cancer, age-specific incidence pattern, 39
Cancer, clonal theory of, 3, 9, 29, 33, 40, 61
Cancer, demographic pattern of, 7
Cancer, early detection of, 219. *See also* Cancer surveillance program
Cancer, early onset of, 82, 139, 216, 219, 228–29
Cancer, site-specific, 82
Cancer of the ampulla of Vater, 228–29
Cancer families. *See* Familial aggregation of cancer
Cancer-family syndrome
actin studies in, 231
age of onset, 223, 225–26, 235
associated clinical features, 228–29
autosomal dominant inheritance of, 82, 86, 141, 223
biological markers of, 230–32, 237
cancer, early identification of, 226
cancer, location of, 99, 223, 228–29
cancer, multiple primary, 223, 225
cancer, risk for family members, 141, 223, 229, 248
cancer, type of, 86, 99, 110–11, 141, 222, 226
cancer surveillance programs, 227, 235–37
carcinoembryonic antigen in, 230
carcinogen avoidance in, 235
differential diagnosis in, 228–29
dominant colorectal cancer in, 114

emotional consequences of, 227, 238
endoscopic assessment in, 226, 232, 236–37
familial clustering of cancer and, 85
family history in, 82, 223–25, 226–27
general information about, 86, 110, 114–15, 248
hereditary colon cancer and, 228–29
and HLA association, 231
inheritance pattern of, 223, 228–29
mixed lymphocyte culture in, 231
osteogenic sarcoma in, 115–16
pedigree analysis of, 224, 226, 229
prolonged survival in, 223, 235
prophylactic surgery in, 227, 232–35, 238
risk of cancer in, 223, 226
skin fibroblast study in, 231–32
and SV40 T antigen, 230–31
treatment in, 226, 232–34

Cancer patient, first-degree relatives of, 96–97
Cancer-predisposing gene, cellular phenotype of
cell growth in medium with low serum, 153
chromosome instability, 150, 213
hypersensitivity to alkylating agents, 152, 213
hypersensitivity to clastogens produced by mutant cells, 152
hypersensitivity to DNA-crosslinking agents, 152, 213
hypersensitivity to ionizing radiation, 151–52, 213
hypersensitivity to ultraviolet radiation, 151–52, 213
immunodeficiency, 151
in vitro cell transformation, 152–53
tetraploidy and hyperdiploidy in cultured cells, 150

Cancer-predisposing genes, and linked traits, 54, 154–56
Cancer-predisposing genetic conditions. *See* Heritable conditions predisposing to cancer
Cancer-proneness, definition of, 4
Cancer surveillance program, 54, 130–31, 218–21, 227, 235–37, 249
Carcinoembryonic antigen, 225, 230
Carcinogen
activation of, 19
classic, 7
exposure to, in familial clustering of cancer, 82, 97
modifying agents of, 14. *See also* Noncarcinogenic modifying factor
as mutagens and chromosome disrupters, 39, 73

Carcinogenesis, dietary factors in, 7, 19, 134
Carcinogenesis, multistage model of, 9–11, 39, 54, 74
Carcinogenesis, two-stage theory of, 9, 29, 40, 44, 67–68, 70, 243
Carcinoid tumor, 107
Carcinoma of the duodenum, 115
Carotid-body tumor, 84
C-cell hyperplasia, 108
Cell growth
anchorage independence and, 11
contact inhibition and, 11
and genes at chromsomal breakpoints, 62
and oncogenes, 34, 65, 87–88
and saline instillation, 14
synchronization of, 164, 176
in transformed cells, 153

Cell membrane protein, 28, 33
Cell proliferation gene. *See* Cell growth; *see also* Oncogene
Cell-surface receptor defect, 86
Cellular oncogene. *See* Oncogene
Central nervous system cancer, 112, 137
Central nervous system dysfunction, 212, 214
Cervical cancer, 110, 198
Chediak-Higashi syndrome, 83, 93
Chemodectoma, 84
Chemotherapeutic agent, sensitivity to, 213, 216, 217
Chemotherapy, 170
Chinese hamster ovary cell, 71, 72
Chromatid exchange
in Bloom's syndrome, 16, 150, 213, 215
in Fanconi's anemia, 150
unequal, 4
in xeroderma pigmentosum, 213, 215

Chromatid segregation, 3, 62
Chromosome, marker, 3, 9, 61, 106, 190
Chromosome, murine, 65, 69
Chromosome banding, methodology and nomenclature, 160
Chromosome breakage, spontaneous
in ataxia-telangectasia, 16, 150, 213
in Bloom's syndrome, 16, 150, 213
and environmentally induced cancer, 15–18
in Fanconi's anemia, 16, 150, 213
in retinoblastoma, 43
in scleroderma, 16
in Werner's syndrome, 150
in xeroderma pigmentosum, 16, 213

Chromosome breakage syndrome. *See* Ataxia-telangectasia; Bloom's syndrome; Fanconi's anemia; Werner's syndrome; Xeroderma pigmentosum, *see also* Chromosome instability

Chromosome breakage syndrome (*continued*)
associated abnormalities in, 16, 150, 211–15
cancer predisposition in, 8, 92, 93–94, 150, 211
complications of, 216–17
diagnosis of, 215–16
handling patient with, 218–21
names of, 16, 92, 150, 211
Chromosome breakpoint
in Burkitt's lymphoma, 30–31, 64–65
in chronic myelogenous leukemia, 30–31
and fragile sites, 155
and oncogenes, 63, 64–65, 66, 155
in renal cell carcinoma, 113–14, 154–55
in retinoblastoma, 41, 155
in Wilms' tumor, 48–49, 155
Chromosome change, in neoplastic cells, 62–66, 186–87
abnormal segregation, 62
change in number, 62, 198, 199, 204
change in structure, 62–65, 198. *See also* Chromosome deletion
chromosome breakage, 62–65
effect on tumor behavior, 186–87, 202–3
gene dosage, 62
marker chromosome, 200–201, 203
nonrandom translocation, 198
position effect, 62–63
primary changes, 186–87
secondary changes, 186–87
significance of, 200
specific breakpoints, 63–65
Chromosome change in human cancer
in acute lymphoblastic leukemia, 171–75, 202, 203, 204
in acute monocytic leukemia, 202
in acute myelomonocytic leukemia, 202
in acute nonlymphoblastic leukemia, 73, 164–71, 202, 203, 204
in adult T-cell leukemia, 203
in Burkitt's lymphoma, 61, 63–65, 187–92, 202
in chronic lymphocytic leukemia, 197–99, 202, 203, 204
in chronic myelogenous leukemia, 60–61, 161–64, 202, 203, 204
correlated with tumor-inducing agents and identical tumors, 73
in Hodgkin's disease, 73, 197
in meningiomas, 198, 200, 204
in myelodyspoietic disorders, 202
in myeloproliferative disorders, 202
in non-Burkitt's lymphoma, 192–96, 202, 203, 204
in non-Hodgkin's lymphoma, 192–96, 202, 203, 204
in ovarian carcinoma, 73
in plasmacytomas, 65
in polycythemia vera, 203
in preleukemia, 202, 203, 204
in prolymphocytic leukemia, 202, 203
in refractory anemia, 203
in renal cell carcinoma, 50, 114
in retinoblastoma, 44, 104
in secondary leukemia, 203, 204
in solid tumors, 198–204
in T-cell lymphoma, 196–97, 202, 203
in Wilms' tumor, 49, 105
Chromosome change predisposing to cancer, 61, 66–68, 85. *See also* Renal cell carcinoma; Retinoblastoma; Wilms' tumor
Chromosome deletion
and hemizygosity, 69, 95
as a mutation associated with neplasia, 4, 32, 39, 51, 85
nomenclature of, 160
in retinoblastoma, 32, 41–43, 85, 95, 155
in tumor progression, 32, 62, 69
in Wilms' tumor, 32, 69, 85, 155
Chromosome disrupting agents, 73
Chromosome duplication, 39, 62
Chromosome gap, 150, 155, 213
Chromosome heteromorphism, 70
Chromosome instability
as a cellular phenotype for cancer-predisposing genes, 150
in the chromosome breakage syndromes, 93–94, 211, 213
in familial retinoblastoma, 16, 43
and fragile sites, 155
and inherited cancer predisposition, 5, 92, 211
in neoplastic cells, 3
Chromosome instability syndrome. *See* Chromosome breakage syndrome
Chromosome inversion, 39, 62, 104, 160
Chromosome mosaicism, 87
Chromosome nondisjunction, 3, 45, 49, 51, 68, 95
Chromosome number, nonrandom change of, 3, 62, 68–69, 73, 150
Chromosome reduplication, 45
Chromosome structure, rearrangement of
in ataxia-telangectasia, 16, 150, 213, 214
in Bloom's syndrome, 213, 214
as a clonal marker, 3, 9, 16
in Fanconi's anemia, 213, 214
in Gardner's syndrome, 115
and insertional activation, 26
and neoplasia, 3
in retinoblastoma, 43
in Wilms' tumor, 150
Chromosome translocation
associated with neoplasia, 26, 60–61, 62–63, 71, 74

balanced, constitutional, 41, 49, 50, 113, 160
in Burkitt's lymphoma, 30–31, 61, 63–65
in chronic myelogenous leukemia, 30–31, 60–61, 161–64, 172
as a mutation, 4, 51
nomenclature of, 160
in renal cell carcinoma, 154–55
in retinoblastoma, 42, 85, 104, 155
unbalanced abnormalities, 60
in Wilms' tumor, 85, 105, 155
Chromosome 1
in acute lymphoblastic leukemia, 174
in acute myelogenous leukemia, 166
in acute nonlymphoblastic leukemia, 166, 167
in acute promyelocytic leukemia, 167
in breast cancer, 198
in Burkitt's lymphoma, 191
in colon cancer, 198
in endometrial cancer, 198
in melanoma, 198
in neuroblastoma, 69, 106, 198
in non-Burkitt's, non-Hodgkin's lymphoma, 194, 196
in retinoblastoma, 44, 204
Chromosome 2, 64, 166, 167, 188–93
Chromosome 3
in acute nonlymphocytic leukemia, 167, 168
in acute promyelocytic leukemia, 167
in chronic myelogenous leukemia, 162
in hypernephroma, 85, 113–14
in nasopharyngeal cancer, 198
in parotid tumor, 198
in renal cell carcinoma, 50, 154–55, 198
in small cell cancer of the lung, 198
in thrombopoietic abnormalities, 168
Chromosome 4, 172–73, 174, 175
Chromosome 5
in acute nonlymphocytic leukemia, 169, 170–71, 177, 204
in bladder cancer, 198
in cervical cancer, 198
in myeloid neoplasms, 69
Chromosome 6
in acute lymphocytic leukemia, 172, 174, 175
in acute myeloblastic leukemia, 168
in acute myelomonocytic leukemia, 167, 168
in acute nonlymphocytic leukemia, 167, 168
and basophilia of the bone marrow, 168
in Burkitt's lymphoma, 190
in chronic myelogenous leukemia, 168
in melanoma, 198
in non-Burkitt's, non-Hodgkin's lymphoma, 194–96
in ovarian cancer, 198
in retinoblastoma, 198, 204
in serous cystadenocarcinoma of the ovary, 198, 200, 201
Chromosome 7
in acute lymphocytic leukemia, 172
in acute nonlymphocytic leukemia, 165, 167, 169, 170–71, 177, 204
in acute promyelocytic leukemia, 167
in ataxia-telangectasia, 214
in Burkitt's lymphoma, 190
in cervical cancer, 198
in chronic myelogenous leukemia, 163
in myeloid neoplasms, 69
in neuroblastoma cell lines, 72
in non-Burkitt's, non-Hodgkin's lymphoma, 196
in preleukemia, 204
Chromosome 8
in abnormal hematopoiesis, 69
in acute lymphocytic leukemia, 172, 173, 175
in acute myelogenous leukemia, 162, 165–66
in acute nonlymphocytic leukemia, 69, 165–66, 169, 171, 177, 204
in B-cell neoplasia, 50
in Burkitt's lymphoma, 50, 64, 188–93
in chronic myelogenous leukemia, 30, 163, 164, 187, 204
in hypernephroma, 85, 113
in parotid tumors, 198
in non-African Burkitt's lymphoma cell line, 162
in non-Burkitt's, non-Hodgkin's lymphoma, 193–96
in polyps of the colon, 204
in preleukemia, 204
in renal cell carcinoma, 50
in salivary gland tumors, 198
Chromosome 9
and *abl* oncogene, 31, 161
in acute lymphocytic leukemia, 172
in acute monocytic leukemia, 162, 167
in acute myeloblastic leukemia, 168
in acute myelomonocytic leukemia, 167, 168
in acute nonlymphocytic leukemia, 162, 167, 168
and basophilia of the bone marrow, 168
in bladder cancer, 204
in chronic myelogenous leukemia, 30–31, 168
and the Philadelphia chromosome, 30–31, 161, 162, 172
Chromosome 10, 167, 174, 175, 190

Chromosome 11
in acute lymphoblastic leukemia, 172–73, 174
in acute monocytic leukemia, 162, 167
in acute myeloblastic leukemia, 166
in acute myelomonocytic leukemia, 167
in acute nonlymphocytic leukemia, 162, 166, 167, 169
in Burkitt's lymphoma, 190
in cervical cancer, 198
in Ewing's sarcoma, 198
in neuroblastoma, 198
in non-Burkitt's, non-Hodgkin's lymphoma, 193–96
in Wilms' tumor, 48–49, 67–68, 85, 87, 105, 155, 198
Chromosome 12, 193–96, 197–99, 204
Chromosome 13, 41–45, 67–68, 85, 87, 104, 155, 198, 204
Chromosome 14
in acute lymphoblastic leukemia, 172, 173–74, 175
in ataxia-telangectasia, 214
in B-cell lymphoma cell lines, 31
in Burkitt's lymphoma, 30, 31, 61, 64, 188, 193
in chronic lymphoblastic leukemia, 197
in Hodgkin's disease, 197
in non-African Burkitt's lymphoma cell lines, 162
in non-Burkitt's, non-Hodgkin's lymphoma, 193–96
in serous cystadenocarcinoma of the ovary, 198, 200, 201
Chromosome 15, 162, 165, 166–67, 169
Chromosome 16, 154, 162, 167–68, 177, 246
Chromosome 17, 162–67, 169, 187
Chromosome 18, 172, 173, 174, 175, 193–96
Chromosome 19, 164, 167, 169, 174, 204
Chromosome 20, 69, 172
Chromosome 21
in acute lymphocytic leukemia, 172, 174, 175
in acute myeloblastic leukemia, 162, 165–66
in acute nonlymphoblastic leukemia, 165–66, 169, 171, 177
in Down's syndrome, 67, 85, 87
in dysregulation of hematopoiesis, 85, 87
in non-Burkitt's, non-Hodgkin's lymphoma, 196
Chromosome 22
in acute lymphoblastic leukemia, 172, 174
in acute nonlymphoblastic leukemia, 204
in Burkitt's lymphoma, 30, 64, 188–93
in Ewing's sarcoma, 198
in meningioma, 112, 198, 204
in neuroblastoma, 198
and the Philadelphia chromosome, 161, 162, 172
in sarcoma, 204
Chromosome X
in acute lymphoblastic leukemia, 174
in acute myeloblastic leukemia, 166
in acute nonlymphocytic leukemia, 166, 169
in Burkitt's lymphoma, 191
in Klinefelter's syndrome, 67, 85, 87
late replication of, 42
Chromosome Y
in acute lymphoblastic leukemia, 174
in acute myeloblastic leukemia, 166
in acute nonlymphocytic leukemia, 166, 169
in chronic myelogenous leukemia, 163
in Klinefelter's syndrome, 67, 85, 87
in mixed gonadal dysgenesis, 67, 85, 87
Chronic leukemia, 107
Chronic liver disease, 93
Chronic lymphocytic leukemia, 136, 197–99, 202, 203
Chronic lymphocytic leukemia, B-cell, 186, 202
Chronic myelogenous leukemia
association with neurofibromatosis, 107
blastic crisis of, 162, 163–64, 172, 203
and c-*abl* oncogene, 24–25, 31, 161
chronic phase of, 161–63, 202
and c-*sis* oncogene, 31
hyperdiploid clones in, 163–64
hypodiploid clones in, 163–64
incidence of abnormal karyotype in, 161
karyotype evolution in, 163–64, 172
median survival in, 161, 163
other karyotypic changes in, 162, 163, 164, 168, 187, 202
and the Philadelphia chromosome, 60–61, 161–64, 172, 187, 202
Philadelphia chromosome negative patients, 162
prognosis of, 161, 163
prognostic significance of karyotypic evolution, 164, 172
risk to relatives of patients with, 136
second Philadelphia chromosome, 163, 164
therapy, 164
Cigarette/cigarette smoke, 7, 8, 14, 19, 134

Clastogenic factor, 152
Clinical evaluation
  of families with cancer aggregation, 142–43
  of patient with cancer, 80–82
  of patient with cancer predisposition 218–21
Clonal chromosome abnormality, 3, 150, 160, 213, 214
Clonal evolution of neoplastic cells, 61, 73, 74
Clonal expansion
  of fibroblasts with chromosome abnormality, 150
  of neoplastic cells, 33, 61, 73
  of T lymphocytes with chromosome abnormality, 150, 213, 214
Clonal selection of neoplastic cells, 33
Clonal theory of cancer. *See* Cancer, clonal theory of
Cockayne's syndrome, 16
Coincidence, in familial aggregation of cancer, 134, 135, 142–43, 144
Colectomy, 232, 233
Colon cancer. *See also* Familial colon cancer
  in adenomatosis of the colon and rectum syndrome, 14
  in cancer-family syndrome, 111, 114, 141
  and dietary factors, 7, 19
  in familial aggregation of cancer, 85, 86, 139
  family studies of, 136
  in Gardner's syndrome, 115
  and Kirsten *ras* oncogene, 24–25
  in multiple polyposis coli syndome, 114, 137–38, 140
  and *myc* amplification, 26
  nonrandom chromosome change in, 198, 204
  and single gene mutation, 140
  in Muir-Torre syndrome, 110, 114
Colon polyps. *See* Polyps of the colon
Colorectal cancer, dominant, 114–15. *See also* Familial polyposis of the colon; Gardner's syndrome; Muir-Torre syndrome; Peutz-Jeghers syndrome
Complementation group, 15, 215
Computed tomography scanning, 107, 237
Concanavalin A, 186
Concordance rate, 123
Congenital hemihypertrophy, 105
Congenital malformation, 91
Consanguinity, 90, 97, 98–99
Constitutional chromosomal abnormalities that predispose to neoplasia, 85, 87. *See also* Down's syndrome; Gonadal dysgenesis; Hypernephroma; Klinefelter's syndrome; Renal cell carcinoma; Retinoblastoma; Wilms' tumor
Contact inhibition. *See* Cell growth
Corneal scarring, 212, 214, 217
Counseling
  of the cancer-prone individual, 100, 218–21
  of families with breast cancer, 130–31
  of families with chromosome imbalance, 41, 49
  of families with hereditary retinoblastoma, 41, 42
Cowan 1 protein, 186
Cowden syndrome. *See* Cowden's disease
Cowden's disease, 108, 246
Crossing-over. *See* Chromatid exchange
Crosslinking agent. *See* DNA cross linking agent
Croton oil, 10–11
Cryptorchidism, 109
Cultural inheritance, 125–26, 128–29
Cutaneous leiomyoma, 108
Cyclophosphamide, 213
Cylindromatosis, 84
Cystic fibrosis, 90
Cytochrome P450 enzyme, 19
Cytogenetic methodology, 160
Cytogenetic nomenclature, 160

Danish twin registry, 98
Daughter-strand-gap-repair deficiency. *See* DNA repair
Debrisoquine-hydroxylation, 19
Delayed mutation, 41
Deletion. *See* Chromosome deletion
Demographic pattern, 7
Dermal hyperpigmentation, 212, 214
Dermatosis, 84, 212, 215
Desmoid, 228–29
Detection of individuals at high risk for cancer, 149. *See also* Cancer-predisposing genes, and linked traits
Diabetes, 212, 216
Diagnosis of a chromosome breakage syndrome, 215–16. *See also* Ataxia-telangectasia; Bloom's syndrome; Fanconi's anemia; Xeroderma pigmentosum
Diallelomorphic gene loci, 19
Diepoxybutane, 152, 213
Dietary factors in carcinogenesis, 7, 19, 134
Differential diagnosis, 228–29
Differentiation, 51
Diffuse gastrointestinal polyposis, 114
Disturbance of sexual development, 83
Diverticulitis, 249
DNA alkylating agent, 152
DNA binding protein, 33, 35

DNA crosslinking agent, 16, 18, 152, 213, 214
DNA damage, 9, 15, 94
DNA damaging agent, 15, 152, 213, 215
DNA interstrand crosslink, 213
DNA repair
  daughter-strand-gap repair, 213
  defects of, 15, 213. *See also* Xeroderma pigmentosum
  deficiency disorders, 83, 92, 94, 151–52, 213
  excision repair, 15, 83, 92, 94, 152, 213
  of gamma endonuclease sensitive sites, 213
  process of, 9, 15, 92, 94, 213, 214
DNA replication, semiconservative, 215
Dolichocephaly, 212
Dominant inheritance, 4, 32, 40, 96, 100
Dominant inheritance of human cancer
  breast cancer, 111, 245–47
  in cancer-family syndrome, 111, 141
  of the central nervous system, 112–13
  embryonal tumors, 104–5
  of the endocrine system, 108–9
  in families, 9, 85
  of the gastrointestinal tract, 114–15, 248–49
  in harmatoma syndromes, 106–8
  in the reproductive system, 109–11
  in the respiratory tract, 112
  site-specific cancer, 82, 99, 245–47, 248–49
  of the skeletal system, 115–16
  of the skin, 111–12
  of the urinary tract, 113–14
Dominant mutation, 32
Dominantly inherited cancer predisposition. *See also* Familial breast cancer; Nevoid basal cell carcinoma syndrome; Renal cell carcinoma; Retinoblastoma; Wilms' tumor
  and aneuploidy, 150
  and cytogenetic change, 54, 103
  and single gene alteration, 97, 101, 124–25
Dominantly inherited disorders predisposing to cancer, 83–84, 86. *See also* Adenomatosis of the colon and rectum syndrome; Nevoid basal cell carcinoma syndrome; Retinoblastoma; Wilms' tumor
Dose response to initiating and promoting agents, 11, 12
Double minute, 44, 71, 72, 106
Down's syndrome, 67, 85, 87
Drug metabolism and activation of chemical carcinogens, 19
Drug resistance, 60, 71
Duplication. *See* Chromosome duplication
Dysgenetic testes, 105, 109
Dysgerminoma, 110
Dyskeratosis congenita, 83, 93
Dysmyelopoietic disorders, 202
Dysplastic nevus syndrome, 84, 86, 112

Early onset of cancer. *See* Cancer, early onset of
Ectodermal dysplasia, 111–12
Embryonal tumor, 104–6, 111. *See also* Gonadoblastoma; Hepatoblastoma; Neuroblastoma; Retinoblastoma; Sarcoma; Teratoma; Wilms' tumor
Endocrine insufficiency, 212, 214, 217
Endocrine system, tumors of, 108–9
Endometrial cancer, 110, 111, 124, 141, 198
Enhancement/enhancer element, 26, 30, 36
env gene, 23
Environmental factors in cancer etiology. *See also* Occupational exposure to mutagen; Sunlight/sun exposure; Tobacco product
  and bladder cancer, 114
  and breast cancer, 129
  effect on susceptibility genes, 122, 128–29
  and esophageal cancer, 115
  and familial aggregation of cancer, 134, 143
  and Hodgkin's disease, 129
  and leukemia, 142, 170–71
  and lung cancer, 129, 140–41
  and X-ray exposure, 142
Enzyme defect, 85–86
E1A protein, 29
Eosinophilia of the bone marrow, 167–68, 203
Eosinophilic differentiation, 168
Epidemiology, *See* Genetic epidemiology
Epidermal growth factor receptor, 34, 36, 65
Epidermolysis bullosa dystrophica, 83, 111, 115
Epigenesis, 9
Epithelial cancer, 137
Epithelioma, 84
Epstein-Barr virus, 29, 129, 186, 188
erb-A oncogene, 24–25. *See also* oncogene
erb-B oncogene, 24–25, 29, 34, 36, 65. *See also* onogene
E-rosette inhibitory factor, 106
Erythema of the face, 212, 216
Erythrocyte catalase, 48–49
Erythroleukemia, 24–25

Esophageal cancer, 84, 110, 115
Esterase D, 41–42, 44–45, 104
Estrogen, 129, 130
Ethyl methanesulfonate, 213
Ewing's sarcoma, 198
Excision repair. *See* DNA repair
Exostosis, hereditary multiple, 84

Familial aggregation of cancer
  approaches to investigation, 4, 122, 134–38
  counseling in, 101, 130
  evaluation of, 4, 99–100, 138–39, 142–43
  in families with cancer-predisposing genes, 83–85, 137–38
  family history in, 122–23, 138–39
  genetic components of, 7–8, 81–82, 100
  incomplete penetrance, 128–29
  kinds of, 139–43. *See also* Familial breast cancer; Familial colon cancer
  multifactorial inheritance in, 99–100, 134
  multiple anatomic sites of cancer, 135, 137–38, 141–43
  nongenetic factors in, 134
  pedigree construction, 123
  preclinical markers of disease, 129–30
  reasons for, 122, 133–34
  risk to family members, 122, 128–29, 131, 136, 143–44
  screening programs in, 130–31
  significance of, 4, 122
  single site cancer, 135–36, 139–41, 143
  twin studies in, 122, 123
  two or three sites of cancer in, 135, 137–38, 141
Familial breast cancer
  age of onset in, 124, 125, 130, 136, 243–45
  anecdotal accounts of, 133
  associated neoplasia, 124–25, 246
  autosomal dominant inheritance of, 124–25, 245–47
  and cancer-family syndrome, 111
  clinical types, 242
  counseling in, 130–31
  cultural factors in, 125–26
  differential diagnosis in, 228–29
  environmental factors in, 125–26
  and estrogen receptors, 130
  and estrogen use, 129
  family history in, 123–25
  family studies of patients with, 123–24, 136
  genetic epidemiology of, 124–25
  genetic hypotheses, 124–25
  genetic marker loci, 126–28, 154
  laterality of tumors, 124, 130, 136, 243–45
  linkage analysis in, 126–28, 154, 246
  mode of inheritance, 228–29
  and ovarian cancer, 154
  pathology of tumors, 228–29, 242
  penetrance of genes, 125, 246
  postmenopausal, 136, 242
  preclinical disease markers of, 129
  premenopausal, 136, 242
  risk determination, analytic risk, 249–50, 252–54
  risk determination, empiric risk, 250
  risk for first-degree relatives, 136, 244–45
  risk for second-degree relatives, 136, 244–45
  risk for women relatives, 123–24, 136
  risk in women without susceptibility alleles, 125
  screening programs, 130–31
  segregation analysis, 250–52, 254
  survival, 130
  tumor multiplicity, 130, 243–45
Familial cancer. *See* Familial aggregation of cancer
Familial colon cancer. *See also* Cancer-family syndrome
  age of onset, 247, 249
  anatomic location and pathology of polyps and/or cancer, 228–29
  associated neoplasms, 248–49
  differential diagnosis in, 228–29
  dominant inheritance of, 248–49, 255 *n*
  mode of inheritance, 228–29
  mutliple primary neoplasia, 247
  nonpolypoid disease, 247, 248–49
  pedigree analysis of, 249
  penetrance, 248, 249
  polypoid disease in, 247
  risk assessment, 248–49, 255 *n*
  tetraploidy in cultured cells, 150
  tumor location, 247, 249
Familial clustering of cancer. *See* Familial aggregation of cancer
Familial intraocular melanoma, 112
Familial polyposis of the colon
  anatomic location and pathology of polyps and/or cancer, 228–29
  associated clinical features, 228–29
  bile duct carcinoma, 115
  colonic neoplasia, 103, 228, 247
  cultured cells from, 115, 231–32
  cytoskeleton study in, 231
  differential diagnosis in, 226, 228–29
  dominant colorectal cancer and, 5, 114
  dominant transmission of, 84, 222, 228
  and hereditary colon cancer, 228–29

Familial polyposis of the colon (*continued*)
  intestinal polyps, 103, 228
  mode of inheritance, 99, 228–29
  penetrance of gene, 103
  predisposition to neoplasia, 5, 84, 222, 247
  prophylactic surgery in, 233
  and site-specific cancer, 140
Familial susceptibility to cancer. *See* Familial aggregation of cancer
Family history
  in evaluation of cancer aggregation, 80–82, 138–39, 142–43
  in familial aggregation of cancer, 134–38, 138–39
Family studies
  in bowel cancer incidence, 136, 140
  in breast cancer incidence, 123–24, 133, 135, 139–40
  in central nervous system tumor incidence, 137
  in epithelial cancer incidence, 137
  in lung cancer incidence, 140–41
  in lymphoma incidence, 137, 141
  in melanoma, 142
  in patients with early age of onset, 138, 141
  in patients with multiple primary neoplasms, 137–38
  in sarcoma incidence, 137, 141
  in stomach cancer incidence, 136
  in uterine cancer incidence, 136
Fanconi's anemia
  abnormalities associated with, 213
  cancer predisposition in, 16, 83, 92–93, 211–12, 217
  cell studies in, 18, 152–54, 213
  as a chromosome breakage syndrome, 16, 17, 92, 93, 211
  chromosome instability in, 150, 213, 214
  clastogenic facor in, 152
  clinical complications of, 212, 217
  diepoxybutane and, 152, 213
  and DNA crosslinking agents, 16, 18, 152, 213, 214
  DNA repair in, 94
  esophageal cancer in, 115
  heterozygotes for, 152
  hypersensitivity to environmental agents, 16, 17–18, 213
  inheritance pattern of, 16, 83
  and mitomycin C, 152, 213, 214
  physician handling of, 218–21
  prenatal diagnosis of, 152
  spontaneous chromosome damage in, 16, 150, 213, 214
  and superoxide dismutase, 18
  treatment in, 217
FBJ osteosarcoma virus, 24–25
Fecal occult blood, 236
Feline leukemia virus, 29
Ferritin, 106
fes oncogene, 24–25. *See also* Oncogene
fgr oncogene, 24–25. *See also* Oncogene
Fibrocystic pulmonary dysplasia, 84
Fibroma, 107, 108, 228–29
Fibrosarcoma, 24–25, 53, 115
First-degree relatives of cancer patients, 96–97
fms oncogene, 24–25. *See also* Oncogene
Focal dermal hypoplasia, 84
Foregut carcinoids, 108
fos oncogene, 24–25. *See also* Oncogene
Founder effect, 99
fps oncogene, 23, 24–25. *See also* Oncogene
Fragile site, 73–74, 155
Freckling, 212, 214–15
Free radical, 18
Fujinami sarcoma virus, 23, 24–25

gag gene, 23
Gamma-irradiation, 213
Gallbladder cancer, 115
Ganglioneuromas, 105, 107
Gardner-Rasheed feline sarcoma virus, 24–25
Gardner's syndrome
  anatomic location and pathology of polyps and/or cancer, 228–29, 247
  associated clinical feature, 228–29
  cancer types associated with, 115
  cellular and chromosomal radiosensitivity in, 115, 151
  colorectal cancer in, 103, 114
  differential diagnosis in, 228–29
  hereditary colon cancer in, 228–29
  mode of inheritance in, 84, 228–29
  multiple primary neoplasms in, 247
  and ovarian cancer, 110
  and polyposis, 84
  and predisposition to neoplasia, 84, 247
Gastric carcinoma. *See* Stomach cancer
Gastrinoma, 108
Gastrocolonic carcinoma, 114
Gastrointestinal adenomatous polyposis, 228–29
Gastrointestinal tract cancer, 114–15
Gene amplification
  of c-*myc* oncogene, 32, 72–73
  cytogenetics of, 71–72
  and methotrexate, 71
  as a mutation, 39
  of N-*myc* oncogene, 46, 66, 72–73
  and oncogene activation, 26, 31–32, 45–46, 73
  and transposition, 72–73
  in tumor cells, 66
  in tumor formation, 39, 45, 62
Gene conversion, 45, 61

Gene copy number, 71–73
Gene dosage, 62, 68–69
Gene mapping, 41, 47, 234, 246
Genetic counseling. *See* Counseling
Genetic epidemiology
  approaches to study, 134–38
  behavioral factors, 122, 125–26, 134
  cultural factors, 122, 125–26, 128–29
  demographic factors, 122
  environmental exposure, 122–23, 125, 125–26, 128–29, 133–34
  of familial breast cancer, 123–28
  genetic factors in, 122–23, 124–25, 125, 133–34, 138–39
Genetic isolate, 99
Genetic marker loci for cancer-predisposing genes
  in familial breast cancer, 126–28,
  fragile sites, 155
  glutamate-pyruvate transaminase, 111, 126–28, 154, 246
  inherited chromosome rearrangement, 154–55
  major histocompatibility complex, 154
  mapping of marker genes, 54
  red blood cell groups, 153
  restriction fragment length polymorphism, 54, 155–56
  for retinoblastoma, 41–42
  serum protein markers, 153
  for Wilms' tumor, 48
Genetic predisposition to cancer. *See also* Cancer-family syndrome; Familial polyposis of the colon
  animal models of, 8
  and interaction with environmental carcinogens, 7–8, 15–18, 46–47, 51–52, 54, 235
  as a primary event, 40
  tissue specific, 53
Genetic susceptibility to environmentally induced cancer. *See also* Fanconi's anemia; Hereditary retinoblastoma; Nevoid basal cell carcinoma syndrome; Xeroderma pigmentosum
  defect in DNA repair ability, 15
  hereditary lack of protective substances, 14, 17–18
  human disorders with spontaneous chromosome breakage, 15–18
  human disorders with susceptibility to an environmental agent, 15–18
  hypersensitivity to induction of DNA damage, 14–15
  inherited initiated characteristic, 14, 18–19, 40. *See also* Adenomatosis of the colon and rectum syndrome
Genital ambiguity, 85
Genitourinary anomalies, 48
Genomic alteration. *See* Mutation
Genomic instability. *See* Chromosome instability
Germ cell suppression, 67
Germ cell tumor, 67, 70–71, 85
Germinal mosaicism, 41
Germinal mutation. *See* Genetic predisposition to cancer
Germ-line chromosome mutation, 66–68, 154–55
Glioma, 107
Glomus tumors, multiple, 84
Glucagonoma, 108
Glucose-6-phosphate dehydrogenase, 9
Glutamate pyruvate transaminase, 111, 126, 154, 246
Glycogen storage disease, 83, 94
Goltz's syndrome, 84
Gonadal dysgenesis, 67, 83, 85, 93, 105
Gonadoblastoma, 67, 85, 87, 105
Growth deficiency, 211–12, 214, 217
Gynecomastia, 67

Hamartoma, 106, 108, 228–29
Hamartoma syndrome, 106–8, 112. *See* Basal cell nevus syndrome; Cowden's disease; Neurofibromatosis; Tuberous sclerosis; Von Hippel-Lindau syndrome
Hamartomatous polyps, 228–29
Haploidy, 62
Hardy-Zuckerman feline leukemia virus, 24–25
Harvey murine sarcoma virus, 23, 24–25, 32
Helper virus, 28
Hemangioblastoma, 107, 109
Hematopoiesis, abnormal, 67, 69, 87
Hemihypertrophy, 48
Hemizygosity, 44, 67–68, 69, 95–96
Hemochromatosis, 83, 86, 94
Hepatoblastoma, 105
Hepatoma, 86
Hereditary adenocarcinomatosis, 85, 248. *See also* Cancer-family syndrome
Hereditary colon cancer, 223, 228–29, 231–32, 249
Hereditary gastrocolonic cancer, 248
Hereditary retinoblastoma. *See* Retinoblastoma
Heritable conditions predisposing to cancer
  and carcinogen exposure, 80, 82
  and cellular oncogenes, 87–88
  constitutional chromosome abnormalities, 85, 87
  dominantly transmitted, 83, 84, 86
  familial clustering, 84, 85, 86
  genetically determined traits associated with cancer, 87
  recessively transmitted, 83–84, 85–86

Heterochromia irides, 105
Heterozygosity
  identification of, 90
  loss of, 44–45, 47, 49, 54, 67–70, 95
Heterozygote, cancer predisposition in, 95
High-resolution banding analysis, 159, 175–77
HLA association. *See* Major histocompatibility locus/complex
HL-60 cell line, 32, 33
Hodgkin's disease
  and cancer-family syndrome, 111
  chromosome changes after treatment, 73
  and Epstein-Barr virus, 129
  karyotypic characteristics of, 197
  and major histocompatibility complex, 87, 129, 154
Homogeneously staining regions, 35, 71, 72, 106
Homologous chromatid exchange. *See* Chromatid exchange
Homozygosity. *See* Heterozygosity, loss of
  on a cellular level, 4, 95
  and consanguinity, 98–99
  via parthenogenesis, 70–71
  and recessive inheritance, 90–91
  and somatic recombination, 4, 95
  at a tumor locus, 203
Host factors, 5
Host resistance gene, 41
Hot spot, 43
H-2 locus, 70
Hydantoin, 106
Hydatidiform mole, 71
Hyperdiploidy, 150, 163, 171, 199, 203
Hypernephroma, 85, 107, 113
Hyperparathyroidism, 109
Hypodiploidy, 163–64, 170, 171–72, 199
Hypospadias, 104

Identification of affected families, 226–27
Identification of gene carriers in affected families, 149. *See also* Cancer-predisposing genes, and linked traits
Immortalization, 39
Immunodeficiency
  in acquired immunodeficiency syndrome, 93
  in ataxia-telangectasia, 83, 91, 212, 214, 216–17
  in Bloom's syndrome, 83, 91, 212, 216
  and cancer predisposition, 4, 83, 91, 93
  as a marker for cancer predisposition, 151
  and organ transplantation, 93
Immunodeficiency-associated diseases, 83, 91, 93, 151. *See also* Acquired immunodeficiency syndrome; Ataxia-telangectasia; Bloom's syndrome; Severe combined immunodeficiency; Wiskott-Aldrich syndrome
Immunoglobulin heavy chain gene
  in Burkitt's lymphoma, 26, 30–31, 63–65, 188, 190–91
  chromosome translocation of, 30–31, 63–64, 173
  and c-*myc* translocation, 26, 30–31, 64–65, 66, 173
  map location of, 63
  in mouse plasmacytomas, 26
  rearrangement of, 101
Immunoglobulin kappa chain gene, 30, 63–65, 66, 188, 190–91
Immunoglobulin lambda chain gene, 30, 63–65, 66, 188, 190–91
In situ hybridization, 155–56
Inbred mouse strain, 8, 135
Inbreeding. *See* Consanguinity
Incomplete penetrance, 128–29
Indicators of cancer predisposition. *See* Cancer-predisposing genes
Infertility, 212, 216
Initiation, 9–14, 18–19, 39, 40, 74
Initiation-promotion model, 9–14, 18–19, 39, 74
Insertional activation, 26, 29–30, 39
Insulinoma, 108
int oncogene, 29. *See also* Oncogene
Interleukin-2, 186
International system for human cytogenetic nomenclature, 160
Intracisternal A-particle, 29, 35
Investigation of familial cancer
  healthy probands, 137
  probands with cancers of many kinds, 135
  probands with cancers of single sites, 124–25, 135–36
  probands selected with a genetic hypothesis, 137–38
Isochromosome, 44, 160
Isonicotinic acid hydrazide, 213

Juvenile polyposis coli, 114, 228–29

Karyotype, 5, 41
  nomenclature of, 160
  prognostic value in leukemia, 159, 161, 163, 164
Keratosis, 84, 212, 215
Kidney abnormality, 214
Kirsten murine sarcoma virus, 24–25, 32, 153
kit oncogene, 24–25. *See also* Oncogene
Klinefelter's syndrome, 67, 85, 87, 111

Knudson's two-hit model. *See* Carcinogenesis, two-stage theory of

Larynx tumor, 110
Late replicating X, 42
Leiomyoma of the skin, hereditary multiple, 84
Leukemia
in cancer-family syndrome, 82, 111
and diagnostic X-rays, 142
and Down's syndrome, 67, 85, 87
in familial aggregation of cancer, 124, 142
family studies of, 97, 136–37
in infancy, 85
in Li-Fraumeni syndrome, 246
in neurofibromatosis, 107
and oncogenes, 24–25
risk to relatives of patients, 136
secondary, 203, 204
in siblings, 97
in twins, 98
Leukocyte alkaline phosphatase, 166
Li-Fraumeni syndrome, 53, 246
Linkage analysis
in cancer-family syndrome, 230
in familial breast cancer, 126–28, 154
of inherited traits and cancer-predisposing genes, 153–56, 230
in retinoblastoma, 41–42
in Wilms' tumor, 49
Linkage associated with genetic cancer
fragile sites on chromosomes, 155
glutamate-pyruvate transaminase, 154
major histocompatibility complex, 154
restriction fragment length polymorphism, 155–56
Lipoma, 84, 108
Lipopolysaccharide W, 186
Liver cancer, 115, 217
Liver disease, chronic, 94
Long terminal repeat, 23, 28, 30, 36
Low serum in medium, 153
Lung cancer
cultural inheritance in, 125–26, 128–29
and debrisoquine metabolism, 19
dominant inheritance in, 112
familial predisposition to, 8, 112, 129, 134, 140–41
genetic susceptibility to, 129, 134, 140–41
in hamsters, 13
and Kirsten-*ras* oncogene, 24–25
marker enzymes and, 19
and occupational exposure to carcinogen, 140–41
in scleroderma, 16
Lymphatic leukemia, 24–25
Lymphocyte. *See* B lymphocyte; T lymphocyte
mitogenic stimulator of, 185–86
Lymphoid cancer
in ataxia-telangectasia, 16
in Bloom's syndrome, 16
and oncogenes, 24–25
Lymphoid leukosis virus, 35
Lymphoma, 24–25, 69, 137, 141. *See also* Burkitt's lymphoma; Hodgkin's disease; Non-Burkitt's lymphoma; Non-Hodgkin's lymphoma

McDonough feline sarcoma virus, 24–25
Major histocompatibility locus/complex, 87, 129, 154, 231
Malsegregation of chromosomes. *See* Chromosome nondisjunction
Mammography, 237
Mapping. *See* Gene mapping
Marker chromosome. *See* Chromosome, marker
Marker locus. *See* Genetic marker loci for cancer-predisposing genes
Medullary thyroid carcinoma, 5, 108
Medulloblastoma, 16, 17, 52–53, 108
Megakaryocyte, 162
Melanin, 94
Melanin spot, 227
Memorial Sloan-Kettering Cancer Center, 14, 94
Mendelian trait, 103. *See also* Dominant inheritance; Recessive inheritance
Meningeal sarcoma, 141
Meningioma, 16, 107, 112, 198, 200, 204
Mental deficiency, 85, 212, 215, 217
Metastatic disease, 106
Methotrexate, 71, 164, 175–77
Methyl-nitro-nitrosoguanidine, 213
Microcephaly, 212, 214
mil oncogene, 24–25. *See also* Oncogene
Mill-Hill-2 virus, 24–25
Mitogen, 185–86
Mitomycin C, 18, 152, 213
Mitotic error, 3, 60
Mitotic recombination
in the chromosome breakage syndromes, 50–51, 69–70, 95
and homozygosity production, 4, 44–45, 49, 68–70, 95
as a mutation, 39
in retinoblastoma, 44–45, 68, 95, 104
in Wilms' tumor, 49, 68
Mitotic spindle abnormality, 60
Mixed leukocyte culture, 231
Moloney murine leukemia virus, 24–25
Moloney murine sarcoma virus, 23
Monosomy, 62, 66
mos oncogene, 23–25, 30, 35, 69, 165. *See also* Oncogene
Mosaicism. *See* Chromosome mosaicism
Mucosal neuroma syndrome, 109
Mucous membrane carcinoma, 249

Muir-Torre syndrome
  cancer found in, 103, 110, 114, 115, 249
  dominant inheritance of cancer, 114, 249
  familial clustering of cancer in, 85, 223
Multifactorial inheritance, 82, 86, 91, 134, 250
Multiple endocrine neoplasia syndrome, 84, 108–9, 141
Multiple exocytosis, 116
Multiple hamartoma syndrome, 84
Multiple neoplasms, 82, 228–29
Multistage model of carcinogenesis. *See* Carcinogenesis, multistage model of
Murine leukemia virus, 29
Murine mammary tumor virus, 29
Murine sarcoma virus-3611, 24–25
Mutagen, 9, 39
Mutant clone, 213, 214
Mutation
  amplification, 39, 45–46
  base substitution, 4, 9, 39
  crossing-over, 95–96
  delayed, 41
  deletion, 4, 9, 39, 41–43, 48–49, 50–51, 95
  by DNA damage 9
  duplication, 39
  germinal, 44, 48, 95–96
  integration of viral DNA, 4, 39
  inversion, 39
  mechanism of, 4, 39
  nondisjunction, 95
  point mutation, 95. *See also* Point mutation
  rearrangements, 9
  recombination, 39, 51
  regulatory gene 32, 45, 86
  of a regulatory gene 32, 45, 86
  somatic, 44, 51, 95–96
Mutation rate, 213
myb oncogene, 23, 24–25, 28, 174. *See also* Oncogene
myc oncogene. *See also* Oncogene
  activation and expression in Burkitt's lymphoma, 25, 26, 30–31, 33, 64–65, 191
  amplification of, 31–32, 72–73
  and avian leukosis virus, 22, 23, 24, 29, 33
  and B-cell lymphomas, 23, 33, 65
  and chromosome translocation (8;14), 30–31, 64–65, 173
  gene product of, 28, 30, 33–34, 65
  insertional activation of, 22–23, 26, 28
  map location of, 30, 69, 173
  and mouse plasmacytomas, 26, 65
  and myelocytomatosis virus, 22–23, 24
  and other viruses, 29, 33
  promoter insertion, 22–23, 26, 29, 33
  translocation of, 26, 30–31, 50, 64–65, 72–73, 173
  and types of neoplasms, 26, 65
Myeloblastic leukemia, 24–25
Myeloblastic leukemoid reaction, 67
Myelocytomatosis virus-29. *See* Avian myelocytomatosis virus
Myelodyspoieteic disorder, 202
Myeloid leukemia, 24–25, 69
Myeloproliferative disorder, 202

Nasopharyngeal cancer, 87, 199
Near-haploidy, 174
Nephrolithiasis, 108
Neural crest cell, 112
Neurinoma, 84, 107
Neuroblastoma
  benign variant of, 105–6
  cell line from, 32, 72
  chromosome studies of, 69, 106, 198
  and E-rosette inhibitory factor, 106
  and fetal hydantoin syndrome, 106
  gene amplification in, 32
  in neurofibromatosis, 107
  and N-*myc* oncogene, 24–25, 32
  nonrandom chromosome change in, 69, 198
  pedigree analysis in, 105
  penetrance of gene for, 105
  and *ras* oncogene, 24–25, 32
  serum ferritin in, 106
Neuroectoderm, 107
Neurofibromatosis
  and central nervous system cancer, 84, 112
  as a hamartoma syndrome, 106–7
  incidence of disease, 107
  incidence of malignancy in, 5, 53, 103
  osteogenic sarcoma in, 116
  penetrance of gene, 103
  and radiation, 53
  types of tumors in, 53, 84, 107
Neurologic deficiency, 212, 214, 215, 216, 217
Nevi, 84
Nevoid basal cell carcinoma syndrome, 15–17, 52–53, 84
NIH 3T3 cells, 27, 32, 66
Nitroquinoline oxide, 213
N-*myc* oncogene, 24–25, 32, 47, 66, 72. *See also* Oncogene
Non-African Burkitt's lymphoma, 162, 173. *See also* Burkitt's lymphoma
Non-Burkitt's lymphoma, 192–96, 202, 203
Noncarcinogenic modifying factor, 12–14
Nondisjunction. *See* Chromosome nondisjunction
Non-Hodgkin's lymphoma, 192–96, 202, 203
Nonhomologous chromatid exchange. *See* Chromatid exchange

Nonintegumentary neoplasia, 217
Nonsister chromatid exchange. *See* Chromatid exchange
Nuclear protein, 28, 33. *See also myc* oncogene
Null allele, 42
Nutrition. *See* Dietary factors in carcinogenesis

Occupational exposure to mutagen, 7, 170–71
Ocular glioma, 107
Oncogene. *See also abl*; *Blym*; *erb*-A; *erb*-B; *fes*; *fgr*; *fms*; *fos*; *fps*; *int*; *kit*; *mil*; *mos*; *myb*; *myc*; N-*myc*; *raf*; *ras*; *rel*; *ros*; *sis*; *ski*; *src*; *yes*
  amplification of, 66, 72–73
  cellular, 22, 24–25, 65–66
  and chromosome rearrangment, 4, 159, 161, 165, 173, 177–78, 204
  and control of cell growth, 4, 34, 65, 87–88
  cooperation between, 33
  and fragile sites, 155
  insertional activation of, 25–26
  linkage analysis of, 128
  mapping of, 63, 155, 204
  polymorphism of, 128
  protein products of, 28, 33–34
  regulatory regions of, 204
  transfection of, 27, 32, 66
  viral, 22–23, 24–25, 65
Oophorectomy, 234
Oral contraceptive, 129
Organ transplant, 93
Osteogenic sarcoma
  and cancer predisposition, 83, 96
  and familial aggregation of cancer, 141
  familial occurrence of, 115–16
  in Gardner's syndrome, 115
  and oncogenes, 24–25
  and retinoblastoma, 15–17, 46–47, 51, 83, 85, 95–96, 219
  and translocation/insertion, 85
Osteoma, 228–29
Ovarian cancer
  and breast cancer association, 110, 124–25, 154, 246
  in cancer-family syndrome, 110, 111
  cultured cells from, 150
  and dominantly inherited disease, 110, 228
  in hereditary adenocarcinomatosis, 248
  nonrandom chromosome change in, 73, 198
  predisposition to, 223–24, 237
Ovarian fibroma, 51–53, 108, 110
Ovarian teratoma, 70–71

Paget's disease, 116
Pancreatic cancer
  in cancer-family syndrome, 111
  in Gardner's syndrome, 228
  hyperdiploidy in cultured cells of, 150
  islet cell tumor, 107, 108, 109
  in multiple endocrine neoplasia syndrome, 108
  site-specific in families, 115
Pancreatic cholera syndrome, 108
Pancytopenia, 214
Papillary thyroid cancer, 108
Papilloma, 108
Paraganglioma, 107
Parathyroid hormone, 108
Parathyroid hyperplasia, 108
Parathyroid tumor, 108
Paris Conference nomenclature, 160
Parotid tumor, 198, 200
Pedigree, 123, 138–39, 140–43, 227. *See also* Family history
Pedigree analysis, 80–82, 96–97, 124–25, 128, 241. *See also* Segregation analysis
Pelvic examination, 237
Penetrance. *See* Gene penetrance
Periampullary carcinoma, 115
Peutz-Jeghers syndrome, 84, 110, 114, 115, 228–29
Phacomatoses. *See* Hamartoma syndrome
Phenocopy, 252
Phenylketonuria, 90
Pheochromocytoma, 84, 107, 108, 109
Philadelphia chromosome
  in acute lymphoblastic leukemia, 172
  associated karyotypic changes, 163–64, 172, 187
  and c-*abl* oncogene, 31, 161
  in chronic myelogenous leukemia, 30–31, 60–61, 161–64, 187
  as a prognostic indicator, 161, 163–64, 172
  second, 163, 164, 187
  specific breakpoint, 31, 161, 162
  typical translocation, 161, 162, 172, 187
  variant translocation, 161, 162, 172, 187
Phorbol ester, 11, 12, 13
Phospholipid pathway, 33–34
Phytohemagglutinin, 160, 186
Pigmented mole, 84
Pituitary tumor, 108
Plasmacytoma, 26, 29, 65, 69
Plasminogen inactivator, 11
Platelet-derived growth factor, 34, 65
Pleiotrophy, 115
Point mutation
  in carcinogenesis models, 4, 39
  in chromosome breakage syndromes, 50–51
  and oncogenes, 26–27, 32
  in retinoblastoma, 43, 51, 95
Pokeweed mitogen, 186
pol gene, 23
Polonium-210, 13–14

Polycyclic hydrocarbon, 10, 13–14, 19
Polycythemia vera, 203
Polygenic inheritance, 134. *See also* Multifactorial inheritance
Polymorphism
  chromosomal, 70
  cytochrome P450 enzymes, 19
  of DNA, 44–45, 127–28
  electrophoretic, 9, 42
  enzyme, 19, 70
  of glucose-6-phosphate dehydrogenase, 9
  of restriction fragment length, 44–45, 49, 70
  in the study of familial breast cancer, 127–28
Polyoma virus, 29
Polyploidy, 62
Polyposis coli syndrome. *See* Familial polyposis of the colon
Polyposis-like condition, 226–27
Polyposis syndrome, 84, 86, 114–115, 137. *See also* Familial polyposis of the colon
Polyps of the colon, 114, 115, 204, 226–27, 228–29
Position effect, 4, 62–63, 65, 69, 71, 103
Preclinical marker of disease, 129
Predisposition to neoplasia. *See* Constitutional chromosomal abnormalities that predispose to neoplasia; Dominantly inherited clinical disorders predisposing to cancer; Familial aggregation of cancer; Recessively inherited clinical disorders predisposing to cancer
Prednisone, 164
Preleukemia, 202, 203
Prenatal diagnosis, 42. 49, 54
Progeria, 16
Progression, 74
Prolactin, 129
Prolymphocytic leukemia, 202, 203
Promoter insertion. *See* Proviral insertion
Promoter sequence, 69
Promoting agent, 10–11
Promotion, 9–14, 18–19, 39, 74
Promyelocytic leukemia, 24–25, 26, 32, 33, 35
Prophylactic surgery, 219, 227, 232–34, 235–36, 250
Prostate cancer, 5, 110, 111
Protease inhibitor, 12–13, 14
Protein A, 186
Protein kinase, 28, 33
Proteolytic enzyme, 11, 12
Proto-oncogene. *See* Oncogene
Proviral insertion, 23–24, 26, 33
Pseudodiploidy, 171
Pseudopolyposis, 228–29
Psoralen plus UV irradiation, 213, 214
Radiation
  cancer induced by, 13–14, 51–54, 142
  cellular transformation by, 39
  exposure to, 7, 134
  and induction of lung cancer, 13–14
  as an initiating agent, 13–14
  and secondary acute nonlymphoblastic leukemia, 170
  sensitivity to, 151
Radiation therapy
  in nevoid basal cell carcinoma syndrome, 17, 52–53, 108
  in retinoblastoma, 15–17, 46, 51–52
Radiomimetic drug, 151
raf oncogene, 24–25. *See also* Oncogene
ras gene family. *See also* Oncogene
  c-*ras*, 33, 69
  gene product of, 33–34
  N-*ras*, 25, 32
  v-Harvey-*ras*, 23, 25, 27, 32
  v-Kirsten-*ras*, 25, 27, 32
Recessive inheritance, 90–91, 96
Recessive inheritance in human cancer
  cancer-predisposing disorders, 83, 91–94, 211
  cancer risk in first-degree relatives, 96–97
  identification of, 90–91
  inbreeding, 98–99
  Knudson's model of, 96
  and multifactorial inheritance, 91
  retinoblastoma model, 95–96, 203
  twin studies, 97–98
  in Wilms' tumor, 203
Recessive mutation, 32
Recessively inherited cancer predisposition
  cancer risk in first-degree relatives, 96–97
  cancer risk in heterozygotes, 85
  the chromosome breakage syndromes, 92, 93–94
  and disturbances of sexual development, 83, 217
  DNA-repair deficiencies, 83, 92, 94, 211–15
  in familial breast cancer, 124–25, 126
  immunodeficiencies, 83, 91, 93, 212
  inbreeding, 98–99
  Knudson's model, 96
  other, 83, 93, 94
  pedigree analysis, 96–97
  retinoblastoma, 83, 95–96
  twin studies, 96–97
Recessively inherited clinical disorders predisposing to cancer, 83–84, 85–86. *See also* Ataxia-telangectasia; Bloom's syndrome; Fanconi's anemia; Werner's syndrome; Xeroderma pigmentosum
Recombination. *See also* Mitotic recombination

in adenovirus infection, 213
in formation of viral oncogenes, 22, 27
Red blood cell group. *See* ABO blood group
Reduplication. *See* Chromosome reduplication
Regulatory gene mutation, 32, 45, 86
Regulatory protein, 32
Regulatory sequence, 23, 28
rel oncogene, 24–25. *See* Oncogene
Renal angiomyolipoma, 107–8
Renal cell carcinoma
age of diagnosis, 50
chromosome breakpoint, 113–14
chromosome 3, 50, 113–14, 154–55, 198
constitutional cytogenetics, 50, 113–14
hereditary factors, 49–50
inherited translocation, 154–55
laterality of tumor, 50
tumor cytogenetics, 50, 113–14, 198
Replication defective virus, 28
Replicon chain initiation, 213
Replicon-fork progression rate, 213
Repressor sequence, 69
Respiratory tract cancer, 112. *See also* Lung cancer
Respiratory tract infection, recurrent, 212, 216
Restriction fragment length polymorphism
mapping of, 155–56
as a marker for cancer genes, 50, 54, 70, 156
in retinoblastoma, 44–45, 70
in Wilms' tumor, 49, 70
Reticuloendotheliosis virus, 24–25, 29
Retinoblastoma
age of onset, 40, 104
amplification in, 45–46
association with other neoplasms, 15, 16, 46–47
breakpoints, specific, 41–43, 67
cancer incidence in relatives, 40, 46–47
cellular and chromosomal radiosensitivity, 151
chromosomal instability in, 43, 151
chromosome 13, 41–45, 85, 87, 95, 104, 155, 203
chromosome 13 deletion, 41–43, 45, 67–68, 85, 104, 155
and esterase D, 41–42, 104
familial heritable tumor, 8, 16, 40–41, 67–68, 95–96, 104, 155
family history, 40–41, 44–45, 104
laterality of tumor, 40–41, 104
mosaicism, 104
multifocal origin of tumors, 204
N-*myc* and, 46
and osteosarcoma, 15, 16, 46, 93, 95–96, 115, 219
other nonrandom chromosome changes, 198, 204
penetrance of gene, 40–41, 104
pleiotrophic effects of gene, 104, 115
predisposition to retinoblastoma, 42, 93
radiation therapy and, 15, 16, 46
regulatory genes and, 45
restriction fragment length polymorphisms and, 44–45, 104
and retinoma, 43
segretation analysis and, 40–41
somatic crossing-over and, 95–96, 203
sporadic nonhereditary tumor, 40–41, 67–68, 95–96, 104
susceptibility to environmental carcinogens, 15–17, 47
suppression, 45
transforming genes and, 45
Retinoic acid, 12
Retinoma, 43
Retroperitoneal sarcoma, 228–29
Retrovirus
acutely transforming, 22, 23, 24–25, 28, 65
nonacutely transforming, 28, 29–30
Reverse transcriptase, 23
Rhabdomyosarcoma, 106, 107
Ring-necked pheasant virus, 29
Risk determination
analytic, 241, 249–50, 252–54
empiric, 241, 250
Risk factors for cancer. *See* Cultural inheritance; Environmental factors in cancer etiology; Genetic susceptibility to environmentally induced cancer; Risk determination
RNA tumor virus. *See* Retrovirus
ros oncogene, 24–25, 28, 34. *See also* Oncogene
Rous sarcoma virus, 23, 24–25, 28

Salivary gland tumor, 198, 200
Sarcoma
in basal cell nevus syndrome, 108
in cancer-family syndrome, 106, 111
childhood, 82
and environmental factors, 106
nonrandom chromosome change in, 198, 204
occurrence in families, 137
in progeria, 16
retroperitoneal, 228–29
as rhabdomyosarcoma, 106
of soft tissue, 246
Schmidt-Ruppin strain of Rous sarcoma virus, 23
Schwannoma, 107, 108
Scleroatrophic dermatosis, 84
Scleroderma, 16
Screening for cancer. *See* Cancer surveillance program

Sebaceous cyst, 228–29
Secondary leukemia. *See* Leukemia, secondary
Segregation analysis, 40–41, 124–25, 128, 223, 250–52. *See also* Pedigree analysis
Selection, 73, 74
Seminoma, 198, 204
Serous cystadenocarcinoma of the ovary, 198, 200, 201
Serum protein marker, 153
Severe combined immunodeficiency, 91, 151
Sex chromosome abnormality, 67, 87
Sex-limited autosomal trait, 109
Sexual development, disturbances of, 83
Shortness of stature. *See* Growth deficiency
Simian sarcoma virus, 24–25
Simian virus 40, 29, 153, 230–31
Single-gene syndrome, 140, 143. *See also* Cancer-family syndrome; Multiple endocrine neoplasia syndrome; Polyposis coli syndrome
Single-gene trait, 83–84
sis oncogene, 24–25, 31, 34, 36, 65–66. *See also* Oncogene
Sister chromatid exchange. *See* Chromatid exchange
Site-specific aggregation of cancer, 135–36, 139–40
Site-specific familial tumor. *See* Familial breast cancer; Familial colon cancer
Skeletal abnormality associated with nevoid basal cell carcinoma syndrome, 17
Skeletal system cancer, 115–16. *See also* Osteogenic sarcoma
ski oncogene, 24–25. *See* Oncogene
Skin cancer
  and albinism, 86, 94
  dominant inheritance of, 111–12
  in nevoid basal cell carcinoma syndrome, 16, 17
  nonmelanotic, 112
  and sunlight exposure, 7, 15
  in xeroderma pigmentosum, 15, 16, 151–52, 217
Skin dystrophy, 111–12
Skin tumor, 9, 17, 108, 110, 11, 249
Small bowel tumor, 110
Small cell carcinoma of the lung, 24–25, 26, 198, 200
Smoking, 7, 8, 14, 19, 125–26, 134
Snyder-Theilen feline sarcoma virus, 24–25
Sodium metaperiodate, 186
Somatic cell hybridization, 155–56, 215
Somatic crossing-over. *See* Mitotic recombination
Somatic mutation theory, 9, 40, 103
Somatic recombination. *See* Mitotic recombination
Soybean trypsin inhibitor, 12–13
Spontaneous chromosome breakage. *See* Chromosome breakage
Squamous cell carcinoma, 15, 53, 111
src oncogene, 28, 33–34. *See also* Oncogene
Stem-line concept, 60
Stomach cancer
  and ABO blood group locus, 87
  in the Bonaparte family, 133
  in cancer-family syndrome, 111
  familial aggregation of, 133
  family studies of, 136
  site-specific, in families, 115
  in Muir-Torre syndrome
Structural protein, abnormality of, 86
Sunlight/sun exposure
  and albinism, 94
  in Bloom's syndrome, 16, 212, 213
  in Cockayne's syndrome, 16
  in nevoid basal cell carcinoma syndrome, 16, 17, 52
  and skin cancer etiology, 7
  in xeroderma pigmentosum, 15, 16, 53, 151–52, 213, 214–15
Superoxide dismutase, 18
Suppression, 10, 12–13, 45
Surveillance program. *See* Cancer surveillance program

T antigen. *See* Simian virus 40
T lymphocyte, 150
Tay-Sachs disease, 90
T-cell growth factor, 186
T-cell leukemia, 24–25, 69, 70, 203
T-cell lymphoma,196–97, 202
Telangectasia, 212
Teratogen, 106
Teratoma, 106, 109, 198
Testicular cancer, 109–10
Testicular feminization, 83, 93
Tetraploidy, 66, 150, 164
6-Thioguanine resistance, 17
Thymidine dimer, 152
Thyroid cancer
  in autopsy material, 5
  in cancer-family syndrome, 111
  in Cowden's disease, 246
  follicular cell, 108
  in Gardner's syndrome, 115, 228–29
  medullary, 5, 108
  papillary, 108
Thyroid function, 129
Tobacco product, 7, 14, 125–26, 134
Torre-Muir syndrome. *See* Muir-Torre syndrome
Transcriptional activation, 22, 26
Transciptionally active chromatin, 73–74
Transduction, 22, 29
Transfection assay, 27, 30, 32, 66

Transformation, cellular, 4, 10–11, 27, 39, 152–53
Transformation, viral, 115
Tranformation protein, 22
Tranforming gene, 45
Translocation. *See* Chromosome translocation
Transposition, 71–73
Trenimon, 213
Triploidy, 66
Trisomy, 44, 62, 66, 69, 87
Trisomy 21. *See* Down's syndrome
Tuberous sclerosis, 84, 107–8
Tumor behavior and karyotypic change
  invasiveness, 186, 187, 195, 200
  metastatic spread, 186, 187, 188, 195, 200, 202
  response to therapy, 186, 187, 188, 195, 200, 202
  selective proliferation, 186
Tumor cytogenetics. *See* Chromosome change in human cancer
Tumor evolution, 68, 74
Tumor forming ability, 10–11
Tumor invasiveness, 60
Tumor progression, 33, 62
Tumor promoting agent, 7, 11–14
Turcot's syndrome, 83, 228–29
Twin studies, 97–98, 105, 123
Two-sage theory of carcinogenesis. *See* Carcinogenesis, two-stage theory of
Tylosis, dominantly inherited, 84, 115

Ulcerative colitis, 228–29, 233
Ultrasound, 237
Unbalanced chromosome abnormality, 60
Unmasking, 41–42
Unscheduled DNA synthesis, 152
Upper airway cancer, 112
Ureteral cancer, 110, 114
Urinary tract cancer, 113–14, 249
Uterine cancer, 110, 136
UV light sensitivity
  in Cockayne's syndrome, 16
  in Bloom's syndrome, 16, 213
  in Gardner's syndrome, 115
  in xeroderma pigmentosum, 15, 16, 17, 151–52, 213, 215

Vincristine, 164
Viral oncogene. *See* Oncogene
Virogene, 22
Virus
  exposure to, 7
  integration of viral DNA, 4.
  prototypes, 24–25
  replication defective, 28
  transducing type, 22, 29
  transformation by, 115
Vitamin A, 12
Von Gierke's disease, 83, 86
Von Hippel-Lindau syndrome, 84, 107, 109, 113
Vulva tumor, 110

Werner's syndrome, 83, 92, 93, 150
Wilms' tumor
  age at diagnosis, 47
  aniridia, 48, 67–68, 85, 155
  associated anomalies, 48–49, 85, 104–5, 155
  association with neurofibromatosis, 107
  association wtih other neoplasia, 43, 48
  chromosome 11, 48–49, 67–68, 85, 87, 105, 155, 198, 203
  constitutional chromosomal imbalance, 85, 87, 155
  deletion, 32, 69, 85, 105, 155
  erythrocyte catalase, 48–49
  familial tumors, 85, 104–5
  gonadal dysgenesis and, 85, 105
  Harvey-*ras* oncogene, 69
  heritable form of, 47–48, 85
  laterality of tumors, 47–48, 104
  linkage to, 48–49
  loss of heterozygosity, 54, 69, 203
  marker loci for, 48, 54
  nonheritable form of, 47–48
  penetrance of the gene, 47, 105
  and retinoblastoma, 43
  risk for offspring, 47–48
  risk for other tumors, 105
  sporadic tumors, 104–105
  tumor cytogenetics, 49, 105, 198, 203
  twin studies, 105
Wiskott-Aldrich syndrome, 83, 91

Xeroderma pigmentosum
  cancer predisposition in, 16, 53, 92, 211–12, 217
  cancer risk for heterozygotes, 85
  cancer surveillance in, 218–20
  cell studies in, 15, 17, 151–52, 213
  as a chromosome breakage syndrome, 15, 16, 92, 93, 211
  clinical complcations of, 212, 217
  clinical features of, 212, 213, 214–15
  and clinical features of, 212, 213, 214–15
  and clinical hypersensitivity to environmental agents, 15–16, 17, 53, 151–52, 213, 214–15
  complementation groups in, 15, 152, 215
  cytogenetic disturbances in, 213
  DNA repair and, 15, 83, 93–94, 152, 213
  general information about, 15, 16, 17

Xeroderma pigmentosum (*continued*)
in vitro sensitivity in, 16, 151–52, 213
inheritance pattern of, 16
physician handling of, 218–221
type of cancer in, 15, 16, 17, 53, 217
variant form of, 83, 93, 213, 215
X-linked genes predisposing to cancer, 124–25
X-ray sensitivity
and ataxia telangectasia, 16, 213, 214
as a carcinogen, 10–12, 142
and Gardner's syndrome cells, 115
and hereditary retinoblastoma, 15–17
and nevoid basal cell carcinoma syndrome, 15–17
and progeria, 16
XY gonadal dysgenesis. *See* Gonadal dysgenesis

yes oncogene, 24–25, 28. *See* Oncogene